GOLD MINING 88

Second International
Conference on Gold Mining

November 7, 8, and 9, 1988
Vancouver, British Columbia, Canada

C. O. Brawner, *Editor*

Advisory Organizations

BC Dept. of Mines and Petroleum Resources
BC Mining Association
BC and Yukon Chamber of Mines

Published by the
Society of Mining Engineers, Inc.
Littleton, CO • 1988

Conference Organizing Committee

Symposium Chairman

C.O. Brawner, P. Eng.

Mining Engineering Department
University of British Columbia
Vancouver, BC, Canada

Advisory Committee

George H. Beckwith

Vice President
Sergent, Hanskins & Beckwith
Phoenix, Arizona

John D. Nelson

Civil Engineering Dept.
Colorado State University
Fort Collins, Colorado

Vic Dawson

formerly Sr. Inspector of Mines
Dept. of Energy, Mines and
    Petroleum Resources

James Madonna

Dept. of Mining Engineering
University of Alaska
Fairbanks, Alaska

Randall J. Scott

Vice President
Pincock, Allen & Holt, Inc.
Lakewood, Colorado

Advisory Organizations

BC Dept. of Mines and Petroleum Resources
BC Mining Association
BC and Yukon Chamber of Mines

C.O. Brawner, *Editor*

C.O. Brawner has spent 35 years in geomechanics and geotechnical engineering, specializing in soil mechanics, landslides, stabilization, earth and tailings dams, and rock mechanics relating to transportation and mining engineering.

He is a graduate of the University of Manitoba and Nova Scotia Technical College.

Brawner spent six years with the British Columbia Dept. of Highways as senior materials and testing engineer. From 1963 to 1978 he was a principal and ultimately president of Golder Brawner and Associates, consulting geotechnical engineers. Since 1978 he has been on the staff of the Mining Dept. of the University of British Columbia teaching geomechanics in mining as well as providing specialist review consulting in the mining and transportation industries. He has served on numerous geotechnical review boards including Suncor, Syncrude Canada Ltd., the North East Coal Program in BC, the State Electric Commision of Australia, Pine Point Mines Ltd., and Trans Alta Utilities.

He has been involved in stabilizing many failures internationally on open pit and underground mines and has carried out extensive research in rock testing procedures, rock slope stabilization, pit slope drainage, blasting, and monitoring instability.

Brawner has been a member of the NRC Associate Committee for Geotechnical Research, vice chairman of the Canadian Advisory Committee on Rock Mechanics, chairman of the CACRM Committee on Mine Waste Embankments, specialist consultant to the Canadian Dept. of Mines for the development of the "Design Guide for Mine Waste Embankments in Canada," specialist technical advisor to the World Health Organization, special technical advisor to the BC Uranium Commission for Uranium Waste, chairman of the CANCOLD Committee on Industrial and Tailings Dams, the Ministers Advisory Committee to CANMET, and program chairman of numerous conferences.

He is author of over 75 technical papers and was supplementary reporter to the Third International Rock Mechanics Conference; Penrose Lecturer for the Geological Society of America; and invited lecturer to over 70 conferences, universities, and engineering institutes internationally, including the United States, Britain, Japan, Germany, Australia, China, and Russia.

Brawner was awarded the Presidents Medal by the Canadian Association of Good Roads, Award of Merit of the Association of Consulting Engineers of Canada, and the Meritorious Service Award of the Association of Professional Engineers of British Columbia. He was selected to receive the Walter Gage Teaching Award at the University of British Columbia.

He was selected a Distinguished Member of SME in 1982 and gave the Daniel C. Jackling Lecture in 1986. In 1988 he was selected a Distinguished Lecturer by the Canadian Institute of Mining and Metallurgy.

FOREWORD

The First International Conference on Gold Mining was very successful, with over 300 delegates representing 26 countries. In addition, exhibitors from 25 companies attended. Many delegates expressed interest in attending a future similar conference. As a result, this Second International Conference has been organized.

Ten themes are included:

- Exploration and Evaluation
- Legal and Ownership
- Feasibility and Financing
- Mining
- Processing
- Heap Leaching
- Placer Mining and Dredging
- Environmental
- Case Examples
- Opportunities in Lesser Developed Countries

The intent of the program is to present a broad overview of gold mining from geology to environmental. For the first time, legal and ownership questions have been included.

A number of representatives of Lesser Developed Countries expressed interest in describing development and investment opportunities in their countries. One complete afternoon session was arranged to fulfill this request. It is a major highlight of the program.

A number of case examples provide a practical contribution.

A truly international flavor has resulted with authors from 12 countries represented to assist in making the Conference a success.

C. O. Brawner
Editor

Table of Contents

1 Exploration and Evaluation   1

2 Feasibility and Financing   93

1

Exploration and Evaluation

Chapter 1

GOLD EXPLORATION AND ORE EVALUATION

Owen E. Owens

Vice-President, Exploration
Cominco Ltd.
Vancouver, B.C.

INTRODUCTION

This paper deals with the main components in exploring for and developing a gold ore deposit. It addresses geology, exploration drilling, ore reserves, and evaluation criteria in an integrated manner to present an overall picture. Gold forms such a small component of a gold ore body, that particular care should be taken in dealing with the data from a developing deposit.

The paper lists the main gold ore deposit types and their character. Their differences, as well as differences to other types of ore deposits, are important to understanding exploration approaches, and problems of ore reserve definition.

Gold exploration has enjoyed a great boom in recent years. Many new ore deposits have been developed but experience shows that definition of gold is not a simple straight forward process and this outline attempts to point out some of the important aspects of ore reserve definition including some pitfalls.

GOLD DEPOSIT TYPES

The Witwatersrand deposits in South Africa produce about 50% of the world's gold. The gold occurs as fine grains in thin beds within gently dipping quartz pebble fan conglomerates usually with associated pyrite grains. It is characterized by rather uniform distribution of gold along thin layers within the conglomerate.

Hemlo, Ontario is a large deposit which also has an unusually uniform distribution of gold. The gold mineralization occurs in relatively regular amounts in a 20-50 feet thick sheared zone in vertically dipping altered acid volcanics and sediments. Calculation of minable ore grade has been effectively carried out by surface drilling alone for the Witwatersrand and Hemlo deposits on account of the relatively regular distribution of gold in these deposits.

This is not the case for the shear and vein hosted gold deposit types. These are usually steeply dipping and have an irregular distribution of gold through the deposit. The ore zones may be only partially mineralized and/or the gold content may vary widely within the zones. Host and wall-rock alteration is usually strong.

The Timmins camp has been the largest Canadian gold producer with production from steeply dipping chloritic shear zones in basic volcanics. It is a typical shear-hosted Precambrian gold deposit. Other examples of shear-hosted gold in basic volcanics are Kirkland Lake, Val d'Or and Yellowknife in Canada or Kalgoorlie in Australia. We think of these deposits as being formed from moderate temperature solutions depositing gold commonly with quartz, carbonate and some pyrite. Because of the irregular distribution of gold in this type of deposit mining experience is normally required to completely establish stope outlines, assay walls, cutting factor needed for high gold assays, and to establish dilution requirements.

The younger Nevada type of lower temperature epithermal deposits is another important gold deposit type. The host rocks are commonly sediments. Gold is broadly distributed in a widespread pervasive alteration zone. Silicification is normal. Average deposit grade is quite variable from greater than a 1/10th of an ounce to economic deposits as low as 3/100ths of an ounce of gold per ton. Structural control is not particularly evident and the limits of the deposit may be quite irregular. The deposits lie in formations which are block faulted and only weakly folded. Mineralization is broadly distributed and a broad drill program of vertical holes is required to outline ore geometry. The deposits usually have a broad horizontal dimension in contrast to the steeply dipping Precambrian type previously mentioned.

New vein deposits are being outlined in Mesozoic sediments and granitic intrusives in western B.C. These are narrow, steep to shallow dipping, pyritic, quartz, calcite veins cutting across weakly to strongly altered sediments and granites. Gold distribution can be very regular for part of the strike length and quite irregular and continuous in other parts. The deposits must be mined underground and minable grade carefully calculated from a combination of cross cutting drill holes and sampling in underground openings all in advance of mining.

The Mother Lode in California is another example of elongated relatively narrow veins, in this case often offset by crosscutting faults. California has had very large gold production, in part from placers derived from the erosion of the primary vein deposits.

The Misima, Likir and other South Sea Island deposits are types of deposits connected with a volcanic vent or in sub-volcanic necks. Gold occurs as disseminated grains and widely distributed fracture filling type of mineralization usually with intense wall-rock alteration. The host rocks, dimensions of the deposit, and distri- bution of values resemble those of porphyry copper deposits. Sometimes, like OK Tedi the gold zone of the early mining phase is a residual zone where the copper of what was once a porphyry copper- gold zone has been leached out by surface weathering.

The Nevada and porphyry type deposits are usually mined in open pits, allowing in-pit selection of ore grade as determined by sampling blast hole drill cutting and selecting those portions of the blast of suitable grade. This control of mill feed grade after the blast is not possible in underground mining. Underground the geologist may visually or with the aid of face sampling guide the miner in preferred directions, but he just doesn't have the opportunity of sending part of individual blasts to the waste pile as does the open pit miner. This difference is of great importance in the application of the concept of cutoff grade which can be effected more reliably in open pits than for ore to be mined under- ground.

EXPLORATION

Each of the gold deposit types calls for a different approach to exploration and particularly different considerations in drilling and definition of the ore reserves. There is however a surprising similarity in the manner in which most recent gold discoveries have been made. Gold is a very stable element. It survives weathering. In many or most of its deposits it leaves a broad physical and chemical halo in rock and resulting soils. It can be picked up in streams by panning or prospecting. Stream silt geochemistry is now of increasing importance. Old fashioned prospecting has located most of the showings which led to the discovery of the world's gold deposits over time. Re-examination of old showings is the basis of most new operations, even today. Extensive rock sampling with atten- dant assaying is still the common method of locating gold. Today soil and silt geochemistry is adding another effective sampling dimension. Geophysics can assist in locating shear structures which may be hosts for the shear hosted type of deposit. IP or EM can help locate the pyrite so commonly associated with gold deposits. However, gold exploration, in contrast to most other metals, seems to respond best to sampling and more sampling, followed by even more sampling, either in a prospecting mode of following creeks or

traversing the area and banging rocks, or sampling soils and silts.
Systematic rock sampling and assaying seems to be most effective to
locate a target and determine the attractiveness of the location or
area.   A significant gold occurrence normally has a broad halo of
low values to help assess the importance of the find or showing.

The search for particularly strong alteration in favourable rocks
is of course an important adjunct to the work.   Strong argillic
alteration with the formation of quartz and sometimes carbonate are
important ore criteria in shear and intrusive hosted deposits.
Gold-bearing solutions usually have a silica component perhaps
leached from other parts of the system.   In any event the deposition
of quartz or fine opaline or cherty varieties are important indicators.   Pyrite is another normal associate and commonly the amount of
gold is proportionate to the pyrite content (or its altered relic).

The newly located or relocated gold discovery might be trenched
which will be helpful in indicating the structural setting and providing assay data.   However it is usually difficult to get a realistic reading on the gold content by surface sampling alone.   The
surface may be irregular, variously weathered with hard and soft
portions, prominent and recessive portions, all making reliable
sampling difficult or uncertain.   In deeply weathered terrain gold
values may be partially leached out; although increasingly we seem
to be coming to the view that while gold is to some extent soluble
in ground water solutions, not much gold, proportionally is, removed
from near surface exposures.   It may be that other elements are more
depleted so gold values stay at realistic values.

To really evaluate a gold showing and get into the excitement of
the chase calls for drilling.   This may be diamond drilling with
full core recovery; percussion drilling with fine cuttings carried
to surface around the drill rods with question regarding gold
recovery; rotary drilling with larger chips and better reflection of
bedrock; or reverse circulation drilling with cutting carried to
surface within the drill rods and thus less chance for loss of gold
on the way up the hole, and less chance for contamination by
abrasion from the side of the hole.   All non-core drilling runs some
risk of contamination by hole wall abrasion.   This is particularly
true in the case of passing through high gold content structures
where drilling abrasion of the wall of the hole may give an exagge-
rated impression of the width of the zone.   In the case of core
drilling great care is needed to get full core recovery in the
mineralized zone.   This is often difficult as the contrast between
hard quartz and softer, more foliated or weathered intervening
materials often makes for core recovery problems.

Drilling from surface allows the trend of the mineralization to
be investigated widely.   It provides the opportunity to determine
the dimensions of the deposit and obtain indications of grade.   It
may be less satisfactory in clearly demonstrating the structural

setting of the deposit in sufficient detail to establish ore continuity. It can and often does provide grade data of somewhat uncertain nature. Simple recovery in the mineralized zone may be less
than 100%. The position of the minable wall of the mineralization
may be uncertain. In any event, the size of the sample is small and
gold by nature is commonly erratically concentrated.

What spacing is advisable for the drill holes? Initially, it is
desirable to be as wide as possible to delineate the deposit,
closing later to define the structure and the gold distribution more
accurately. Quite closely-spaced drilling is normally needed to
define structure, grade and dimensions, perhaps 50 or 25 meter
intervals (and even 8 m underground to define stope limits). Multi-
phase drilling of the zone has the advantage of allowing average
grade comparisons of each phase to help assess the reliability of
grade determination.

In the case of shear and vein hosted deposits underground
openings in ore and closely spaced underground drilling is usually
required to establish the structural control of the mineralization,
variability of the gold content within the deposit, wall rock
competency, and to more accurately determine ore reserves.

TREATMENT OF ASSAY RESULTS

There are several important considerations in interpreting drill
results for gold which are worth thinking about for a moment. What
does one do with particularly high values? These really excite us
and drive the process ahead but do they present special problems?
What do we do with near zero values within that portion of the
structure being considered for mining? Where does one place the
wall or limit of the particular drill intersection being considered?
In the case of wall rock limit, the situation may be straight
forward if specific rock type, identifiable to the miner, is taken
as the ore unit. However, in the case of underground mines, problems will arise if, for grade reasons, part of this unit is
excluded in the grade calculations because the grade is low. This
is, of course, called an assay wall situation and for underground
mines it is a major consideration fraught with problems in outlining
ore reserves and determining the grade of the ore body which can be
mined practically. If assay walls are used they should not be fed
into a computer and left to the program to determine minable grade.
The outlined ore must be returned to ore sections and plans and be
shown to be a structurally or geometrically continuous and a practical body for mining. If there appear to be questions in this regard
the assay wall situation has to be reconsidered. In my view, the
preferred practice is to establish the physical outline of the ore
body first, then to apply geostatistics to determine its grade,
probably with some alteration of the process back and forth in
complicated cases.

What about those particularly high grade gold values or what is
often called nugget effect? Perhaps, the drill sample only caught a
rich grain or a rich narrow vein. The question is whether this
concentration of gold extends far enough to represent a significant
volume of rock, for instance half way to the next hole, so its value
can be taken without further consideration, or perhaps alternatively
that there are so many of these high-grade occurrences that they can
be taken as typical of the deposit. Experience shows this is only
rarely the case. Nugget effects are common in most vein deposits
and their erratic nature is such that they should not be assumed to
extend a significant distance out from the hole. Neither should
adjacent low values be automatically assumed to be outside the
ore body. Another nearby or adjacent drill hole might locate a high
value in the lower grade part of the ore structure. The usual
approach is to cut particularly high values to some lower number to
reduce nugget effect distortions on the overall average. Otherwise,
they can produce a significant but unrealistic effect on grade.
Arriving at a procedure to handle the cutting question is difficult.
Given good sampling and reliable assaying, the most productive technique to gain insight into the distribution of grades is to plot
drill holes with individual assays, on sections. This, together
with geology, is useful in indicating those risky situations where
the grade of a large volume may be unduly influenced by a single
high-grade value.

Plotting frequency distribution of all gold values obtained in
the ore body is useful, particularly if there is a lot of data.
Gold deposits commonly exhibit strongly skewed distributions. If
this is the case, cutting the high-grade tail is a frequent
practice. The level to which to cut high values is a question which
can only be answered finally with mine experience in that particular
mine. Our practice is to plot frequency distribution to determine
skewness. If distribution is strongly skewed, it is necessary to
cut the high-grade end of distribution, usually to 2.5 to 5 times
the uncut mean.

ORE RESERVES

Let's turn to the most positive aspect of the exploration process, the calculation of the ore reserve itself. This is where we
get a chance to see how it all adds up and we can eventually even
apply some dollar values.

The principles used in the calculation of reserves in a gold
deposit are of course similar to those for other deposits. The
objective is to provide the most reliable estimate of the asset by
using all relevant data. Since ore deposition is a natural geologic
process the relationship between lithology, alteration and ore
limits is very significant. The ore should be placed in its geological context whenever possible.

The evolution of geostatistics has proven to be helpful particularly the development of variograms, which permit the continuity and variability between drill holes to be quantified, as well as treating the nugget effect. However, in spite of much research and the development of many elegant and innovative mathematical weighting techniques, (disjunctive kriging, universal kriging, indicator kriging, lognormal kriging, inverse distance cubed or squared, and multi-gaussian techniques) no method has emerged which is clearly superior for the variability so common in gold deposits. Experience is that as long as some method is used to treat high and low grade samples, include the effect of adjacent grade information, and importantly, care is given to identify geological and minable continuity, then the modern mathematical approaches are usually effective.

When dealing with a large system where significant tonnages of ore of various grades exist, it is often possible to consider mining at different cutoff grades. In this case an understanding should be developed of the tonnage, grade and geometry of ore available at various cutoffs, and in the case of open pit deposits, the effect of cutoff grade upon the strip ratio. In these cases, it is worthwhile to invest the time in developing a clear picture of the geometric and geological effect of the cutting process in order that a reliable grade-tonnage relationship can be developed.

In the case of narrow vein deposits the most common pitfall in estimating reserves, is the temptation to exclude low-grade samples, and correlate only the high-grade samples between intersections. Only rarely is it possible to mine a high-grade section out of a vein, and even more rare to be able to identify these situations at the exploration stage. It should also be realized that high values are often illusory as just adjacent rock is often much lower in grade.

One other certainty in a miner's life is dilution. Dilution comes in many forms:-

(a) Addition in smoothing ore outlines in the mine planning process.
(b) Overbreak in the stopes due to weak ground.
(c) Overbreak due to poor stope control.
(d) Low grade or waste getting into the ore pass to make up tonnage.
(e) Backfill contamination during pillar recovery.

The effect of dilution is always to reduce grade (and profits). Only rarely does something happen in mining which has a positive effect on the grade.

Determining the specific gravity and variations in specific gravity is important in arriving at the tonnage factor. There can be
significant errors in tonnage where weathering, moisture or mineralogical variations within the deposit are not adequately investigated. Assays and most density determinations are done on dried
samples, but it's not unusual for highly altered rocks to contain
10-20% moisture by weight, even in a dry climate.

The reliability of ore reserve estimates depends on the quantity
and understanding of the data used in the estimate. Thus ore
reserves should be quantified in some way on the basis of information available for the calculation. Public reporting of reserves is
best served by reporting minable reserves according to the classification system of "measured", "indicated" and "inferred" ore.

The securities agencies and engineering associations have accepted definitions of ore reserves which should be followed. These can
be summarized as follows:-

Measured (Proven)

Material for which tonnage is computed from dimensions revealed
underground or in drill holes and for which the grade is computed
from sampling so spaced and the geological character so well defined
that the size, shape and mineral content are established accurately. Normally the term should be limited to those reserves at a
mine which can be projected from one or more exposed faces on the
basis of operating results.

Indicated (Probable)

Material for which tonnage and grade are computed partly from
specific measurements and partly from projections for a reasonable
distance on geological evidence. Normally there should be sufficient information and accuracy to form the basis of a mine production forecast.

Inferred (Possible)

Material for which quantitative estimates are based largely on
geologically character and widely-spaced sampling. This reserve
category usually involves limited drilling and application of geological projection. Data is insufficient to support a mine production
forecast.

There is an increasing tendency to see gold reserves reported as
"ounces" with no mention of tons, grades, or statement regarding the
degree of confidence in the numbers. There may be no indication
given regarding the portion of the contained ounces which will be
mined. This type of reporting is very questionable.

The practice of reporting "geologic reserves" is widespread as well. Sometimes the reason is the requirement for prompt disclosure necessitates that announcements be made before preliminary mine planning can be carried out. In other cases there is considerable information but a final decision on mining is not yet possible. On the other hand operators and project promoters have an obligation to their shareholders and the investing public that the assets of the company are fairly and accurately reported. Ore reserves are, by definition, material which calculation indicates can be mined at a profit. In the case of early stage calculation of reserves assumptions have to be made on mining method and costs and the reserves drawn up on that basis. The term geological reserves is becoming a useful term where minability and profitability are unclear but I do not believe it should be used for a body or tonnage that is not outlined as a minable unit.

Reporting reserves with a stated or implicit accuracy, e.g. "measured ore has an estimation variance of +/- 10%" was a concept embraced by geostatisticians and accountants a few years ago, but seems to have less reverence in the gold mining scene today. The main reason is that it is difficult to accurately quantify the accuracy without considerable data from production experience.

It is important to the operator and to the investor that ores with differing metallurgies are not consolidated under a general heading "Reserves" without further explanation.

EVALUATION

Success in surface drilling and the location of a continuous mineralized body leads to an evaluation in economic terms. A first preliminary evaluation can usually be carried out by comparison with costs and approaches developed at similar situations elsewhere. When the results suggest attractive economic possibilities more detailed work and metallurgy should be investigated. Samples may be obtained for preliminary metallurgical studies from coarse assay sample rejects. These studies indicate the appropriate gold treatment process and likely recovery, providing necessary data for a first approximation of the economics of a mine.

Underground investigation provides the necessary understanding of ore grade, tonnage, continuity of ore grade, ore structure, wall rock competency and larger samples for more complete metallurgical work. The work will provide larger and better samples for more extensive metallurgical work to define the process, recovery, grinding work index, operating costs and capital costs for a gold extraction plant. In instances of a broad ore structure large diameter core, broadly spaced through the deposit, can substitute in place of rock mined underground for metallurgical testing. The underground work also provides more accurate information on which to

select the mining method, estimate pre-production underground
development costs and time to prepare the mine for production.

Ore reserves, mining and metallurgical data give the basis to
work out the profitability of a prospective operation. Location,
access, employee housing, power and environmental considerations
round out the data necessary for an economic evaluation often called
a feasibility study. Such a study may take many forms dependent on
the size and capital cost, degree of profitability, profitability
risk, need to raise outside funding, environmental and permitting
requirements. There is no one series of conditions involved in the
term feasibility study. Rather the process requires sufficient
definition of the project (scope) and sufficient costing and economic evaluation to satisfy the owners, and often government, on the
project's viability and environmental compatibility.

CONCLUSION

The paper has attempted to bring together the series of features
involved in finding and establishing a gold ore deposit. It
develops the premise that ore reserve development and estimation is
not a matter of routine or mathematical treatment of the data but a
procedure which involves assumption and judgment about the geological and geometrical nature of the deposit, ore limits, treatment of
ore values, mining and metallurgical considerations in an integrated
manner. The paper has covered a wide variety of subjects briefly to
focus on the interrelationship of considerations and point out
special concerns in the case of gold where so much value occurs as
such minute and usually invisible portions of the ore mined. While
the features mentioned are of particular relevance to gold deposits,
most of the considerations apply to other mineral deposits as well.

Chapter 2

LATERITIC GOLD GROWTH AND NUGGET FORMATION
WITH CASES FROM MATO GROSSO STATE, BRAZIL

RAO, BHASKARA A.*
BARROS, JORGE G.C.*
ADUSUMILLI, MARIA S.*
MENDONCA, AUGUSTO F.**

*DEPT. OF GEOSCIENCES, UNIVERSITY OF BRASILIA
BRASILIA, BRAZIL
**EXPLORATION GEOLOGIST
BRASILIA, BRAZIL

ABSTRACT

Lateritic gold has been the object of study only for a very few years, though nuggets have been mentioned in earlier works even as early as 1984 by Hartt in Brazil.

In Mato Grosso State, lateritisation is intense, as is common in all tropical forest areas of the world. Exploration has revealed the presence of recoverable "nugget gold" of variable sizes (30 - 200 gms) but frequently of good size (0 - 40 gms), especially near Cuiaba and Alta Floresta regions. In both of them the host system is a lateritised breccia indicative of a fault or shear zone breccia, though in different lithological domains. The Cuiaba occurrences are phyllitic and show a profile as follows, from top to bottom: 1) Compact ferralite crust. 2) Oolitic to pisolitic compact crust (+ some gold). 3) Mottled concretionary laterite crust. 4) Lateritised "eluvial material" (gold nuggets). 5) Mottled patchy to banded laterite. 6) Banded lateritised phyllite with lenticular to banded goethite. 7) Altered red goethitic phyllite, with quartz veins (+ fine gold). 8) Unaltered grey phyllite with quartz veins (Protore). The Alta Floresta-Juruena River occurrences are volcano-sedimentary and show a profile as follows: 1) Deep brown siltitic, argillaceous and arenaceous material with organic matter (+ gold). 2) Nodular and concretionary laterite, compact and indurated, with arseno-phyrite in quartz fragments (gold nuggets). 3) Band with silty-clayey matrix and arenaceous in nature. 4) Bedrock meta-arcosian arenite. 5) Volcano-sedimentary with rhyodacites and andesitic basalts, with intercalations of arkosic and greywacke nature. Breccias and quartz veins, and granites associated in different portions (Protore).

The gold growth is attributed to: 1) Disseminated gold in the protores. 2) Initial enrichment due to granite and/or quartz veins, and hydrothermal activity. 3) Development of shear and/or fracture zones; gold mobilisation and segregation and/or formation of small ore shoots. 4) Chemical alteration and lateritisation, mostly active and dominating in the weak shear zone, forming iron hydroxides, carrying and redepositing and/or enriching gold in such zones, also through colloidal form.
5) The nuggets thus formed show a linear arrangement in their distribution pattern. 6) Release of such gold nuggets due to desintegration of laterites forming eluvium, colluvium or soil. This could or might have suffered the reincidence of lateritisation, as evidenced in some places (paleo-surfaces); or has contributed to the alluvial gold placers.

INTRODUCTION

The three International Seminars on Lateritisation Processes and laterities (Trivandrum, 1979; Sao Paulo, 1982; Tokyo, 1985) have given a new impetus to the study of laterites in greater detail. The more attractive part, however, is being reserved to indurated laterites or laterite crusts with gold. They are being the object of study of the authors and some reference need to be made to some interesting results.

Laterites are classified after their chemical components by Schellmann (1979), their mineralogy (Aleva, 1981) and their structures (Bhaskara Rao, 1983). The structures are attributed to the laterite crusts which are defined as: Lateritic crusts are indurated compact coverings on any bedrock type, with a deep brown colouration indicative of the abundance of geothite, either dominating all over or as intergranular locking material between rock fragments, or bands or nodules constituted of varying mineralogy. Such crusts are usually autochtonous but could be allochtonous, either way indicative of bedrock below or nearby. Studies are normally concentrated on in situ crusts (Bhaskara Rao, 1985 a). The guide horizons for lateritic gold are:
1) pisolitic or oolitic crusts on the ultramafic to mafic sequences, both volcanic or intrusive, and parts of greenstone belts (Bhaskara Rao, 1985 a); 2) lateritised gravel bed (Bhaskara Rao, 1985 b); 3) lateritised colluvium in tropical rain forest regions, and over volcanogenic sequences both of intermediate and acidic nature (Bhaskara Rao et al, 1988); and 4) lateritised eluvium on metasedimentary sequences, and in wet forest regions (Bhaskara Rao and Adusumilli, 1988 - communicated).

GOLD DISTRIBUTION

The distribution of gold seems to be more versatile than ever
before, as a consequence of new and intense investigations all
over the world in search of this noble metal. Several new
geological environments are being considered as hospitable for
gold. Of them, certainly some are more prominant than the others
(Table 1).

In resume the greenstone belts, volcanogenic and
volcano-sedimentary sequences, quartz veins, granitic stocks
intruding into the above sequences, lateritic formations and
placers are the most economic of the host systems for gold.
Studies on the transport of gold through solutions rich in Cl, CO
and S and the consequent paragentic associations as guides for
exploration continue to stimulate the search for gold, in other
geological environments. Since greenstones are constituted of a
sequence of volcanic-intrusive-sedimentary formations, a case of
exploration guides envisaged for greenstones is of interest
(Table 2; modified after Bhaskara Rao & Divakara Rao, 1988).

Table 1: GEOLOGICAL ENVIRONMENTS AND GOLD DEPOSITS

1. GREENSTONE BELT UNITS, specially volcanic suites both
   komatiitic and tholeitic, acid to intermediate; clastic,
   chemical and biogenic sedimentary formations, policyclic;
   granites, K-Si rich hydrothermal forming shear zones and lode
   quartz veins.

2. GRANITIC ROCKS including dykes, stocks, and their
   neighbouring environments; and pegmatites.
   STRUCTURAL TRAPS as shear zones and fractures surrounding the
   granites and their fillings by quartz veins.

3. SKARN TYPE deposits, normally related to granite-granodiorite-
   monzonite clan; either Mg or Ca type.

4. GOLD SILVER VEINS, stockworks, pipes and silicified bodies in
   deformed volcanic formations (greenstones included).

5. VEINS, SHEETED ZONES and SADDLE REEFS in folded and fractured
   sedimentary terrains; and also replacement bodies in
   fractured and/or chemically favourable rocks.

6. GOLD-SILVER VEINS, STOCKWORKS and SILICIFIED ZONES in mixed
   volcano sedimentary sequences which have been tectonically
   deformed, and intruded by granite.

7. DISSEMINATED and STOCKWORK GOLD-SILVER deposits in igneous
   instrusive, volcanic and sedimentary rocks.

Table 1: GEOLOGICAL ENVIRONMENTS AND GOLD DEPOSITS (Continued)

8. QUARTZ PEBBLE CONGLOMERATES WITH GOLD and associated
    mineralisations.

9. LATERITIC GOLD DEPOSITS, specially on greenstone belt units,
    volcanogenic sequences, volcano sedimentary formations,
    metasediments (phyllites) and in general on shear zones.

10. REGOLITH TYPE deposits in arid and sub-arid regions,
    containing residual gold.

11. PLACER DEPOSITS, specially fluvial type in cratonic and
    mobile belt areas.

Source: Special Supplement on Gold to Mining Journal,
        Sept. 12, 1986, with some additions by the authors.

Table 2: EXPLORATION GUIDES FOR GREENSTONE BELT UNITS AND GOLD

Lithotype	Controls	Nature of Gold
Laterites	Shear zone Fault zone Enrichment	Nuggets Fine to invisible.
Hydrothermal K-Si rich granite	Endo and Exo- fracture systems. Shear zones. Lode quartz veins.	Coarse, crystalline and ore shoots.
Acidic and intermediate suites*	Axial planes of tight folds. Fracture and	*Fine to disseminated.
Komatiitic to tholeitic basalts with minor ultramafics**	shear zone fillings, bor- dering the granite. Alteration by the $Si-CO2-S$ systems.	**Concentra- ted/segregated in some phases.

Table 2: EXPLORATION GUIDES FOR GREENSTONE BELT UNITS AND GOLD
Continued

Lithotype	Controls	Nature of Gold
Intra continental clastic, chemical and biogenic sediments.	Axial planes of folds, fracture/ shear zones.	Coarse to fine grained.
Tonalite-trondhjemite-granodiorite intrusives (Polyphase migmatities)	Contact zones and related fracture fillings. Tight fault zones.	Coarse grain + fine grain.

Adopted from: Bhaskara Rao & Divakara Rao, 1988.

Table 3:   LATERITE TYPES WITH GOLD OCCURRENCES

Host Rock	Laterite Crust Type (Profile)
GREENSTONE BELT	Breccioid laterite Concretionary laterite Nodular laterite Pisolitic laterite Transition zone (saprolite) Bedrock
MAFIC TO ULTRAMAFIC VOLCANOGENIC SUITE	Lateritised gravel bed Nodular laterite Mottled laterite Pisolitic laterite Altered bedrock (saprolite) Bedrock
ACID VOLCANOGENIC SUITE	Lateritised colluvium Nodular lateritic colluvium Nodular laterite Pisolitic laterite Altered bedrock Bedrock

Sources: Publications of the author & collegues.

Some interesting aspects are to be inferred from Table 2. They are:
1) The volcanic activities that prevailed permit the normal disseminated distribution of gold in the flows.

2) Intrusives like the TTG (Tonalite-trondhjemite-granodiorite) or GG (granodiorite-granite) emplacements favour the mobilization of the disseminated gold and permit its concentration at the borders or in weak zones, like fractures and/or shears.

3) The process of desintegration and sedimentation, which alternates with the volcanism, has a capacity to deposit the detrital gold in a disseminated form in the sediments.

4) The evolution of the older to younger greenstones, include a set of activities like: tectonism, intrusions, erosion and sedimentation, volcanism, and diapiric or sheet like intrusions, which constantly have "mobilised" or "remobilised" or "re-remobilised" the original gold and other gold which was brought by the activities. Such an evolution certainly could favour, occasionally, formation of segregations or concentrations or small to big ore shoots.

5) The repeated tectonic events that are documented in the evolution of these sequences, are again favourable loci for the concentration or segregation or formation of ore shoots.

6) The chemical weathering, which is very dominant in warm climates resulting in lateritisation with more than one phase in its evolution, is an instrument for transportation, or dissolution and reprecipitation of gold. When concentrations or segregations or ore shoots are already existing in the shear or fracture zones, lateritisation could be a factor that could add to the existing gold and thus form nuggets.

Thus, the process of lateritisation seems to be the enriching factor and nugget gold builder.

LATERITE TYPES WITH GOLD

During these studies on laterite crusts, profiles of laterite crusts which have gold are observed. A resume of them is included in Table 3.

It is seen that the characteristic of the profiles is practically the inclusion of pisolitic and nodular varieties. Yet the difference which calls the attention is the overlying laterite type. In case of greenstone belt units it is often a breccoid laterite; in case of mafic to ultramafic volcanogenic suite it is lateritised gravel bed; and in case of acid volcanic suite it is lateritised colluvium. All these contain quartz, either as pebbles or as fragments, though the fragments are more characteristic of breccoid types and colluvium indicating the presence of quartz veins in the bedrock.

All have one typical characteristic as well. Very rarely a nugget is known in the pisolitic type, though this is the preferred gold repository in laterite system. The gravel bed, colluvium and sometimes the nodular varieties do have fine gold, consistently.

The importance of Table 3 is in the following information, specially in tropical, sub-tropical and sub-arid regions of the world where lateritisation is a dominating process.

1)   The bedrock type and its chemistry is important in the relative abundance of participation of goethite in the process of lateritisation and induration.

2)   Lateritisation is certainly much more active on Fe-rich units in any system, to form laterite crusts.

3)   Lateritisation is, further, a dynamic process in regions/localities/zones where fracturing, shearing or faulting is a dominating structural feature.

4)   Such zones, already considered as possible hosts of relative concentration of gold or other minerals, are natural loci of accumulation and growth of dissolved and migrating gold through lateritisation process.

5)   Thus, laterite crusts, grown in certain environments and in favourable structural features, are extremely useful as exploration guides in volcanic, volcano-sedimentary, sedimentary and metamorphic suites, and igneous environments.

CASES FROM MATO GROSSO STATE:
PHYLLITE-LATERITE-NUGGET GOLD MODEL

The Mato Grosso State in Brazil is known for gold occurrences and deposits specially in the laterites, and in different parent rock environments.

Cuiaba, the capital of Mato Grosso State, is privileged with terrain where after each rainfall people find small nuggets in the gulleys and drained lands. All these parts are covered by a lateritic soil, often to a depth of over 3 m and further as very indurated and compact material forming crusts. Several prospects are known in the neighbourhood of the city, and some of them are still very active.

The Cuiaba occurrences have the following characteristics, from the observations made in Jatoba, Casa de Pedra and other deposits.

1)   The host protore is phyllites, which are extremely fine grained and sericitic, belonging to Cuiaba Group, where several units have been identified and of which Units 3 and 5 are considered to be hosts of gold. They are intercepted by quartz veins, often with very high frequency, and with centimetric thickness.   Two preferential directions are noted for the veins. The fertile veins are with N 70 E - S 70 W direction, and the sterile veins have the direction N 30 E - S30W. The fertile veins are semi-transluscent, and often are not milky. They are also very much fractured. The sterile veins attain thickness of about 3 m and are often milky.

A guide for the fertile quartz veins also is the presence of pseudomorhosed siderite resulting in geothite, with good rhombohedral forms; and also streaks and/or powdery black tourmaline either as bands or filling the cavities (Table 4).

2)   The phyllitic rocks, when constituted of disseminated fine grained pyrite are considered as ore, with recoverable tenors of gold.   Such pyrite is usually of less than a mm size, and is totally altered in the zones where the rock is considered as ore.

Where the size of pyrite increases, and specially in contact with the quartz veins occasionally attaining even 60 mm, the distribution of gold is not considered directly related or indicative of free gold.   Further, sometimes, totally pseudomorphosed pyrite is seen to have some grains of gold inside the crystal.   If such gold is as inclusions in pyrite, or has been released due to alteration, is not known.

3)   Intense structural deformations including shearing and faulting, resulting in bands of ultramylonites.

4)   Surficial chemical weathering resulting in intense lateritisation, showing different phases of development with a characteristic lateritised eluvial material called "lateritised alluvium" (Bhaskara Rao & Adusumilli, in Press) where the concentration of the metal is favoured. This phase is indicated also as a paleo-erosion surface which suffered lateritisation. Successive stages of surficial alteration and deposition resulted in the oolitic and concretionary laterites, and a stage of dehydration and oxidation formed the ferralite covering.

5)   The process, as a result of descending and circulating
meteoretic   and   sub-surface waters and hydroxide formation,
mobilised   and enriched zones below the laterites, with some gold,
with tenors often greater than those in the protores.

6)   The same process of dissolution and reprecipitation of gold
favoured the deposition and growth of gold in the vugs of rock
crystal in the lode quartz veins.

7)   The mobilisation and deposition of gold and its constant
growth has permitted, often, development of nuggets.

Thus, the laterite profile here can be described as constituted
of the following stages (Table 5).

VOLCANO SEDIMENTARY-LATERITE-NUGGET MODEL

Located NNE of Alta Floresta town along the banks of Rio Juruena
in North of Mato Grosso State, are occurrences of nuggets in
laterite cover (Barros, 1987: Unpublished report).

Geologically the region is in the domain of Guapore Craton.
Stratigraphically, from base to top, is the following sequence:

* Xingu Complex: Early to Middle Proterozoic, of polymetamorphic
basement of the amphibolite-granulite facies, constituted of
gneisses, migmatites and anatectic granites.

* Uatuma Group: Irari Formation: Middle to Late Proterozoic of
volcanics, acid to intermediate in nature with arcosic and
greywacke intercalations.   Rhyodacites and andesitic basalts
predominate over the meta-arcosian sandstones.   Breccias and
quartz veins are also recorded.

Sub-volcanic Unit, Late Proterozoic, is represented by granites.

Intrusive Unit, Late Proterozoic, is constituted of diorites,
gabbros and diabases.

* Lateritic cover, Tertiary to Quaternary, with alluvial, eluvial
and colluvial material; extensive development of laterites all
over the area.

The structural feature that marks the area is a "shear zone" in
the E-W direction. The fracture systems NE-SW and NW-SE are seen
through the drainage pattern. The Northern part of the area is
characterised by a typical recumbent anticline, closing towards
West.   The Southern flank of the structure composed probably of
two meta-arenites is limited by faults resulting in a "breccia
band" of 10 x 0.5 km, mineralised in gold.

Table 4:   FERTILE AND STERILE QUARTZ VEINS IN PHYLLITE REGIONS:
MATO GROSSO STATE

CHARACTER	FERTILE QUARTZ VEINS	STERILE QUARTZ VEINS
Trend	N 70 E	N 30 E
Nature of Quartz		
a) Transparency	Semi-transparent to Semi-transluscent	Transluscent
b) Colour	Colourless, grey Brownish when covered by geothite	Milky white
Thickness	Thin, up to 80 cm	Up to 300 cm
Associated Minerals		
Carbonates	Rhombohedral siderite (+ ankerite ?) pseudomorphosed to goethite	---
Sulphides	Pyrite cubes (varying sizes) pseudomorphosed.	Not common
	Arsenopyrite	---
Borates	Tourmaline inducing black banding or as streaks; black powdery masses in cavities.	---
Relict Structures	Skeletal platy structures in vugs, represented by oxides (ilmenite ?) or altered sulphides (pirhotite ?)	Skeletal platy structures in vugs, represented by oxides (ilmenite ?) or altered sulphides (pirhotite ?)
Fracturing	Extremely fractured resulting in brittled fragments	Good fracturing with larger blocks
Hydrothermal Alterations	Frequent clay mineral pockets	Sparse argillisation

Table 5: LATERITIC PROFILE OVER PHYLLITES IN MATO GROSSO STATE

TYPE	NATURE	PRESENCE OF GOLD
Ferralite: Compact crust with pisoli-layers.	Not always present; very expressive when present.	---
Pisolitic and oolitic, compact crust.	Always present.	Gold as fines and as small (to medium) sized nuggets.
Mottled concretionary (of cm concretions) laterite crust.	Not always present.	---
"Lateritised Eluvium: with quartz fragments and blocks.	Thick layers and very expressive. Very compact, indurated.	Rich zone; also with gold nuggets.
Mottled patchy to banded laterite crust.	Very compact, with gibbsite and clays, occasionally with concretions.	---
Banded lateritised phyllite	Banded geothite with the schistosity of phyllite and alternating with phyllite.	---
Altered geothitic phyllite	Or alterite, with quartz veins, usually thin and in swarms, with a red to red brown colouration.	Presence of gold, usually transforming it to ore.
Unaltered grey green phyllite.	With quartz veins and dominating evidences of shear zones, and intense tectonism.	Protore, normally uneconomic.

In this band, with about 5 m thickness, the following horizons are detected from bottom to top:

* Bedrock: Meta arcosian arenite of Iriri Formation.

* A band with greenish grey colouration, arenaceous in character, with silty-clayey matrix. quartz occurs as granules, cobbles and blocks, with meta-arenite blocks of the bedrock. This is the principal auriferous horizon with thickness up to 1.5 m.

* Laterite horizon, with nodules and concretionary types (Nodular lateritic crusts) in a matrix of silt and clay, and grains of quartz. The quartz fragments show association of arsenopyrite. This ochre coloured laterite is the auriferous zone, and shows some organic material, with a thickness of 0.5 - 2.0 m.

* Laterite crust with siltitic, argillic and arenaceous material, with a deep brown colouration and thickness of 1.0 m. Rarely cobbles of quartz are seen. Organic material is frequent. This laterite zone is also auriferous, though the tenors are low.

The chemical analyses revealed mineralisation in the entire sequence of rocks, with disseminated gold. In some preferential zones, such as the intersections of faults, gold occurs in significant tenors and rarely as small grains (or nuggets). The placers are also mineralised and are being worked for a decade. The average tenor of the area of 1.0 gm/m , with very fine granularity for gold.

The lateritic cover in the brecciated zone where angular fragments and blocks of quartz with sulphides, such as arsenopyrite, are seen as the richest band. The tenor reaches here up to 12 gm/m . The presence of nuggets is very common, and sizes weighing up to 90 gm are obtained, not infrequently.

These nuggets show a variety of physical aspects. Some are massive with planar surfaces, some are like an aggregate of fine crystals, and others are with a spongy aspect showing evidences of accretion and mosaic growth with some twins. The granules of gold, often, have different colours, probably indicative of differentiated composition, such as the content of silver which reaches up to 8% in some cases.

EVOLUTION OF GOLD NUGGETS

Gold growth in laterites is still not yet well understood. It has, undoubtedly, several processes involved to some extent or other, such as: 1) detrital; 2) dissolved, transported and precipitated; 3) authigenic grown; 4) residually enriched, and 5) other mechanisms.

Some photographs with detailed legends are included illustrative of the occurrences and the nature of nuggets from Mato Grosso State (Photos 1 - 8).

The evolution of the nugget gold is conceived through the following:

1) Presence of disseminated gold in the protores, such as phyllites and volcano (acid to intermediate) - sedimentary sequences, with possible relative abundance in some faciological variations.

2) Initial mobilisation, concentration and/or addition due to quartz veins resultant of tectono-hydrothermal activity and/or granite emplacements.

3) Development of shear and/or fracture zones, and brecciation. Mobilization of gold and segregation and/or formation of small ore shoots.

4) Chemical alteration and lateritisation on the bedrocks, dominating and active in the weak shear zones. Generation of iron hydroxides and dissolution of gold in the bedrock. Migration, redeposition and enrichment of gold in such zones forming grains and nuggets; also as colloidal deposition.

5) General distribution pattern along linear zones or restricted to brecciated bands, in both cases indicative of shear zone enrichment as another important factor for nugget formation, besides lateritisation.

6) Erosion, desintegration and release of such nuggets of laterites forming eluvium, colluvium and soil. Reincidence of lateritisation as indicated by paleo-surfaces covered by pisolitic-oolitic or ferralitic laterites. Also contribution to the formation of alluvial gold placers.

Photo 1: A view of the phyllites with swarms of quartz veins, intensely fractured, showing two preferential directions. Some meta arenites as blocks are also recorded. Locality: Road Cutting, 25 km from Cuiaba to Casseres.

Photo 2: Banded phyllites with fillings of iron hydroxide as a passage to overlying laterite cappings. White patchy clays are recorded, indicative of argillic alterations. The rock is extracted as ore when in contact with fertile quartz veins. Locality: Cascalheira de Prefeitura Mine, Pocone Town.

Photo 3:   Blocks of sterile concretionary mottled and patchy laterite crusts, with centimentric concretions and frequent clays and gibbsite. Locality: Casa de Pedra Mine 35 km NE of Cuiaba.

Photo 4:   "Lateritised eluvium" with angular fragments of quartz of varying sizes and constituted of matrix with irregular distribution of grains, fragments of quartz, indurated by colloidal goethite with or without nodular or banded forms. Locality: Cascalheira da Prefeitura Mine, Pocone Town.

Photo 5:   A view of the extensive laterite crust covering, already exploited and beneficiated, with a relict (2 meters thick) of original surface and profile. The area that has been covered and is now being reworked (washing of tailings and beneficiation, with recovery of small nuggets of very pure deep yellow gold) is of a few sq. km. The level now exposed is the mottled concretionary laterite, highly compact, and/or the upper leveld of altered phyllite with quartz veins showing powdery tourmaline and pseudomorphosed siderite. Locality: Jatoba Mine, 25 km SE of Cuiaba.

Photo 6: Compact pisolitic aggregate passing to compact ferralite crust. Locality: Jatoba Mine.

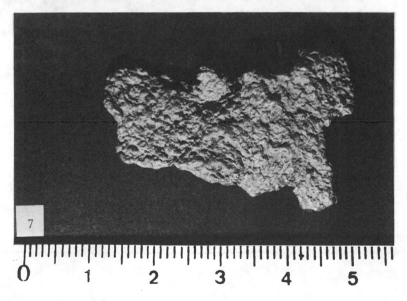

Photo 7: Gold nugget of 51.3 gm with 8% of silver, obtained from Rio Juruena, Alto Floresta District. The surface is rugged, granular to spongy in aspect, suggesting a probable growth via accretion. The sample is from manually dug pit in the lateritised eluvionary horizon, over a shear zone.

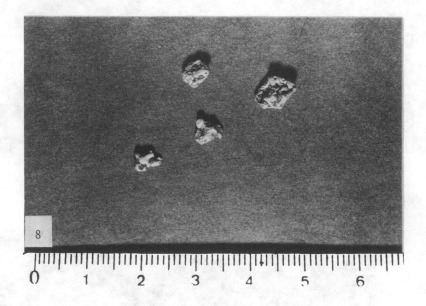

Photo 8:   Small nuggets of 0.5 gm each, from Jatoba Mine. All of them still show
relicts of red brown lateritic aggregated material.

ACKNOWLEDGEMENTS

Thanks are due to: Geologist Jose da Silva Luz, Director of
DNPM, Cuiaba Office; Mr. Iguatemy Mendonca Filho, Director of
Mineracao Manati Ltda, Cuiaba, for providing facilities and
making the trip to Cuiaba possible for the first two authors.
Special thanks are due to Geologist Nilson Batista de Sousa of
DNPM for the field programme, visits to the deposits and for
discussions.

REFERENCES

Aleva, G.J.J., "Laterite versus Saprolite" IGCP Newsletter II,
Hyderabad. June (1981). p 30-34.
Bhaskara Rao, A., Depositos Minerals : Ambients de sua Formacao e
Modelos. Texts, Restricted Circulation, Univ. Brasilila. (1983)
202 pp.
Bhaskara Rao, A., "Guide to Horizons for Gold Mineralisation in
Lateritic Crusts"d. in: International Seminar on Laterites,
Tokyo. Proceedings. Ed. Y. Ogurs (1985 a), p. 119-128.
Chemical Geol., (1987) Vol. 60, p.293-298.
Bhaskara Rao, A., "Lateritic Gravel Bed: A New Guide Horizon for
Lateritic Gold?" in: International Seminar on Laterites, Tokyo.
Proceedings. Ed. Y. Ogura. (1985 b). p. 111-118. Chemical
Geol., (1987) Vol. 60, P.287-291.
Bhaskar Rao, A., and Adusumilli, M.S., "Lateritic Gold: Option
for Small Mine Development". Small Mines Development in Precious
Metals. Soc. Mining Engineers, AIME, Littleton (1987) p.29-36.
Bhaskara Rao, A., and Adusumilli, M.S., "Lateritised Eluvium: A
New Guide Horizon for Lateritic Gold, with a case from Mato
Grosso State". Communicated to World Gold 89, Reno, Nevada. Oct.
(1989).
Bhaskara Rao, A., Barros, J.G.C., and Adusumilli, M.S.,
"Lateritic Gold Project". Lateritisation Processes, Proceedings.
2nd. Int. Sem. on Late. Proc, Univ. Sao Paulo. (1983). p.159-176.
Bhaskara Rao, A., Borges, M.R., Adusumilli, M.S., "Lateritised
Colluvium: A Guide Horizon for Lateritic Gold in Tropical Rain
Forest Regions". Bicentennial Gold 88, Melbourne. Extended
abstracts, Poster Session, Geol. Soc. Australia, Abstract Series
No. 23 (1988). p.565-569.
Bhaskara Rao, A., and Divakara Rao, V. "Distribution, Migration
and Concentration of Gold in Greenstone Belts." Bicentennial gOLD
88. Melbourne. Extended Abstracts, Poster Session, Geol. Soc.
Australia, Abstract Series. No. 23 (1988),.
Carvalho Melo, S., "Indice d'Or de Jatoba, Cuiba, Mato Grosso,
Brassil." Rapport Stage CESEV, Ecole Nat. Sup. de Geol. Applique
et de Prospection Miniere, Nancy (1983), 44 pp.
Michel, D., "Concentration of Gold in situ laterites from Mato
Grosso". Mineral. Deposita. 22, p. 185-189.

Schellmann, W., "Considerations on the definition and classification of laterities". <u>Lateritisation Processes,</u> Proc. Int. Sem. of Lat. Proc., Trivandrum (1979). p.1-10.

Chapter 3

GOLD & THE PLATINUM GROUP - WAYS TO THE RIGHT ANSWER

by Ken Bright

Senior Geochemist
Bondar-Clegg & Co.

ABSTRACT

This paper is an overview of approaches to assessment of ore grade for gold and the platinum group at the exploration stage of a project. Proper sampling techhniques must be complemented by good assaying. In turn, correct assaying has its roots in thoughtful, systematic sampling and sample preparation procedures.

While intricacies of the subject matter could occupy a book, we will comment on acceptable sampling approaches, outline a lab handling scheme, and discuss preferred analytical methods.

Pitfalls previously encountered by the mining and exploration industry will be highlighted with emphasis on a perceptive understanding of the nature of ore occurrence. Lastly, focus will be centered on reliable assay approaches.

INTRODUCTION

Exploration groups are now looking for, discovering, and mining or planning to mine almost every conceivable ore occurrence of gold. Platinum group metals are principally being mined in placers and large, zoned ultrabasic intrusives. Future production may also come from hydrothermally re-worked rocks near zones of deep-seated, ophiolitic intrustion or from volcanogenic/sedimentary sources with large-scale secondary overprints.

In order to evaluate these deposits, an understanding of the

nature of ore occurrence and its geochemical expression is most helpful, but we often do not have, or think we cannot have, that luxury
in exploration. Before a production decision is to be made, we must
know how a precious metal occurs, and its grade, tonnage, and
amenability to economic recovery systems. Cross checking by reliable
evaluation procedures is a must, but in the final analysis it is the
first analysis that counts.

Meaningful assays begin in the field long before the assayer produces a number. The purpose of this paper is to first identify a
series of alert points in sampling and analysis, then to critically
review analytical approaches, and finally to highlight a preferred
scheme for routine assay of Au and the Pt Group.

ORE ASSESSMENT - EN ECHELON EVALUATION CHECKPOINTS

After checking for quality of geology and meaningful, organized
surface sampling, one must give considered attention to the following
ore evaluation checkpoints:

Drilling (sample taken by the drill)

Drill Site Sampling (sample cut by man from the drill return)

Assay Sampling (procedure used by lab to further cut the sample
prior to assay)

Assaying (chemical or physical extraction of the specific noble
metal(s) sought)

Engineering (projection of geological sample logs and assay values
into the earth and extrapolation of this data into a mine plan)

Feasibility Studies (mining and metallurgical recovery,
environmental, and economics)

The whole of an ore assessment program can be deeply affected by
superficial treatment of any of the above areas of endeavor. Pitfalls
in establishing ore grade and its mineable potential are many.
Correct assessment is predicated on systematic, thorough application of solid fundamentals - less on high tech schemes of
evaluation.

Cognizance of Pitfalls - by Illustration

Awareness of areas of pitfall is of great help. The following
extracts from case histories illumine focal points for trouble.

The Drilling Pitfall - A property in Montana was drilled by two
groups. Group I used down hole hammer and encountered an erratic 4.2

ppm Au over 16 m, using Lab A. Group II used dry reverse circulation and measured 2.7 ppm Au over 20 m, using Lab B.

The spotlight was first on the assay labs. However, cross check of pulps and rejects substantiated two general ranges of values. Ore had been encountered in silica-rich blobs and stringers in a quartz--carbonate horizon surrounded by low value zones of intense alteration. Each group respectfully challenged the other's drilling procedures.

Were the down hole hammer results biased by preferential capture of hard, coarser siliceous material? Or were lower value alteration products subtracted into dust or water running down the road?

Were the reverse circulation results diluted by soft gouge sucked into the hole, or was there insufficient air pressure to bring coarse, free-milling gold to the surface?

Old workings allowed accessibility to some of the holes. The area surrounding the void created by drill penetration was bulk sampled and assayed, screening for metallics. These assays and later work revealed a continuous horizon grading 14 ppm Au over 6 m and < 1.3 ppm over 14 m.

Another example of drill sampling bias happened during exploration of a Basin and Range property conducted by a major Au company. A rotary drill chewed through several tens of meters containing free Au of sub-economic grade. A flatly dipping siliceous ledge was encountered and assay values for the ledge interval intersections were of economic interest.

It was later found that as the rotary bit went through the softer Au-bearing rock that small amounts of Au were being concentrated in the hole until the harder ledge was encountered. Then the concentrated residuum along with true values in the ledge were sampled as one interval. This drill sampling problem cost the company in excess of one million dollars.

These two case history excerpts resulted in biased samples submitted for assay. The solution lies in an understanding of the nature of potential ore occurrence, and of drilling systems being used - plus on-site alertness to problem potential.

The Drill-Site Sampling Pitfall - Drill-site sampling is considered important by everyone. However, this author has observed geologists taken right out of the play as necessities stole their presence or their mind from drilling oversight.

As a consequence, boys have been seen sampling by inserting a shovel into a muck pile. Expensive dry splitters have been used by drillers so as to overrule bias considerations in favor of maximizing footage per day. Wet sludge has been decanted in favor of a coarse, heavy fraction that would dry more easily. Mineralization has been missed because core was never assayed. Extremely small splits were taken because of the size of the bags on hand.

In each of these cases, human factors distorted methodical science or dangerously assumed ore continuity. Solution of the problem lies in thoughtful preparation and dedicated alertness. Representative assays can only come from representative samples.

The Assay Sampling Pitfall - All-important assay values are based on a "critical mass" selected by the assayer, usually less than 1% of the sample provided by drill site sampling. The mathematics of samp- ling theory and resultant computer software is highly developed. By performing sampling studies, one can determine a "critical mass" for every step of sample reduction. However, there are complications to perfect assay sampling:

1. There may be major variables from one sample to the next, and

2. Exploration and production efficiencies usually do not allow for the time, money, and after the fact intrusion of sampling studies.

Consequently, empirical practicality and input from sampling theory need be blended into a routine with flexible turnouts that allow for a concomitant check program. As in discussion of other areas of pitfall, scientifically descriptive assay sampling is based on a logical understanding of the nature of ore occurrence and a methodical scheme of treatment. Often, a geologist may adequately undertake his site work, but overlook or fail to communicate with his lab in the area of assay sampling prior to assay. Below are more cases-in-point.

1. A placer concentrate was treated as a rock. A 4 kilogram sample was split (error #1) into a 0.5 kilogram sample. The result- ant split was pulverized (error #2). Some of the noble metal was smeared onto grinding surfaces.

Result: a. Non-representative sample
            b. Values low-biased
Solution: Follow alternate procedure (tabling + amalgamation + Cyanide) suitable for placer cons. No pulverizing.

2. Crushing and grinding equipment was not properly cleaned between sample batches. A virgin exploration project was processed after high grade material. Action (staking) was taken based on contaminated pulps (pulps were checked, but not rejects).

Result: High-bias contamination
Solution: Good clean-up procedures. Check rejects.

3. Routine splits of 250 gm of -180 mm (-10 mesh)were taken and

assayed without further check for several months. Checks on reject material were finally taken during "final" assessment of the project and values did not check.

Result: Necessity of a complete re-evaluation and much wasted effort.
Solution: Early check of reject splits and consequent custom sub-sampling prior to assay split.

The Assay Pitfall - In consideration of assaying pitfalls, these abstracted cases alert us to potential error points:

1. A ½ assay ton notation was not recognized by an assayer. A .1 reading was entered as a 1 by a computer operator.

Result: Value X 5 high bias.
Solution: Supervisory check for human error.

2. An auto sampler/instrument was programmed to by-pass human error, but an intake progressively clogged between standards and this was not picked up by automated printout.

Result: Some values were too low.
Solution: Recognition/correction of program input/output signals. Monitoring of fail-safe checkpoints.

3. A 10 gm sample was leached and read instrumentally by a competent, but geochemically inexperienced chemist who did not remove major elements from the leach solution.

Result: Au and Pt values were calculated to be high.
Solution: Recognition of interferences in method used.

4. Fire assay values on a carefully homogenized pulp were 30% higher at lab 1 than by Cyanide or acid leach at two other labs.

Result: Lower values were taken as accurate because two name labs agreed, although the Pb fusion (fire assay) values were later proven correct.
Solution: Choice of a total metal procedure.

5. Platiniferous samples were "fire-assayed" for Au and Pt by two labs. Lab 1 ran both elements by atomic absorption off of a silver bead dissolved in aqua regia and diluted with water. Lab 2 ran the Au in like manner, but also corrected for matrix; Pt was fused and analyzed separately using additives specifically designed for Pt collection during fusion and its isolation during measurement. Later check proved higher Au values to be 10% high (Lab 1), while higher Pt values (Lab 1) were up to 20% low.

Result: Only partially reliable assays at Lab 1.
Solution: Matrix correction for Au; use of appropriate

collectors/buffers during Pb Fusion/instrumental
measurement for ore grade Pt.

The Engineering Pitfall & The Feasibility Pitfall are very real. As
they span a second phase of ore assessment, this paper will not
discuss them except to note that the most common areas of error are:

1. Assumption of ore continuity between data points.

2. Failure to cut high grade intercepts or allow for practical
mining dilution factors.

3. Assumption of bulk leach recoveries based on pilot samples
that may represent the ore   not the heap.

The above examples illustrate pitfalls. What, then are the ways
to the right answer of assessing ores for Au or the Pt group?

A SIMPLIFIED FUNDAMENTALS CHECKLIST

Sampling

Respecting Drill Sampling & Drill Site Sampling, note these positive
check points:

1. Do your homework and be prepared.

2. Be well organized in advance.

3. Watch what is happening; keep good chip logs.

4. Personally oversee splitting and bagging of the sample.

When it comes to Asssay Sampling:
For Lodes -

1. Ensure that your laboratory:
    A. Crushes the entire sample to at least -1.80 mm (-10 mesh)
        prior to splitting.
    B. Retains sample integrity when using fine grinding
        equipment.

2. Take additional reject splits once ore is encountered, and
quickly have them checked.

3. If ore zones prove to be erratic, conduct sampling studies to
better document mineral type and particle size. Use this data to determine the optimum, practical sample size and mesh size to be used.

4. Employ custom sub-sampling if required.

5.   If coarse -0.15 mm (-100 mesh) native metal is encountered, screen out metallics and assay separately, using a procedure that selectively releases the majority of metallics.

For Placers - make a careful gravity concentrate.   Use fluid convection systems for extremely fine $-.075$ mm (200 mesh) noble metal recovery.   Treat the entire sample.

The following <u>Analytical Flow Chart</u> (Fig. 1) for Au by J.P. Saheurs of Bondar-Clegg, with minor input from this author, illumines a scheme of approach to assay sampling and consequent analytical procedure.   The sample preparation section is also applicable to the Platinum Group Elements.

Assaying

After the <u>Analytical Flow Chart</u>, Table 1 outlines common methodology and technique for Au, noting salient features; Table 2 outlines the same for the Pt Group.

ANALYTICAL FLOW-CHART

Fig. 1

Copyright Bondar-
Clegg & Co., 1987

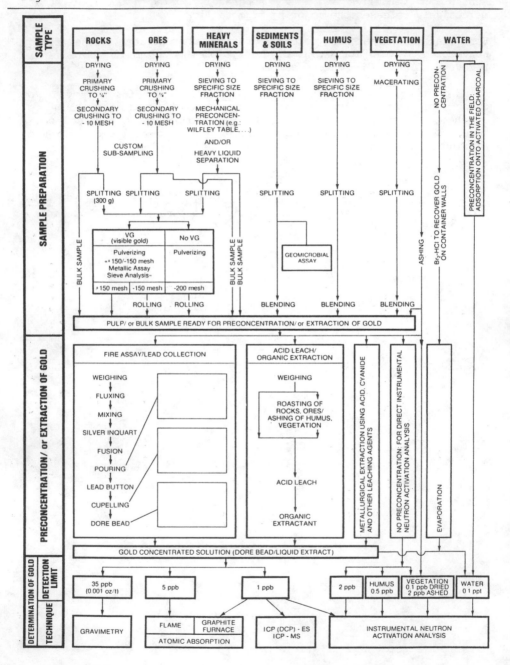

Table 1
METHODOLOGY & TECHNIQUE — AU

Applicable Method	Notes
(a) Pb Fusion (fire assay)	Gravimetric, Atomic Absorption, Inductively Coupled Plasma ($\pm$ Atomic Fluorescence), or Neutron Activation endpoint.
	Requires: Individual approach, close monitoring.
	Good for all geological samples — total metal from any known earth material.
(b) Acid Leach & Extraction	Atomic Absorption or Inductively Coupled Plasma ($\pm$ MS/AFS) endpoint. Matrix problems must be eliminated or corrected for. Technique important. Less confidence overall for total Au.
(c) Amalgamation	Gravimetric endpoint. Particle cleanness/encapsulation may hinder efficiency. Tails can be leached by cyanide.
(d) Cyanidation	Gravimetric or instrumental end-point. Problems from slimes, Cu Zn depression, etc. Expected recoveries may be approximated.
(e) Direct Neutron Activation	Light element interferences must be overcome. Particulate Au may not report quantitatively.
(f) Mass Spec	Too cumbersome for routine work
(g) Charged Particle	5–500 ppb detection limit; some interferences.
(h) Emission Spectrography X-Ray Fluorescence Others	Routine accuracy/detection limit difficult to achieve. May have special applications

Table 2

METHODOLOGY & TECHNIQUE — PT GROUP

Applicable Method	Notes
(a) NiS Fusion	Neutron Activation or other instrumental endpoint on 50 gm sample.
	Slower and more expensive.
	All 6 Pt metals.
	After 5 years of use, the method still has problems and is undergoing further development.
(b) Pb Fusion	Gravimetric, Atomic Absorption, Inductively Coupled Plasma ($\pm$ MS/AFS), or Neutron Activation endpoint.
	Good for all geological samples.
	Requires: Individual approach and close monitoring.
	Pt, Pd, Rh reliable if appropriate collectors/buffers are used.
(c) SnO_2 Fusion	Gravimetric or instrumental endpoint.
	Requires hotter fusion.
	All 6 Pt metals (in research)
(d) Acid Leach Emission Spectrography Direct Neutron Activation Whole Rock Fusion Dry Clorination	Questionable except for custom work, or as a step in assay of high grade ores. Many problems; complicated interferences.
(e) Charged Particle	Uses small sample (0.3 mm layer)
	Some interferences.
	Especially good for Os, Ru.

In the realm of our experience, we would offer the following preferred scheme for routine evaluation:

I. Gold

Lode Ores

A. Assay of multiple splits rather than a single larger sample
because one can monitor homogeniety and assess values being
averaged.

B. Pb Fusion (fire assay). Atomic absorption or neutron
activation endpoint for values of less than 5 ppm.
Gravimetric endpoint for values greater than 5 ppm.

C. Acid leach or cyanide leach for "leach specific" Au
extraction.

Placers

A. Gravity concentration, amalgamation, and cyanide leach of
almagamation tails. Fire-assay free gold concentrates.

B. If assay of cyanide tails demonstrates appreciable values,
identify mode of occurence, and if warranted design a tails
beneficiation circuit.

II. The Platinum Group Elements

Lode Ores

A. Platinum, Palladium, Rhodium - Pb fusion with appropriate
collectors/buffers.

B. Osmium, Iridium, Ruthenium - Ore grade material may be
assayed by wet chemistry after custom leach and fusion.

C. All six Pt group elements - NiS fusion with an instrumental
endpoint shows excellent promise pending method
modification.

Placers

A. Gravity concentration; assay concentrates as above.

B. If tails contain appreciable values, identify mode of
occurence, and if warranted design a tails beneficiation
(normally grinding, flotation, and acid leach) circuit.

CONCLUSION

In this paper we have reviewed the points in sampling and routine analysis where trouble is most likely to occur. In both the field and the lab, a perceptive understanding of the nature of ore occurence is the first ingredient to the right answer in assessing ore for Au or the Pt group. Good organization and control are the framework for systematic, thorough sample treatment.

High tech schemes can often provide welcome resolution to specific roadblocks in noble metals evaluation. Such schemes may then be involved in the ore assessment process. However, each technological application is tied to, and its answers are no better than, the accompanying fundamentals of sampling and assay.

Alert, intelligent application of solid fundamentals can make the difference between a geologist who is in command of his ore and one who is constantly backtracking in his search for the right answers.

While no revolutionary processes have been presented in this paper, we trust some useful information has been gleaned that will catalyze the reader's own thinking, leading to confidence in his or her work and the service groups chosen. Always maintain an ongoing alertness and commitment to thoroughness. Right answers for Au and the Pt group will be the reward.

SELECTED BIBLIOGRAPHY

Beamish, F.E., and Van Loon, J.C., 1977, Analysis of Noble Metals, Academic Press, p. 344.

Boyle, R.W., 1979, The Geochemistry of Gold and its Deposits (together with a chapter on geochemical prospecting for the element), Geol. Surv. Can. Bull. 280, 584 pp.

Buchanan, L.J., 1981, "Precious Metal Deposits Associated With Volcanic Environments of the Southwest," Relations of Tectonics to Ore Deposits, Arizona Geol. Soc. Digest, Vol. 24, pp. 237-262.

Bugbee, E.E., 1940, A Textbook of Fire Assaying, Colorado School of Mines Press, Golden, p 314.

Bumstead, E.D., 1984. " Some Comments on the Precision and Accuracy of Gold Analysis in Exploration." Proceedings Australas. Inst. Min. Metall., no. 289: pp. 71-78.

Burn, R.G., 1984. " Factors Affecting the Selection of Methods of Gold Analysis." Min. Mag. May 1984: pp. 468-475.

Cabri, Louis J., 1986, "Factors Affecting the Recovery of PGE's," in the seminar, "What Canadian Exploration Geologists Should Know About Platinum," sponsored by the Toronto Geological Discussion Group, 13 Nov.

Clifton, H.E., Hubert, A.E. and Phillips, R.L., 1967. Marine sediment sample preparation for analysis for low concentrations of fine detrital gold. U.S. Geol. Surv. Circ. 545.

Clifton, H.E., Hunter, R.E., Swanson, F.J. and Phillips, R.L., 1969. Sample size and meaningful gold analysis. U.S. Geol. Surv. Prof. Paper 625-C.

Conn, H.V., 1979, "The Johns-Manville Platinum-Palladium Prospect, Stillwater Complex, Montana, U.S..," The Canadian Mineralogist, Vol. 17, pp. 463-468.

Cook, D.J. and Rao, P.D., 1973. " Distribution Analysis and Recovery of Fine Gold from Alluvial Deposits." Univ. Alaska, Fairbanks, AK, Mineral Industry Res. Lab. Rep. 32.

_____ , 1982, Drilling & Drilling Management, procedings of Northwest Mining Association short course, Nov. 30-Dec. 2, 1981.

Garrett, R.G., 1983. " Sampling methodology." In: R.J. Howarth (ed) Statistics and Data Analysis in Geochemical Prospecting, Handbook of Exploration Geochemistry 2, Elsevier, Amsterdam, pp. 83-110.

Giusti, L., 1986. "The Morphology, Mineralogy, and Behavior of 'fine-grained' Gold from Placer Deposits of Alberta: Sampling and Implication for Mineral Exploration." Can. J. Earth Sci., Vol. 23, pp. 1662-1672.

Glover, J.E. and Groves, D.I. (eds), 1979. Gold Mineralization. University Extension, University of Western Australia, 106 pp.

Hafftig, J., Haubert, A.W., and Page, N.J., 1980, "Determination of Iridium & Ruthenium in Geological Samples by Fire Assay and Emission Spectrography," U.S. Geol. Surv. Prof. Paper 1129 G, 4 p.

Haffty, J., Riley, L.B. and Goss, W.D., 1977. "A Manual on Fire Assaying and Determination of the Noble Metals in Geological Materials," U.S. Geol. Surv. Bull. 1445.

Harris, J.F., "Sampling and Analytical Requirements for Effective Use of Geochemistry in Exploration for Gold." In: A.A. Levinson (ed), Precious Metals in the Northern Cordillera. Proceedings of a symposium held April 13-15, 1981, in Vancouver, B.C. The Assoc. of Expl. Geochemists, Special Publication no. 10, Rexdale, Ontario, 214 pp., pp. 53-67.

Henley, K.J., 1975, "Gold-ore Mineralogy and its Relation to

Metallurgical Treatment," Minerals Sci. Engng. 7:289-312.

Krason, Jan, 1985, "The Redox Interface Precious Metal Deposits –
Exploration Objectives," in the symposium "Organics & Ore Deposits,"
Denver Region Exploration Geologists, April.

Macdonald, James, 1987, "The Platinum Group Element Deposits:
Classification and Genesis," Geoscience Canada, Vol. 14, No. 3,
Sept., pp. 155-166.

Mertie, John B., Jr., 1969, "Economic Geology of the Platinum
Metals," U.S. Geol. Sur. Prof. Paper 630, 120 pp.

Murphy, G.J., 1982, "Some Aspects of Sampling in Terms of Mineral
Exploration & Mine Geology," in Symposium on Sampling and Analysis
for the Minerals Industry, IMM (London)

Nichol, I., 1983, "Sample Representativity with Particular Reference
to Gold."   In A.S. Joyce and I. Nichol, 1983, Geochemical
Exploration.   Workshop course 232/83, Adelaide April 11-22, 1983.
Australian Mineral Foundation, Glenside, S.A., 483 pp., pp. 355-378.

Ottley, D.J., 1983.   "Calculation of Sample Requirements and the
Development of Preparation Procedures for Reliable Analysis of
Particulate Materials." Nev. Bur. Mines & Geol. Rep. 36: 132-144.

Puddehpatt, R.J., 1978, The Chemistry of Gold. Elsevier, Amsterdam,
274 pp.

Romberger, S.B., 1986. "Disseminated Gold Deposits." Geoscience
Can. 13:23-31.

Riddle, C., 1983, "Analytical Methods for Gold." In A.C. Colvine
(ed) The Geology of Gold in Ontario," Ont. Geol. Surv. Misc. Paper
110, 278 pp., pp. 272-278.

Saheurs, J.P. G. and Massie, C., 1985, "L'analyse de l'or en
Prospection Geochimique: Problemes, Solutions et Exemples." In: M.
Baumier et al. (eds), La geochimie d'exploration au Quebec –
Seminaire d'information 1985. Gouvernement du Quebec, Ministere de
I'Energie et des Ressources, DV 85-11, pp. 159-162.

Scoates, Jon, 1986, "The Bird River and Fox River Sills, Manitoba,"
in the seminar, "What Canadian Exploration Geologists Should Know
About Platinum," sponsored by the Toronto Geological Discussion
Group, 13 Nov.

Todd, S.G., Keith, D.W., and LeRoy, L.W., 1982, "The J-M
Platinum-Palladium Reef of the Stillwater Complex, Montana: I.
Stratigraphy & Petrology," Economic Geology, Vol. 77, pp. 1454-1480.

Tredoux, M., and Sellschop, J.P., 1981, "The Determination of All the

PGE and Gold in Small Bushveld Samples by Charged-Particle Activation-Analysis," presented at the Third International Platinum Symposium, Pretoria, So. Africa.

Wang, W. and Poling, G.W., 1983, "Methods for Recovering Fine Placer Gold." _CIM Bull._, Vol. 76, no. 860:47-56.

Wilson, W.E., 1982. "The Gold-containing Minerals," _Min. Record_, Vol. 13: 389-400.

Chapter 4

SHALLOW HYDROCARBON ACCUMULATIONS, MAGMATIC INTRUSIONS,
AND THE GENESIS OF CARLIN-TYPE, DISSEMINATED GOLD DEPOSITS

by Judith M. Ballantyne

Consulting Geochemist, Ballantyne Geochemistry
Salt Lake City, Utah

ABSTRACT

Carlin-type, disseminated, sediment-hosted gold deposits can be better described as hydrocarbon-associated gold deposits. A new genetic model for these deposits involves the intrusion of one or more small bodies of magma into a shallow hydrocarbon reservoir or area of hydrocarbon-enriched rocks, imposing a convecting geothermal system on the local groundwater. Pyrite, commonly associated with the organic material, provides a substrate for deposition of gold and other trace elements in an environment rich in dissolved sulfide. Both diagenetic and hydrothermal pyrite may act as gold depositional sites. Thermal cracking of hydrocarbons, catalyzed by illite, subsequently generates an oxidizing environment by depleting hydrogen. As oxidation increases, arsenic sulfide minerals and barite are stabilized at first. Increasing hydrogen depletion converts sulfide to sulfate, generating acid oxidizing conditions which: alter pyrite to hematite; continue the deposition of sulfates; and ultimately develop late stage acid-sulfate assemblages containing minerals such as kaolinite, alunite, and jarosite. Gold released from the oxidizing pyrite is redeposited as free gold, and perhaps remobilized.

INTRODUCTION

Up until now, a comprehensive genetic model for the occurrence of disseminated, sediment-hosted gold deposits has been elusive. A genetic model provides a framework for understanding the spatial distribution of ore relative to other features in a hydrothermal system, and allows development of more effective exploration strategies. A behavioral model is particularly useful because it transcends the specific features of a particular deposit. Using a behavioral model the explorationist can evaluate each deposit as a separate entity, with its distinct set of structural, lithologic and hydrologic characteristics.

Previous genetic models for sediment-hosted gold deposits, summarized by Percival et al. (1988), have noted the similarities to geothermal systems, and the importance of shallow intrusions as heat sources. Evidence for the shallow (< 2 km) depth of formation of these deposits is summarized by Bonham (1985). A genetic link between hydrocarbons and gold deposition has been proposed by Radtke and Scheiner (1970), Fournier (1985), Berger (1986), and Tafuri (1987). Tafuri (1987) proposed that the activated carbon and asphaltene at the Mercur deposit are degradation products of mobile hydrocarbons such as petroleum, which circulated in the gold-depositing hydrothermal system.

Any genetic theory for Carlin-type, disseminated gold deposits must include the following features, summarized from Percival et al. (1988):
- Carbonate and silty carbonate rocks as the predominant hosts, although shales and other rock types host some deposits.

- Carbonaceous material in parts of the deposit, and removed by oxidation from other parts.

- Intrusive rocks in the vicinity, often of unknown association to gold mineralization.

- A reduced ore zone and a hydrothermally oxidized zone, the latter sometimes, but not always, overlying the reduced zone. The reduced zone is absent in some deposits.

- Very fine grained, usually less than micron sized, disseminated gold, occurring (i) in unoxidized zones as coatings on pyrite, adsorbed on carbonaceous or organic material, and associated with arsenic sulfides, and (ii) in oxidized zones as free gold associated with quartz, clays or iron oxides.

- Combined structural and lithologic localization of ore: along near-vertical faults and fractures, and within favorable lithologies.

- Arsenic associated with the gold, usually in disseminated pyrite, and also as realgar, orpiment, and other sulfosalt minerals.

- Illite and kaolinite present in host rocks for ore.

- Extensive decarbonation in some areas.

- Barite and silica as additional introduced hydrothermal products.

THE MODEL

Geologic Setting

The geologic occurrence and the mineralogic and geochemical features of Carlin-type gold deposits can be explained by a situation in which one or more small bodies of magma intrude a shallow region of hydrocarbon-rich rock, setting in motion a hydrothermal system. Depending on the hydrologic environment at the time of intrusion, the aqueous fluid may be connate basin brine or local meteoric groundwater. The geochemical behavior of this type of gold-depositing hydrothermal system exhibits important differences from typical hot springs systems because of the nature of the host rocks and the presence of the hydrocarbons.

Hydrocarbon Accumulations

Most of the known sediment-hosted gold deposits occur in the Western U.S. (Percival et al., 1988). Structurally, many deposits occur beneath major thrust planes, exposed by erosion through the lower plate (e.g. Carlin and Cortez, Nevada), or in rocks immediately above the thrust plane (Roberts, 1960; Percival et al., 1988). Some occur in the crests or paleo-crests of anticlines or antiformal structures (Percival et al., 1988). Mercur, Utah, is a previously unrecognized example of a deposit located on the crest of an anticline. At Mercur, the apparent position of the gold deposits on the eastern flank of an anticline is an artifact of Basin and Range tilting. The gold deposits, intrusions of two ages, and the more than 12 km long, intermittent zone of jasperoid called the Silver Reef, all occur on the eastern flank of the Ophir anticline, about 1000 to 1500 m east of the anticline crest, where beds currently dip about 15° east (Tafuri, 1987). Rotating the Oquirrh mountain range back the 15° eastward tilt reported by Tafuri (1987) to be due to Basin and Range faulting, aligns all of these features along the original crest of the anticline.

The structural locations described above for the gold deposits are also those typical of hydrocarbon reservoirs. Hydrocarbons tend to accumulate in the higher portions of anticlinal and antiformal structures, in permeable reservoir rocks beneath an impermeable cap. Shales are typical cap rocks. At Mercur, the Long Trail Formation shale overlies the gold deposit and is largely unmineralized (Tafuri, 1987).

The concentration of gold deposits in eroded "windows" through the Roberts Mountain thrust in Nevada appears related to the coincidence that the topographically higher portions of a thrust plane are those beneath which hydrocarbons accumulate, and are also eroded through and exposed earlier than topographically lower regions.

One feature of hydrocarbon accumulations that is important to gold deposition is the presence of pyrite. Pyrite and organic matter are intimately related, the pyrite commonly occurring as microcrystalline aggregates termed framboids (Tissot and Welte, 1984). In kerogen concentrates the abundance of pyrite may range from 0% to more than 50% (Tissot and Welte, 1984). Pyrite at Carlin occurs as euhedral, subhedral, and framboidal crystals, associated with organic material (Radtke et al., 1980). Euhedral and framboidal diagenetic pyrite, and hydrothermal pyrite and marcasite, often associated with organic material, occur at Mercur (Jewell and Parry, 1987; Tafuri, 1987).

Another important feature of hydrocarbon accumulations is the presence of hydrogen sulfide, which promotes arsenic and gold deposition on the pyrite, as discussed later.

Intrusions

Small bodies of felsic to intermediate igneous rocks, in the form of plutons, dikes, and sills, are associated with most of the known sediment-hosted gold deposits (Percival et al., 1988). More than one stage of intrusive activity may have occurred, as in the case of Mercur, Utah. Unless evidence is presented to the contrary, it seems reasonable to ascribe the heat source driving each hydrothermal system to these igneous rocks. Modern, gold- and silica-depositing geothermal systems are normally associated with igneous activity.

At Mercur, the Porphyry Hill granodiorite, dated at 36.7 m.y. (Moore, 1973) occurs about 2 km north of the mine, while the Eagle Hill intrusive rhyolite (dated at 31.6 m.y., Moore, 1973) occurs on the southern edge of the mine (Tafuri, 1987). The rhyolite is topographically and stratigraphically higher than the ore zone, and would have intruded any hydrocarbon accumulation in the anticline crest.

The intrusive rocks in sediment-hosted deposits are variably altered, and may or may not contain gold mineralization (Percival et al., 1988). Alteration and mineralization of intrusive bodies is a function of their position within the hydrologic system relative to hydrothermal fluid conduits. At Getchell, Nevada, alteration of dikes is confined to mineralized areas, which are closely associated with faults (Berger, 1985). At Mercur, the Eagle Hill intrusive rhyolite is probably associated with the gold-depositing geothermal system, but is weakly altered and only rarely mineralized (Guenther, 1973; Shrier, 1988, personal communication). The presently exposed portions of the rhyolite may have been above the water table at the time, or separated from the hydrothermal system by an aquiclude like the Long Trail Formation shale.

Effects of Magma Intrusion into a Hydrocarbon Accumulation

The intrusion of a felsic or intermediate magma at 600° or 700°C into a hydrocarbon reservoir or hydrocarbon-rich region will produce the following results:

-   Heating of the groundwater or brine in the aquifer associated with the hydrocarbon accumulation, and development of a hydrothermal system in which fluid flow is controlled by permeable structures. The hydrothermal system developed will have similarities to fracture-dominated geothermal systems, with the greatest fluid flux occurring along faults and dilatant zones in folded rocks. Such dilatant zones are important ore controls at a number of deposits (Percival et al., 1988). The system will discharge to the surface through permeable features like fractures and breccia pipes. Hot springs-like gold deposits (see Buchanan, 1981) may be formed in these permeable structures.

-   Explosive breaching of the system may occur, if it is sealed, because of a pressure rise due to:
    -   vaporization and boiling of aqueous fluid in the vicinity,
    -   heating and expansion of the associated non-condensable gases, and
    -   production of additional gases due to reactions with carbonate rocks and hydrocarbons.
    The explosion of the reservoir is likely to take the form of gas and hydrothermal eruptions along fractures, sometimes forming breccia pipes in the process. If the reservoir had been breached prior to intrusion, gases may not escape as explosively, although hydrothermal eruptions are likely to be involved.

-   Dissolution of silica from surrounding rocks, and subsequent deposition in upper, cooler, outflow regions of the hydrothermal system. Sealing of the upper and outflow portions of the system by silica, and also by carbonate deposition, is likely.

-   High carbon dioxide gas pressures due to carbonate dissolution will maintain temperatures below 300°C (Henley, 1985), because the onset of boiling above this temperature will convert additional heat to steam. Release of non-condensable gases after rupture of the hydrothermal system may account for the higher fluid inclusion temperatures measured in oxidized assemblages at Carlin and Mercur by Radtke (1985) and Jewell (1984).

-   Onset of boiling in the upper levels of the hydrothermal system. Boiling horizons, and the vertical depth over which boiling occurs will depend on the reservoir permeability and the temperature and pressure distribution. Boiling will promote further carbonate deposition. Repeated boiling, sealing and rupture of the hydrothermal system may take place.

- Precipitation of arsenic on the pyrite that is associated with the organic material. Both diagenetic and hydrothermal pyrite may be involved.

- Adsorption of gold onto pyrite surfaces and organics. Boiling may or may not cause gold precipitation, depending on the concentrations of gold and sulfide in the fluid (Reed and Spycher, 1985).

- Thermal cracking of larger hydrocarbon molecules, forming lighter hydrocarbon and later methane and other gases as temperatures rise. This process is catalyzed by the clay minerals present, particularly the illitic minerals (Hunt, 1979). Hydrogen required by the cracking reactions may be supplied either directly or indirectly from the hydrothermal fluid.

- A rise in the oxidation state of the hydrothermal system due to the depletion of hydrogen gas as cracking reactions occur. Loss of methane and hydrogen due to boiling also increases oxidation (Reed and Spycher, 1985).

- Generation of hydrogen sulfide from the thermal cracking of hydrocarbons. Organic matter in carbonate sequences is particularly rich in sulfur (Tissot and Welte, 1984), and hence is capable of producing large amounts of hydrogen sulfide. Generation of hydrogen sulfide takes place at relatively high temperatures during late catagenesis, along with methane production (Tissot and Welte, 1984).

- The oxidation of dissolved sulfide to sulfate as dissolved hydrogen is depleted, generates the acid sulfate conditions that produce the alteration assemblage associated with the oxidized ore in these deposits. Pyrite becomes oxidized to hematite, and minerals characteristic of advanced argillic alteration, such as kaolinite, jarosite, and alunite, may develop in some areas.

- Neutralization of hydrogen ions in decarbonation reactions in regions where interaction of the acid fluid with the host rocks occurs. In some cases, armoring of fracture margins by mineral deposition or organic material may inhibit fluids from reacting with host rocks.

- Precipitation of arsenic as realgar and orpiment as conditions become increasingly oxidized due to hydrogen depletion, but still within the stability range of aqueous sulfide species (Heinrich and Eadington, 1986).

- Deposition of barite and other sulfate minerals as sulfate concentrations increase, and their solubility products are exceeded.

- Oxidation of pyrite and sulfides, releasing adsorbed elements such as gold and arsenic, which may then be re-adsorbed onto the resulting iron oxides, or transported to other depositional sites.

- Eventual destruction of hydrocarbons as oxidation progresses, particularly in the hotter parts of the deposit. Conversion to lighter, fugitive molecules, and oxidation of the remaining carbonaceous material may both contribute to removal of hydrocarbons.

DISCUSSION

Hydrocarbon occurrences in hot spring discharges are known from five areas in and near Yellowstone National Park, Wyoming (Love and Good, 1970). All are associated with hot springs and vents, and contain abundant sulfur. Four of these areas are in volcanics, but all are believed to be underlain by sedimentary rocks (Love and Good, 1970). The authors conclude that the hydrocarbons are extracted from the sedimentary rocks at depth and transported to the surface by the hydrothermal solutions. These hot springs may be the surface expression of a presently forming, sediment-hosted gold deposit.

An actual hydrocarbon reservoir may not be necessary for Carlin-type, disseminated gold deposits to form, but the importance of the organic material is paramount, to provide a pyrite substrate, and the geochemical conditions necessary for gold deposition, and for later hydrothermal oxidation. Methane occurs in fluid inclusions at Carlin (C. A. Kuehn, personal communication, cited by Ilchik et al., 1986), providing evidence for catagenesis at dry gas temperatures (Tissot and Welte, 1984). Oxidation is important because it frees gold from the pyrite and makes the ore amenable to cyanide leaching.

Some deposits, such as Carlin and Mercur, contain remobilized hydrocarbons, (Radtke et al., 1980; Jewell and Parry, 1987; Tafuri, 1987), while others, such as Alligator Ridge, Nevada, do not (Ilchik et al., 1986). Heating remobilizes hydrocarbons, a fact that is utilized in enhanced oil recovery. The nature of the original hydrocarbon material and the thermal history of the hydrothermal system both affect whether or not remobilization occurs. The initial distribution of organics may be more important than that of the remobilized material. These remobilized hydrocarbons, unless associated with pyrite or arsenic minerals, may bear little relationship to ore.

Cracking of hydrocarbons may not be the only way in which oxidizing conditions are generated in the hydrothermal system. Lowering of water tables due to insufficient recharge of the hydrothermal reservoir, can produce a steam-heated zone that is oxidizing if open to the atmosphere. However, the higher temperatures recorded for fluid inclusions in acid-sulfate assemblages, and the addition of organic material to oxidized rocks at Mercur (Jewell and Parry, 1987; Tafuri, 1987), argue against oxidation in such an environment. In addition, the close parallel between the lowest hydrogen index isopleth and the lower boundary between oxidized and carbonaceous rock at Alligator Ridge, Nevada (Ilchik et al., 1986, Fig. 13), supports the relationship between cracking and oxidation.

In the Carlin and Cortez deposits the gold in unoxidized ore occurs mainly in arsenian pyrite (Wells and Mullens, 1973), with little or none in the carbonaceous material. In the oxidized ores both gold and arsenic occur with Fe oxides in pseudomorphs after pyrite. A positive correlation between gold and arsenic content of pyrite, observed at both deposits (Wells and Mullens, 1973), suggests similar behavior of gold and arsenic. At Carlin, the presence of mercury and antimony on pyrite surfaces along with the arsenic and gold (Radtke et al., 1972) suggest that they deposited at a similar time and in a similar manner to the arsenic.

The solubility of arsenic in hydrothermal fluids is a function of temperature, partial pressure of hydrogen sulfide, oxidation state and pH (Ballantyne and Moore, 1988). High concentrations of hydrogen sulfide, likely in these systems for reasons described below, reduce arsenic solubility in geothermal reservoirs, where reactions with pyrite control arsenic solubility (Ballantyne and Moore, 1988). Ballantyne and Moore (1988) proposed a paired oxidation/reduction reaction to account for the distribution of arsenic in pyrite in geothermal reservoirs. They suggested a similar type of reaction might account for the low temperature gold deposition on pyrite shown experimentally by Jean and Bancroft (1985). Such a reaction is also possible in the disseminated gold deposits. The gold may adsorb on pyrite surfaces by reduction at the surface and destabilization of the transporting complex.

Orpiment and realgar are the stable arsenic-bearing minerals under more oxidizing conditions than cause arsenic deposition on pyrite (Ballantyne and Moore, 1988; Heinrich and Eadington, 1986). At Carlin the association of orpiment and realgar with rocks enriched in hydrocarbons and with barite in veins near the bottom of the zone of oxidation (Radtke et al., 1980) is consistent with a fluid containing both sulfate and sulfide. At Mercur, Tafuri (1987) describes a disseminated sulfide suite containing pyrite, marcasite, barite, orpiment and gold, and an associated vein mineral suite with the same mineralogy except for a lack of pyrite. Jewell and Parry (1987) note the common occurrence of organic material with this assemblage. In general, arsenic concentrations limited by orpiment solubility are high, on the order of tens of mg/l at neutral pH's and temperatures of 100° to 150°C (Ballantyne and Moore, 1988). Lowered pH, due to sulfide oxidation, sharply decreases orpiment solubility, promoting deposition (see Ballantyne and Moore, 1988).

In the oxidizing environment, inflowing, gold-bearing fluids could deposit gold due to destabilization of the gold bisulfide complex, as suggested by Reed and Spycher (1985). At Carlin and Cortez, gold and arsenic are commonly associated with Fe oxides after pyrite (Wells and Mullens, 1973). In oxidized rocks at Carlin, gold is associated with quartz, clays and iron oxides (Radtke et al., 1980). An alternative hypothesis is that acid, oxidizing fluids release the previously deposited gold by destroying its substrate of pyrite and carbonaceous material. The released gold may be partly remobilized or may remain as free gold at its original site of deposition. The association of gold with silicification is not addressed in this paper.

Relationships between structure, lithology, fluid chemistry, location of magma intrusion within or adjacent to the hydrocarbon accumulation, depth, and the hydrologic system will all affect the morphology of the gold deposit produced and the distribution of alteration mineralogy and trace element chemistry within it. Overprinting of earlier features as the system waxes and wanes, or due to multiple thermal events, is to be expected.

CONCLUSIONS

The association of Carlin-type, disseminated gold deposits with hydrocarbons, in structural situations characteristic of petroleum reservoirs, suggests that the presence of the hydrocarbons is fundamental to the formation of these types of deposits. A new genetic model for these deposits involves the intrusion of one or more small bodies of magma into a shallow hydrocarbon reservoir or area of hydrocarbon-enriched rocks, imposing a convecting geothermal system on the local groundwater. The sequence of events generated in this environment can explain the mineralization and hydrothermal alteration observed in these gold deposits. Oxidation caused by thermally induced hydrocarbon cracking, frees gold that was originally deposited on pyrite, making the ore amenable to cyanide leaching.

On the scale of an individual prospect or deposit, success in exploration can be improved if exploration strategies stress the importance of understanding hydrothermal system behavior. The behavioral model described in this paper provides a new framework for interpreting sediment-hosted gold deposits. In an exploration program, relationships between gold mineralization, hydrocarbon distribution, alteration assemblages, and fluid-bearing structures, can be used to infer the paleohydrology and thermal history of the hydrothermal system. This information can then be used to develop new exploration targets for both unoxidized, and the readily leachable, oxidized ore.

Exploration for Carlin-type, sediment-hosted deposits should focus on areas where igneous intrusions occur in rocks suitable for petroleum reservoirs. The importance of utilizing the knowledge of petroleum explorationists in a mineral exploration program is emphasized.

ACKNOWLEDGEMENTS

I am grateful to geologists at the Mercur Mine, Utah, for several field trips in recent years, that initiated my interest in developing a genetic model for the hydrocarbon-associated deposits discussed in this paper. Thomas Faddies of Getty Mining Company, and Barrick Mercur Mines, Inc. geologists Tracy Shrier, Steven Kerr and Robert St. Louis have provided stimulating discussions about Mercur geology. Discussions with Lee M. Allison and James W. Bunger have been informative. The ideas presented here, however, are mine. I wish to thank Jeffrey B. Hulen and Duncan Foley for their constructive reviews of the manuscript, and Dianna Lehmann-Goodwin for her editorial comments.

REFERENCES

Ballantyne, J. M., and Moore, J. N., 1988, "Arsenic Geochemistry in Geothermal Systems," Geochimica et Cosmochimica Acta, Vol. 52, pp. 475-483.

Berger, B. R., 1985, "Geological and Geochemical Relationships at the Getchell Mine and Vicinity, Humbolt County, Nevada," Discoveries of Epithermal Precious Metal Deposits: Case Histories of Mineral Discoveries, Vol. 1, Hollister, V. F., ed., Society of Mining Engineers, A. I. M. E., pp. 51-59.

Berger, B. R., 1986, " Descriptive Model of Carbonate-Hosted Au-Ag," Mineral Deposit Models, Cox, D. P., and Singer, D. A., eds, United States Geological Survey Bulletin 1693, p. 175.

Bonham, H. F. Jr., 1985, " Characteristics of Bulk-Minable Gold-Silver Deposits in Cordilleran and Island-Arc Settings," Geological Characteristics of Sediment- and Volcanic-Hosted Disseminated Gold Deposits - Search For an Occurrence Model, Tooker, E. W., ed., United States Geological Survey Bulletin 1646, pp. 71-77.

Buchanan, L. J., 1981, "Precious Metal Deposits Associated with Volcanic Environments in the Southwest," Relation of Tectonics to Ore Deposits in the Southern Cordillera, Dickinson, W. P., and Payne, W. D., eds, Tucson, Arizona, Geological Society Digest, Vol. 14, pp. 237-262.

Fournier, R. O., 1985, "Silica Minerals as Indicators of Conditions During Gold Deposition," Geological Characteristics of Sediment- and Volcanic-Hosted Disseminated Gold Deposits - Search For an Occurrence Model, Tooker, E. W., ed.,

United States Geological Survey Bulletin 1646, pp. 15-26.

Guenther, E. M., 1973, The Geology of the Mercur Gold Camp, Utah, M.S. Thesis, University of Utah, 79 pp.

Heinrich, C. A., and Eadington, P. E., 1986, "Thermodynamic Predictions of the Hydrothermal Chemistry of Arsenic, and Their Significance for the Paragenetic Sequence of Some Cassiterite-Arsenopyrite-Base Metal Sulfide Deposits," Economic Geology, Vol. 81, pp. 511-529.

Henley, R. W., 1985, "The Geothermal Framework of Epithermal Deposits," Geology and Geochemistry of Epithermal Systems, Berger, B. R., and Bethke, P. M., eds, Reviews in Economic Geology, Vol. 2, Society of Economic Geologists, pp. 1-24.

Hunt, J. M., 1979, Petroleum Geochemistry and Geology, Freeman, San Franscisco, 617 pp.

Ilchik, R. P., et al., 1986, "Hydrothermal Maturation of Indigenous Organic Matter at the Alligator Ridge Gold Deposits, Nevada," Economic Geology, Vol. 81, pp. 113-130,

Jean, G. E., and Bancroft, G. M., 1985, " An XPS and SEM Study of Gold Deposition at Low Temperatures on Sulfide Mineral Surfaces: Concentration by Gold Adsorption/Reduction," Geochimica et Cosmochimica Acta, Vol. 49, pp. 979-987.

Jewell, P. W., 1984, "Chemical and Thermal Evolution of Hydrothermal Fluids, Mercur Gold District, Tooele County, Utah," M.S. Thesis, University of Utah, 77 pp.

Jewell, P. W., and Parry, W. T., 1987, "Geology and Hydrothermal Alteration of the Mercur Gold Deposit, Utah," Economic Geology, Vol. 82, pp. 1958-1966.

Love, J. D., and Good, J. M., 1970, "Hydrocarbons in Thermal Areas, Northwestern Wyoming," United States Geological Survey Professional Paper 644-B, 23 pp.

Moore, W. J., 1973, " A Summary of Radiometric Ages of Igneous Rocks in the Oquirrh Mountains, Northcentral Utah," Economic Geology, Vol. 68, pp. 97-101.

Percival, T. J. et al., 1988, "Physical and Chemical Features of Precious Metal Deposits Hosted by Sedimentary Rocks in the Western United States," Bulk Minable Precious Metal Deposits of the Western United States, Schafer, R. W., et al., eds, Symposium Proceedings, April, 1987, The Geological Society of Nevada, pp. 11-34.

Radtke, A. S., 1985, "Geology of the Carlin Gold Deposit, Nevada," United States Geological Survey Professional Paper 1267, 124 pp.

Radtke, A. S., and Scheiner, B. J., 1970, "Studies of Hydrothermal Gold Deposition (I). Carlin Gold Deposit, Nevada: The Role of Carbonaceous Materials in Gold Deposition," Economic Geology, Vol. 65, pp 87-102.

Radtke, A. S., et al., 1972, "Chemical Distribution of Gold and Mercury at the Carlin Deposit, Nevada," [abstract], Geological Society of America Abstracts With Programs, Vol. 4, No. 7, p. 632.

Radtke, A. S., et al., 1980, "Geology and Stable Isotope Studies of the Carlin Gold Deposit, Nevada," Economic Geology, Vol. 75, pp. 641-672.

Reed, M. H., and Spycher, N. F., 1985, "Boiling, Cooling and Oxidation in Epithermal Systems: A Numerical Modeling Approach," Geology and Geochemistry of Epithermal Systems, Berger, B. R., and Bethke, P. M., eds, Reviews in Economic Geology, Vol. 2, Society of Economic Geologists, pp. 249-272.

Roberts, R. J., 1960, "Alinements of Mining Districts in North-central Nevada: Geological Survey Research, 1960," United States Geological Survey Professional Paper 400-B, pp. B17-B19.

Rye, R. O., et al., 1974, "Stable Isotope and Lead Isotope Study of the Cortez, Nevada, Gold Deposit and Surrounding Area," Journal of Research of the United States Geological Survey, Vol. 2, No. 1, pp. 13-23.

Tafuri, W. J., 1987, Geology and Geochemistry of the Mercur Mining District, Tooele, Utah, Ph.D. Dissertation, University of Utah, 180 pp.

Tissott, B. P., and Welte, D. H., 1984, Petroleum Formation and Occurrence, Second Edition, Springer Verlag, New York, 699 pp.

Wells, J. D., and Mullens, T. E., 1973, Gold-Bearing Arsenian Pyrite Determined by Microprobe Analysis, Cortez and Carlin Gold Mines, Nevada," Economic Geology, Vol. 68, pp. 187-201.

Chapter 5

GEOLOGY AND PROCESSING PROBLEMS OF PRIMARY
GOLD DEPOSITS OF NORTHEAST BRAZIL

José Salim and T.V. Subrahmanyan

UFRN - Departamento de Geologia
59.000 - Natal(RN) Brazil

ABSTRACT

The occurence of primary stratiform and vein gold
deposits in the Precambrian rocks of Northeastern Brazil is
discussed. The stratiform deposits are related to clastic
sediments (quartzites and conglomerates). The veins are
associated with quartz and sulfides and are structurally
controlled by shear zones.

The gold was recovered for a long time by small miners.
Developmental programmes for large scale mining are now
envisaged. The recovery problems and gold economics are
also discussed.

INTRODUCTION

Gold occurences in NE Brazil (fig. 1) are known since
the last century (Williamson, 1904). Initially the
deposits were discovered and developed during 1940's.
However, major developments took place during the
prospecting of W, Nb, Ta, Li, Be and other base and
strategic metals for industrial needs during the second
world war.

Gold deposits are in the Precambrian shield and in recent
weathering covers. Primary ores are mainly gold-sulfide
quartz veins, with a few occurences in calcsilicate and

58

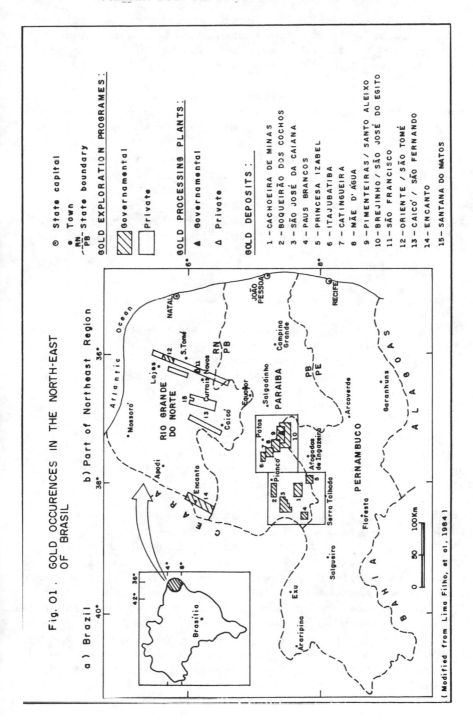

Fig. O1. GOLD OCCURENCES IN THE NORTH-EAST OF BRASIL

a) Brazil   b) Part of Northeast Region

State capital
Town
RN
PB — State boundary

GOLD EXPLORATION PROGRAMES:
Governamental
Private

GOLD PROCESSING PLANTS:
Governamental
Private

GOLD DEPOSITS:

1 — CACHOEIRA DE MINAS
2 — BOQUEIRÃO DOS COCHOS
3 — SÃO JOSÉ DA CAIANA
4 — PAUS BRANCOS
5 — PRINCESA IZABEL
6 — ITAJUBATIBA
7 — CATINGUEIRA
8 — MÃE D'ÁGUA
9 — PIMENTEIRAS / SANTO ALEIXO
10 — BREJINHO / SÃO JOSÉ DO EGITO
11 — SÃO FRANCISCO
12 — ORIENTE / SÃO TOMÉ
13 — CAICÓ / SÃO FERNANDO
14 — ENCANTO
15 — SANTANA DO MATOS

(Modified from Lima Filho, et al, 1984)

amphibolite host rocks. Old placer deposits are associated
with clastic beds of the Proterozoic schist belts. Recent
secondary gold is panned from river and colluvial placers
and residual and lateritic hard mantles.

These deposits being smaller in reserves, did   not
attract mining companies for a long time and consequently
were the targets of small scale miners. As a result of the
price of gold and technological. developments in processing
many abandoned and once considered uneconomic deposits have
come into operation with the participation of both domestic
and multi-national groups. Emphasis is also laid on
tapping new reserves and therefore, major developmental
activities are envisaged. Along with the prospecting and
mining activities also arise problems of gold processing,
in particular the treatment of gold associated with
refractory sulfide ores.

GEOLOGICAL SETTING

The regional geology shows a wide shield covered by
phanerozoic sediments named the Borborema Province (Almeida
et al, 1977). This shield comprises an Archean gneiss
basement and Proterozoic metamorphic schist belts. Both are
intruded mainly by granites and cut by shear zones and
transcurrent lineaments (fig. 2).

Among the belts, two are important in the context of
this paper: Seridó and Cachoeirinha. Their pile comprises
transitional/platformal clastic and chemical sediments,
subordinate volcanic flows, and thick orogenic turbidites
(fig. 3). Their structure is very complex with at least
three phases of folding, and metamorphism and granite
intrusions. The earliest phases develop isoclinal and
recumbent folding, thrust and transposition structures
related to tangential tectonics. The latest phases
overprint the earliest ones with normal folding, although
overturned and recumbent styles are also reported (Jardim
de Sá, 1984a). The earlier metamorphic events were linked
to a lower Proterozoic orogeny (Transamazonic Cycle, 2,1-
2,0 Gy), and the latter to a upper Proterozoic orogeny
(Brazilian Cycle, 700-500 My), as discussed by Jardim de Sá
(1984b) which conforms to a policyclic evolution for these
schist belts.

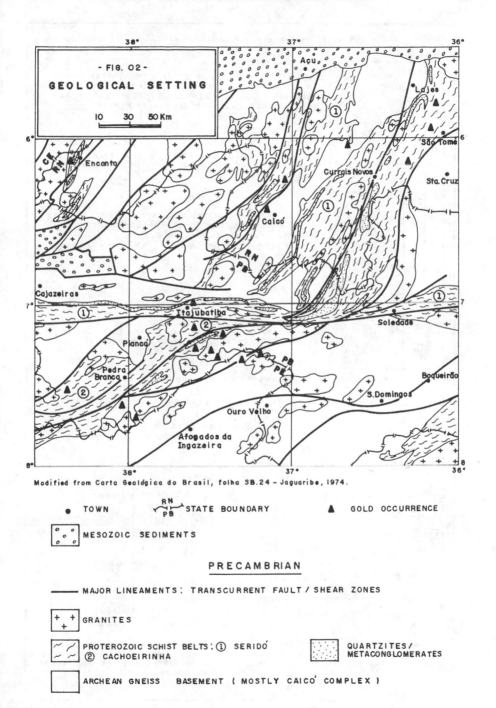

Modified from Carta Geológica do Brasil, folha SB.24 – Jaguaribe, 1974.

● TOWN   ⟍RN↑PB⟍ STATE BOUNDARY   ▲ GOLD OCCURRENCE

[o o o / o o] MESOZOIC SEDIMENTS

PRECAMBRIAN

——— MAJOR LINEAMENTS: TRANSCURRENT FAULT / SHEAR ZONES

[+ + / +] GRANITES

[∿∿] PROTEROZOIC SCHIST BELTS: ① SERIDÓ   [∴∴] QUARTZITES/
② CACHOEIRINHA   METACONGLOMERATES

[ ] ARCHEAN GNEISS BASEMENT (MOSTLY CAICÓ COMPLEX)

Fig. 03 : Regional stratigraphic setting

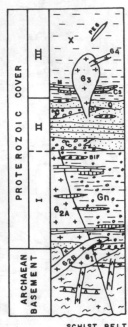

III-MAINLY ALLUMINOUS SCHIST (X) WITH INTERCALATION OF
QUARTZITES (Q), BIF, MARBLES (M) AND CALCSILICATES/
AMPHIBOLITES (Cs) INTRUDED BY GRANITE DIAPIRS
(G₃), DYKES (G₄) AND PEGMATITES (PEG).

II-QUARTZITES, QUARTZ-MUSCOVITE SCHIST AND META-
CONGLOMERATES.

I-MAINLY BIOTITE-GNEISSES (Gn) WITH INTERCALATIONS
OF SCHISTS, QUARTZITES, METACONGLOMERATES, BIF,
MARBLES, CALCSILICATES, VOLCANICS.
GRANITIZED (G₂ₐ) AND INTRUDED BY AUGEN-GNEISSE
SHEETS (G₂ᵦ).

BASEMENT : METASSEDIMENTS/VOLCANICS (∼∼) INTRUDED
BY ACID (++) AND BASIC (ʌʌ) DYKES AND TONALITES (G₁).
INTENSILY AND EXTENSIVELY GRANITIZED (∼+∼+).

SCHIST BELTS: CACHOEIRINHA (III); SERIDÓ (I + II + III).

Marbles (M)	Quartzites (Q)	BIF
Calcsilicate amphibolite (Cs)	Conglomerates (C)	Granites
Schist	Gneiss	

Fig. 04 : Secondary gold deposits

a) Recent secondary

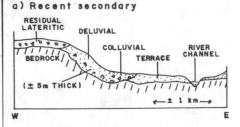

RESIDUAL
LATERITIC   DELUVIAL
                    COLLUVIAL   RIVER
BEDROCK   TERRACE   CHANNEL
(± 5m THICK)
←— ± 1 km —→
W   E

b) Old secondary

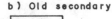

PROTEROZOIC COVER
III — SERIDÓ
II — EQUADOR
I
— SABUGI
ARCHAEAN BASEMENT

EXEMPLES OF RECENT PLACERS ARE FROM
"ENCANTO" DEPOSITS (SEE FIG.1 / Nͬ 14)
SPECIALLY THE DELUVIAL/COLLUVIAL ONES.

THE THREE COARSE CLASTIC LEVELS
OF THE SCHIST BELTS COLUMN ARE
IMPORTANT REEFS FOR PROSPECTING
OF OLD PLACERS.

GOLD DEPOSITS

Gold deposits are known for a long time and are described since the beginning of this century (Williamson, 1904; Rolf, 1946). Systematic geological prospecting also is recent (Lins & Cheid, 1981; Lima Filho et al, 1984; Lins 1984; Santos, 1984). The deposits may be classified into four types, as described below:

Recent and old Placers

Recent placers (fig. 4a) are panned from slope detrital fan and stream sands and gravels, mostly near the primary deposit sites. Gold is also recovered elsewhere from alluvial tantalite placers in streams that cut pegmatites. These deposits, however, are not rich in gold content (less than 1 ppm).

Old placers (fig. 4b) occur in coarse clastic beds of the schist belts and are not yet mined, (only the recent colluvial types). Systematic investigations are underway in these areas (Araujo et al, 1988).

Quartz Veins

Gold-sulfide quartz veins are the principal primary gold deposits that are known today. Several small occurences may be observed regionally (fig. 5) and are described as controlled by shear and fracture zones cutting the schist belts (Salim, 1984a; Souza et al, 1986).

The São Francisco Mine is one of the best examples (Melo & Souza Jr., 1986). The gold-sulfide (pyrite, chalcopyrite, arsenopyrite, galena) veins are hosted in ductile shear zones that cut biotite-schists. This mine has been a small prospect for 40 years and currently is producing 25 kgs/of gold/month since the middle of 1987. There are two kinds of ores: the oxidized and the unaltered. Presently only the oxidized ore is mined by open pits which has depths of 25m. It has proved reserves of 450.000 tons with an average gold content of 2 ppm. The unaltered ore will be explored at a later stage by underground mining methods.

Fig. 05 : Primary gold deposits in quartz veins

a) São Francisco Mine (n. 11 , fig. 01)

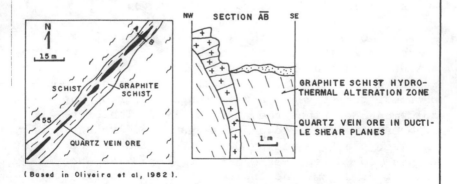

ANDALUZITE, CORDIERITE, ALMANDINE, BIOTITE SCHIST WITH MILONITIC ASPECTS (PHYLONITES).

TRACES OF FIRST TRANSPOSITION FOLIATION ($S_{1,2}$)

METAMORPHIC QUARTZ VEINS

S_3 TRACES OF SHEAR (S/C) OVERPRINTED FOLIATIONS ($S_{3,4,5}$)

HYDROTHERMAL QZ-VEIN ORE (Au AND Fe / As / Cu / Pb SULFIDES).

HYDROTHERMAL ALTERATION ZONE (OXIDATION ARGILIZATION)

SE   SECTION   NW

3 cm

b) Farias Mine (n. 01 , fig. 01)

N

15 m

SCHIST   GRAPHITE SCHIST

55

QUARTZ VEIN ORE

NW   SECTION $\overline{AB}$ SE

GRAPHITE SCHIST HYDROTHERMAL ALTERATION ZONE

QUARTZ VEIN ORE IN DUCTILE SHEAR PLANES

1 m

(Based in Oliveira et al, 1982).

GOLD CONTENT IN VEIN : < 1 TO 9,8 ppm ; AVERAGE : 3 ppm

Gold in Calcsilicate Rocks

These are main host rocks of scheelite and molybdenite of the Northeastern Brazil Tungsten Province(Salim, 1984b). However some deposits have also gold associations in the calcsilicate sequence, as in the Itajubatiba prospect(fig. 6a). This gold association is poorly known and needs special attention because the Tungsten Province has about 20,000 km^2 and more than 700 recorded scheelite occurences.

MINING, PROCESSING AND ECONOMIC ASPECTS

Small Prospects

In small mining districts panning is the traditional method for gold recovery. The dry climates of North-East Brazil restricted the use of this method to placers of some abundant heavy minerals (Tantalite and Cassiterite). Thus gold was a secondary target and was recovered as a by-product from these alluvial prospects. Now with attractive gold prices increasing interest is evinced in gold placers.

The gold recovered by small miners mostly comes from the primary associations (mainly the quartz veins). The ore is mined and processed by hand methods/or partly mechanized. After crushing and grinding the ore is subjected to gravity separation in a 12-15º inclined wooden box where the inclined surface is reinforced with a jute cloth.   The particles retained on the cloth are collected and panned to obtain a concentrate which is subsequently recovered   by amalgamation. The average gold recovery varies from 40-60%.

Organized Mining

Gold mining activity is now being organized in the region. The São Francisco Mine treats about 20.000/mo. of secondary oxided ore with a production of 25kg of gold/mo. The processing involves the cyanidation followed by carbon adsorption. Future expansion plans include the processing of the oxidized gold ore. The viability of this type of ore is to be tested by CPRM (a governamental agency) in the Itapetim Mine (N. 10, fig. 1). The plant consists essentially of a gravity circuit with facilities to carry

FIG. 06 : PRIMARY GOLD DEPOSITS IN CALCSILICATE ASSOCIATION

a) Section in Itajubatiba Mine (N⁰ 6, Fig. 01)

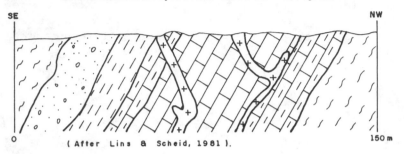

(After Lins & Scheid, 1981).

`+ +` `+`	GRANITIC PEGMATITE.
	DOLOMITIC MARBLES.
	CALCSILICATES AND AMPHIBOLITES. THE GOLD ORE, SPECIALLY THE MORE AMPHIBOLITIC TYPES.
	GRAPHITIC SCHISTS (GARNET AND SILLIMANITE),WITH PYRITE AND CHALCOPYRITE.
	GARNET / AMPHIBOL QUARTZITES.
	GNEISSES, WITH QUARTZITES INTERCALATIONS.

This sequence seems to the transitional red bed / carbonate / evaporite sequence (Sabkha Model of Renfro, 1974).

b) SABKHA SEQUENCE

	EVAPORITE.
	DOLOMITIC MARBLES.
	METALLIFEROUS.
	REDUCED PELITES.
	OXIDED CLASTICS.

c) METAMORPHIC RESULTING

? ?	(BARITE BEDS).
	DOLOMITIC MARBLES.
	CALCSILICATE /AMPHIBOLITES.
	GRAPHITE / PYRITE SCHISTS.
	QUARTZITES.

out pilot scale leaching tests. Besides, the complex ores associated with several sulfides are under investigation to develop suitable techniques for improving the recoveries (Souza et al, 1985).

In developing Countries where the environmental controls are not rigid, the methods to adopt are the traditional processing techniques which are both cost effective and flexible. As seen from the gold production figures (Tables I and II), a major contribution comes from placer mining districts, where the methods are essentially gravity and amalgamation. The Brazilian Government estimates that during the last 11 years there is a loss of US$ 26,91 billion from the following: 317 tons by illegal trading, 1.038 tons as losses in mining, and 519 tons as losses in beneficiation (Gomes, 1988). It is expected that by organized gold mining and processing, these losses can be avoided and may even bring in revenues in

ECONOMIC ASPECTS

The domestic gold production during the last 20 years (Table I) shows an increase only from 1980, before which period it remained stagnant or even a decrease during the 1970's. It is interesting to note that the industrial participation dominated until the 1980 and is surpassed by production from small miners working in placer prospects predominantly in the Amazon region (Table II). The small miners gold production, as per the government's projections represents only 20% of the real production and considerable quantities are illegally smuggled out (about 50 tons or US$ 650 million in 1987).

CONCLUSIONS

The North-East Brazil gold production mostly comes from old occurences and the prospecting work also has been dedicated to the economic evaluation of the same old deposits. As seen, the possibility of new gold discoveries are promising. Thus the region needs systematic studies to define clearly the geological guides and controls for the gold deposits.

The lack of proper management and organizational skills

TABLE I - DOMESTIC PRODUCTION OF GOLD(Kgs), 1966-1990

Year	Companies	Small Prospects	Total	Company Participation
1966	6.142	1.260	7.402	83.0%
1976	3.718	1.204	4.922	75.0%
1986	9.348	14.776	24.124	38.8%
1987	13.142	22.659	35.801	36.7%
1988(*)	26.000	-	-	-
1989(*)	30.000	-	-	-
1990(*)	34.000	-	-	-

Source: DNPM, AFTER GOMES(1988)
(*) Projection.

The gold production of North-East Brazil is smaller (about 2,5 tons/year) in comparison to other regions in Brazil, particularly the Amazon region (Table II).

TABLE II - STATE WISE GOLD CONCENTRATE PRODUCTION(Kgs) FROM BRAZILIAN SMALL PROSPECTS, 1987.

STATES	AMAZON REGION	OTHER REGIONS
Pará	13.346	-
Mato Grosso	4.972	-
Rondônia	3.902	-
Amapá	528	-
Roraima	448	-
Goiás	-	423
Amazonas	276	-
Maranhão	-	123
Outros	-	1.257
T O T A L	23.372	1.703

Source: DNPM, AFTER GOMES(1988).

with respect to gold mining and processing lead to large
loss of gold and revenues (U$ 26,9 billon or 1/4 of
Brazilian external debt!). It is very important, therefore,
to systematically organize prospecting for new reserves,
mining and processing of proven reserves.

REFERENCES

Almeida, F.F.M. et al., 1977, "Províncias Estruturais Bra-
    sileiras", Atas do VIII Simp. Geol. Nordeste, pp. 363-
    391 (portuguese).

Araújo, J., Salim, J. and Melo Jr., G., 1988, "Prospecção
    para ouro e outros metais em Lajes e São Tomé(RN)".
    XXXV Cong. Bras. Geol., (to be published) (portuguese).

Gomes, D.R., 1988, "Produção de ouro volta a crescer", Bra-
    sil Mineral, março, 52, pp. 15-20 (portuguese).

Jardim de Sá, E.F., 1984a, "Geologia da Região do Seridó:
    Reavaliação de dados", Atas do XI Simp. Geol. Nordeste,
    pp. 278-296 (portuguese).

Jardim de Sá, E.F., 1984b, "A evolução proterozóica da Pro-
    víncia Borborema", Atas do XI Simp. Geol. Nordeste: pp.
    297-316 (portuguese).

Lima Filho, C.A., Oliveira, J.L. and Brito, A.L.F., 1984,
    "Mineralizações auríferas no Nordeste Oriental", DNPM,
    série Geologia, 4, pp. 198-231 (portuguese).

Lins, C.A.C. and Cheid, C., 1981, "Projeto Ouro de Pernambu
    co e Paraíba", CPRM, unpublshed report (portuguese).

Lins, C.A.C., 1982, "Mineralização Aurífera de Itajubatiba
    (PB) - Caracterização geológica preliminar", Anais do
    XXXII Cong. bras. Geol., 3, pp. 886-889 (portuguese).

Lins, C.A.C., 1984, "Mineralizações Auríferas nos Estados
    de PE, PB, RN", Atas do XI Simp. Geol. Nordeste, pp.
    452-464 (portuguese).

Melo, J.F. and Souza Jr., E., 1987, "Geologia da Faixa Aurífera de São
    Francisco, Currais Novos(RN)". Unpublished report. (portuguese).

Oliveira, J.L., 1982, "Projeto Ouro Vale do Piancó-PB". CDRM, Unpublished report, pp. 125 (portuguese).

Renfro, A.R., 1974, "Genesis of evaporite-associated stratiform metalliferous deposits: a sabkha process", Economic Geology, 69, pp. 33-45.

Rolf, P.A.M.A., 1946, "Bismuto, Cobre e Ouro na Borborema", DNPM/DFPM, avulso nº 75, pp. 36 (portuguese).

Salim, J., 1984a, "Ouro nos micaxistos do Complexo Seridó (RN-PB)", XI Simp. Geol. Nordeste, Resumos (portuguese).

Salim, J., 1984b, "Origem das mineralizações de tungstênio no Seridó(RN-PB): fatos e especulações", Ciências   da Terra, 10, pp. 13-16 (portuguese).

Santos, E.J. et al., 1984, "Geologia e pesquisa do Distrito Aurífero de Itapetim(PE)", Atas do XI Simp. Geol. Nordes te, Bol. 9, pp. 465-473 (portuguese).

Souza, Z.S., Subrahmanyam, T.V., and Aguiar, A.P., 1985, "Caracterização do Minério de Ouro da Região de Lajes (RN)", XI Encontro Nacional Tratamento Minérios Hidrome talurgia, vol.1 pp. 225-234 (portuguese)

Souza, Z.C., Araújo, J.G. and Carvalho, O.O., 1986, "Geolo gia e Controle de Mineralizações Auríferas entre Lajes e São tomé, região do Seridó(RN): tópicos preliminares", Atas do XI Simp. Geol. Nordeste, Bol. 9, pp. 169-182, (portuguese).

Williamson, E., 1904, "Geologia das Regiões Auríferas da Paraíba e de Pernambuco", Rev. Inst. Arch. Geog. PE, 11, pp. 110-118.

Chapter 6

GEOSTATISTICAL ESTIMATION OF GOLD RESERVES: A CASE STUDY

by Dominique Francois-Bongarcon, PhD

Head of Geomathematical Services
City Resources (Canada) Limited

ABSTRACT

This paper presents a step by step vulgarized review of
the applications of geostatistics to the estimation of the
mineable, recoverable reserves of the Cinola gold deposit
in British Columbia.

INTRODUCTION

The Cinola gold deposit is located in the Queen
Charlotte Islands, British Columbia (Figure 1). It is
presently owned by City Resources (Canada) Limited,
Vancouver, B.C. The deposit has been known to exist since
1971 and a considerable amount of exploration data has
been gathered through the years. Today, at the final
feasibility stage, as much as 37,600 m of drilling
information is available as well as data from mapping and
sampling 460 m of underground workings. As to the size of
the Cinola deposit, it ranks as one of the 25 largest gold
deposits presently available in the world with a total of
approximately 100 tonnes of gold contained "in-situ", more
than one half recoverable by an open pit mining operation.

The Cinola reserve study constitutes a typical
methodological case study for geostatistical estimation of
gold reserves of classical hot spring epithermal gold
deposits. The approach to the estimation of its reserves,
as presented here, draws from modelling methodologies
recognized as standard geostatistical methods today in the
North American gold mining industry. The concepts used
are simple to understand and implement accounting directly
for the specific geology of the deposit. It is not
believed however, that a usable, realistic grade model of
selective mining reserves, suitable for such local

71

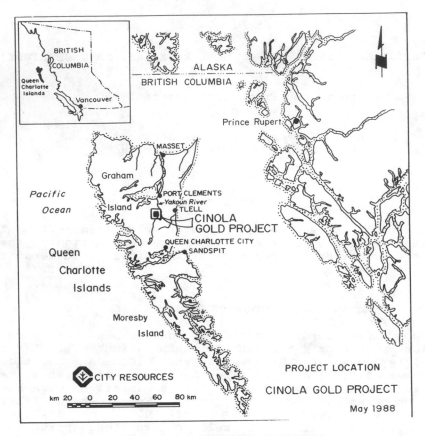

BRITISH COLUMBIA

Queen Charlotte Islands

Vancouver

ALASKA
BRITISH COLUMBIA

Prince Rupert

MASSET

Graham

Pacific

Ocean

PORT CLEMENTS
Yakoun River
TLELL

Island

CINOLA
GOLD PROJECT

QUEEN CHARLOTTE CITY
SANDSPIT

Queen

Charlotte

Islands

Moresby

Island

CITY RESOURCES

km 20   0   20   40   60   80 km

PROJECT LOCATION

CINOLA GOLD PROJECT

May 1988

Figure 1   DEPOSIT LOCATION

calculations as open pit outline determination, mine
scheduling, and grade control can be achieved using non-
geostatistical methods. (Hand-calculated, cross sectional
reserves, however, were built as well at Cinola, as a
means of double checking some of the results of the
geostatistical modelling procedure.)

METHODOLOGICAL STEPS OF THE GEOSTATISTICAL STUDY
OF THE RESERVES OF THE CINOLA GOLD DEPOSIT

The final mineable reserve statements of the Cinola
Gold Project have been derived from a computerized grade
block model developed by use of a three step modelling
procedure.

In the first step, the geology was studied in close collaboration with the project geologists. The geological features controlling the mineralization such as events of mineralization, and modes and mechanisms of metal precipitation were then coded into the block model. Application of these geological features created a series of gold grade populations and statistically homogeneous zones.

In the second step, the gold grades were projected into 30 m x 30 m x 6 m blocks using geostatistical methods and the above mentioned geological coding. This "big block" model was then processed to derive global tonnage-grade curves of the in-situ reserves (at the level of the entire deposit) recoverable by a mining operation including selection against cutoff grade of mining units much smaller than the 30 m x 30 m x 6 m blocks.

This model was compared to hand-calculated reserves, as a double check, before being used in the third step. The third step being the completion of the final geostatistical model of local, recoverable, and finally mineable, reserves and tonnage-grade curves, assuming a given optimal pit outline, a calculated mining cutoff grade and a 5 m average spacing for the grade control blasthole assays.

The next sections present a step by step description of these successive tasks, outlining the methodology that was followed rather than the computational details and results.

GEOLOGICAL CONTROLS OF THE MINERALIZATION

The validity of each single step of the modelling procedure conditions the validity of all further steps. This gives a critical importance to the very first step of the work: the analysis of the geological theory of the deposit which is relevant only if done in terms of modelling parameters.

General Geology of the Cinola Deposit

The details of the deposit geology are described elsewhere. In short, the deposit genesis is believed to have followed the following phases: fault scarp from tectonic uplift along the Specogna Fault, which forms the western boundary of the deposit; alluvial fan from erosion of the scarp; intrusion of the sediments by a highly fluidized rhyolite unit ascending along the Specogna

Fault.   This intrusion initiated the movement of meteoric
water   and   the   development   of   a   hot-spring system,
including vertical (fault and fractures) and lateral
(along   and   through   sediment   beds)   migration   of
hydrothermal   fluids.   A   complex   series   of   metal
deposition, silica sealing, and explosive eruptions events
have progressively increased the quantity of precipitated
gold in a disseminated, pervasive form as well as in the
form of highly mineralized fracture-related quartz veins,
and in the form of a large hydrothermal breccia unit
dipping along the Specogna fault.

Controls of the Gold Mineralization

    As   a   result   of   all   the   above, and as in most gold
deposits, the host lithologies themselves bear very little
weight in controlling the grades and are thus not relevant
to the modelling methodology.

    Conversely, the genesis of the orebody gives clear
indications that there exist several populations of gold
values in the data, and probably more than one zone to
consider and model separately.

    The present step consisted of two parallel tasks: close
examination   of   the   geological   interpretation with
correlated gold assay values in cross sections, and basic
analysis of the data using simple statistical and
geostatistical tools such as cumulative frequency plots on
log-probability   paper,   histograms,   coefficients   of
variation and variograms.   These tasks were complemented
with a similar analysis of the underground workings, which
played an invaluable role at Cinola in understanding the
geometrical features of the highest grade population (the
gold from the highly mineralized veins and fractures).

    It   should   be   emphasized   here   that   these   basic
statistical tools are no panacea.   They help guide the
visual observation (and vice-versa) but may sometimes be
inconclusive by themselves. At that extremely
interpretative stage where most methodological decisions
have to be made, the recipe is to use every relevant bit
of information, numerical or qualitative, to pin down the
structure   of   the   geological   controls   of   the
mineralization.

    At Cinola, this resulted in the definition of two main
populations of grades (a pervasive background and a
discontinuous, vein controlled, high grade population) and
three homogeneous zones or "sub-deposits" (Figure 2):
Zone 1: the dipping breccia units with the neighboring

rhyolite and stockworks, where the background material
bears medium-high gold grades and the mineralized veins
have very high grades and an average dip angle of 50
degrees; Zone 2 West: the western part of the sediments,
with medium-grade background and high-grade, subvertical,
densely distributed veins surrounded by a halo of higher
grade background gold mineralization; Zone 2 East: the
eastern part of the sediments where the rock has been
argillized, the grades are lower on average, the
subvertical veins are more sparse and no halo effect can
be observed. All of the mineralized veins are 0.10 m to
0.80 m wide, and continuous over several tens of metres in
a plane direction.

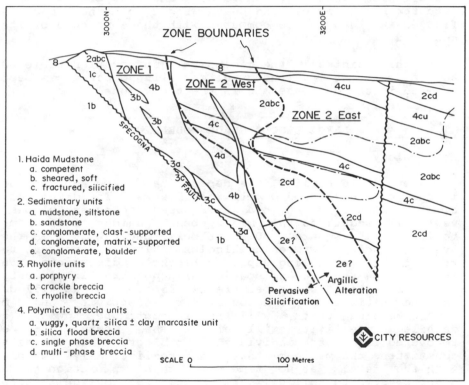

Figure 2 LITHOLOGIES AND ZONES

Once these zones have been separated, and only then,
the basic statistical parameters start showing precise
indications of homogeneity. In particular, in each zone,
the high grade population from the mineralized veins can
then be identified on the cumulative frequency plots and

separated from the rest of the gold values simply by using
a threshold value on the gold grade (the threshold value
changes with each zone). Also, the background grade in
each zone presents a well defined lognormal-like histogram
with coefficients of variation close to 1.0. Finally the
variograms show different systems of ranges and
anisotropic directions in each zone, as expected.

Incorporating the Geology into the Block Model

All data can then be flagged in the database with the
zone and the population it belongs to. Cross sections
and/or plan views of the zone boundaries are digitized
into the block model and each block of the model is
flagged in turn in the computer with the zone code of its
centroid.

At this point all the geological information believed
to be relevant to the grade modelling process has been
entered in the database and in the block model.

LARGE BLOCKS GRADE MODELLING

Methodology

As suggested earlier, the deposit has been represented
in the computer as a 3-D matrix of blocks. The dimensions
of the blocks at Cinola have been chosen as follows:
vertically, equal to the mining bench height, i.e. 6 m;
horizontally, square, 30 m blocks, because 30 m is the
average spacing between drillholes. It is important to
realize that modelling separate blocks of dimensions much
smaller than the average spacing of the data in order to
get "more detail" would be mere illusion. What we will do
instead, ultimately, to account for selectivity on 5 m
square mining units, will be to estimate how such a
detailed selectivity will affect the reserves to be
recovered from each individual 30 m block. This actually
constitutes the maximum that can realistically be done
with the data: at no time shall this exploration grade
model be able to tell us where exactly each ore unit is to
be found, although the final model will tell us with a
reasonable accuracy how many of these units will be above
cutoff grade within any domain covering a few 30 m x 30 m
x 6 m blocks, and what the grade of this ore will be.

In this step, however, we need only to estimate the
gold content of each 30 m block. In order to properly
account for the geology, a very simple two step indicator-
type kriging procedure has been used at Cinola: Each 30 m

block is considered a mixing of the two grade populations, "background gold" and "vein gold" (Figure 3).   Three

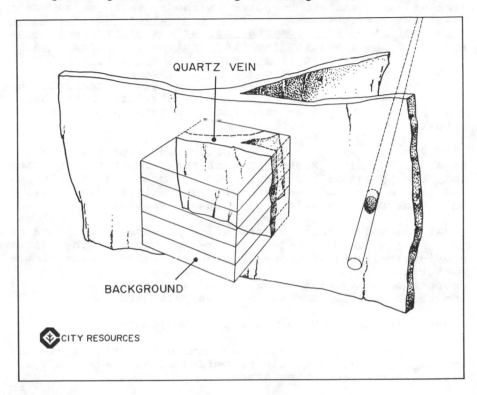

QUARTZ VEIN

BACKGROUND

CITY RESOURCES

Figure 3   TYPICAL MINERALIZED BLOCKS

variables are   thus   estimated for each block: the background grade; the grade of the vein material; and the proportion between 0. and 1. of this vein material in the block.   The latter variable is obtained from the data by flagging with the value 1. any assay belonging to the vein population, i.e. any assay above the threshold value defined for each zone in the previous section, and with the value 0. any data below this threshold.   This 0./1. variable is called by geostatisticians the "indicator" of the vein values. Its interpolated values will   range from 0. (=0% vein) to 1. (=100% vein). The three variables are projected into the blocks using kriging, which requires the calculation and interpretation of variograms.   The literature about how to do it is abundant, and the corresponding details are not the object of this paper. Besides, although it is to be preferred, kriging is not mandatory.   Provided the variograms have been properly

interpreted, acceptable projections of the grades or
indicator variables can be achieved by properly using
simpler and faster computer methods such as inverse-
distance-type methods. <u>The recognition of the three
zones, of the two separate populations of grades in each
zone, and the resulting two-step indicator approach is of
far more impact on the results than the particular
algorithm used to interpolate the variables.</u> However,
skipping the variogram calculation and interpretation
i.e., choosing a completely non-geostatistical approach,
has shown many times to lead to totally unacceptable
results.

Of course, each zone is processed separately, as a
separate deposit. Blocks in each zone are estimated using
only data previously coded as belonging to the same zone,
and the background and vein grades are interpolated using
only background or vein grade assays, respectively.

At the end of this interpolation process, each block of
the block model that belongs to one of the three
previously identified zones has an estimated value of the
background grade, the vein grade and the corresponding
proportion of veining. The estimated overall block grade
can be obtained by combining these estimated figures.

Global Recoverable Reserves and Comparisons

As already mentioned, the mining selection against a
cutoff grade will actually be performed on 5 m x 5 m x 6 m
blocks at the production stage.

Although the model is not yet in its final form,
preliminary in-situ, global recoverable reserves taking
full account of this future selection level can be
obtained now from the block model. Several simple ways of
transforming the global results for mining selectivity
exist. The objective however, is always the same: obtain
an estimated frequency histogram of the gold grades of the
mining units in the deposit. Cutoff grades can then be
used to truncate the histogram and therefore derive the
recoverable reserves and grade-tonnage curves. Assuming,
as can be observed in practice, that the shape of the
histograms of the assay data, small mining units or large
blocks is the same lognormal-like shape, and that
remembering these histograms all have the same mean and
different variances, then, the dispersion variance is the
only missing parameter. Geostatistics allows us to
calculate the average dispersion variance of the mining
units from the variograms. Once this is done, the
histogram of the previously estimated grades of the 30 m

blocks in the deposit can be distorted using an affine
correction of the variance around the mean, i.e. an affine
transformation of each block value (Z) that preserves the
mean (m) and changes the variance (V) to (V'):

$$Z' = (Z - m) \times SQRT(\frac{V'}{V}) + m$$

as was done for Cinola.   Or one can play with fitted
lognormal distributions.   In all cases, it is recommended
to distort or fit estimated <u>block grades rather than the
original assay values</u>, in order to avoid any problems
related to data clusters and to odd histogram artifacts.

Besides providing us with preliminary recoverable
figures, this global histogram distortion technique
allowed us to make the estimated reserves as comparable as
possible to an existing hand-calculated estimate in spite
of the already built-in selectivity of the latter.   Such
comparisons are recommended as a double check before
proceeding further: One should expect discrepancies, but
the order of magnitude and the sign of such discrepancies
can usually be predicted and explained, so that any
unexpected result in the comparison may be the indication
that something went wrong and needs immediate correction.

MODELLING OF LOCAL RECOVERABLE FIGURES

After satisfactory global reserves comparisons, as well
as any other possible verifications have been performed,
the block model can be transformed to provide local
recoverable reserves, which is the prediction of how many
tons of ore at which grade is likely to be ultimately
obtained from each block of the block model.

Methodology

In nature, this final problem has the very same nature
as estimating the global recoverable reserves, as in the
previous section, but the distributions of interest are
now the histograms of the 5 m mining units within each 30
m block.   The mean of such a histogram is the previously
estimated block grade, but no local histogram is available
to be "distorted" as previously. In addition, one should
consider the additional following difficulties:

.   the variogram-derived dispersion variance of the
     grade of mining units within a block, although
     available, is an average value on the entire
     deposit.   It needs to be customized to local

variability conditions, including the famous
"proportional effect" that makes gold values more
variable in richer areas than in low grade areas,
but also local geological features that may add or
subtract to the variability.

.   Because the mean grade of the 30 m block is only an
estimate of the true mean, the reserves that can be
calculated for a set of 30 m blocks having the same
estimated grade, differ from the reserves that could
be calculated for those same blocks if we knew their
exact grades. The difference is a calculable
statistical bias that needs correction.

.   Once an experimental histogram or a theoretical
distribution has been selected for either variance
distortion on distribution fitting, steps should be
taken to make sure that the best representativity or
fit is obtained for the part of the distribution
likely to be used, i.e. the part above the probable
cutoff grades.

The method chosen at Cinola to calculate the local
recoverable reserves and at the same time solve all these
difficulties is known as the "lognormal shortcut" and has
been described in many geostatistical publications and
textbooks. In short, the distribution of the mining units
in any single block is assumed to follow, and is fitted
with, a lognormal distribution. This assumption is widely
confirmed in practice for gold deposits, and holds
particularly true at Cinola, where the method was applied
only to the background grades, the vein material being
added afterwards to the selected reserves. The local
variance customization was studied by quantifying the
variability of the assay data available in a moving
neighborhood centered on each 30 m block, in order to
derive local corrective factors for the "average" variance
calculated from the variograms. The bias correction was
performed by adding the estimation variance of the block
grade to the dispersion variance of the mining units, an
approximation that can be shown to be adequate and has
strong intuitive justifications. Lognormal distributions
fitted to the global histogram of the original assay data
also helped design an additional corrective factor for
improving the fitting of the distributions above the
practical range of the cutoff grades, rather than
achieving the best overall fittings. Finally, the
resulting, corrected dispersion variance was stored in the
block model for each block.

These apparently complex manipulations can be traced back to basic, common sense concerns such as best fitting above cutoff grade, higher variability in richer areas, influence of local geological features, well observed shape-permanency of frequency distributions of blocks of different sizes, etc. The above does not pretend to be a recipe to follow, but rather suggests a general approach for solving that category of problems in a simple and controlled way.

Results

As mentioned earlier, the above "lognormal shortcut" methodology was applied to the background grades only, allowing us to estimate the tons and grade of "background" ore above any given cutoff grade.

The high grade material from the veins was added afterwards to the reserves selected above the cutoff grade, in different ways in the various zones. As a matter of fact, at Cinola, the geology indicates that the blastholes will most likely miss the subvertical veins in the argillized zone (Zone 2 East), but will detect them systematically in the breccia area, where they dip. In the central zone (Zone 2 West), these veins will be missed most of the time, but mined any way thanks to the halo effect in the background mineralization.

Based on these additional assumptions it is easy to use the block model to sum up the reserves recoverable when selecting 5 m x 5 m mining units over 6 m benches against any given cutoff grade. Approximate formulae allow us to quickly calculate the tons and grade above any cutoff grade from a lognormal distribution of grade known by its mean and (corrected, local) variance. The cutoff grade can therefore be changed without changing the block model, which makes all detailed sensitivity analyses possible and easy.

Finally, because it provides reserves that are both local and recoverable, the obtained block model is suitable for detailed engineering, mine scheduling, pit outline optimization, and will still be used at the production stage as a guide to the grade control process.

CONCLUSION

The author thanks City Resources (Canada) Limited for allowing the publication of this paper. The Cinola Gold deposit summarizes most of the features that can be found in primary gold deposits, and therefore its estimation covers most methodological situations. Again, the simple methodological choices made at Cinola, as presented above, are not the only possible ones, but they have the merit of using concepts that are easy to control and to relate to the physical aspects of the reality. In the complex case of epithermal gold, where geological complexity too often takes its toll, that controllability of the concepts used is a major guarantee of success.

Chapter 7

AZTEC – THE AUTOMATIC GOLD ANALYSER

C.J. Sharland[1], M.E.A. Robertson[1] and V. Roze[2]

[1] Interlect Instruments Limited, Honiton, Devon, England.
[2] Intermetrix Systems., Mississauga, Ontario, Canada.

Current Techniques - Fire Assay

Fire assay of gold has been used for several thousand years and is a well known technique to the traditional assayer. Very few assayers wish to use the method but to date there has been no viable alternative. Fire assay is expensive in labour, time, capital and running costs. Workers can be exposed to hazards and uncomfortable working conditions such as lead fumes, heat stress, dust and eye strain. The high capital cost of putting up the building and equipping it, together with running costs which increase annually at a prohibitive rate makes the technique less attractive each year.

The main requirement in most assay laboratories is to maximise throughput economically. This is difficult with fire assay as it is impossible to automate. Each sample for analysis is handled by at least 12 people and each person represents a potential source of error. Record keeping and report generation has to be done manually which is both slow and prone to errors. Replacement of staff requires a considerable training period and with a large turnover of personnel this presents yet another problem.

The presence of silver and dross in the gold sample affects the accuracy at low gold values.

X-ray fluorescent analysis

Two methods of XRF instrumentation have been used with limited success. The first type is a laboratory instrument employing L X-rays. Small samples of a few grams, which are not representative,

and have a large sampling error, have to be finely ground and
prepared and presented to the instrument. Sensitivity is low due
to geometry and self absorption of the L X-rays within the sample
and interferences are a major problem. The other instrument is a
hand held device which is placed against the face of the stope. An
isotope is used to excite the K X-rays which are detected by a
small germanium detector cooled by liquid nitrogen. Errors occur
due to varying geometry of an uneven stope face and sensitivity is
low due to the low signal to background ratio obtained with an
isotope.

The AZTEC has been designed in collaboration with major mining
houses to greatly improve the method of gold assay. Four years of
intense development have provided an instrument that fulfils the
requirements of the modern assay laboratory. See Fig. 1.

Gold is a difficult metal to analyse due to the low levels
encountered and the very high sensitivity required to measure
these low levels. The AZTEC is a computer operated system
comprising X-ray unit, detector, robotic sample changer and
associated electronics. The basis of AZTEC is the excitation of
gold K X-rays by high energy X-rays. The energy spectrum from a
typical gold sample with uranium present is shown in Fig 2. The
immediate advantages of working at the higher energy K levels are;
the ability to read out larger, more representative samples; high
sensitivity; and, lower interference from other elements. The high
energy X-rays are produced from a compact generator which is water
cooled by a chiller unit suitable for use in many varied climates.
Many innovative design features have been incorporated to achieve
the criteria of measuring gold in ore from zero to 10 000 g/t, with
a precision of +/- 0.5 g/t in 100 secs., or +/- 0.25 g/t in 400
secs., in the range 0 - 10 g/t.

A novel high resolution detection system was critical to
achieve the precision at the low end of the measurement range. The
detector is cooled by liquid nitrogen from an integral cryostat
which is automatically topped up as required from a large dewar
alongside the instrument.

The sample changer is computer controlled to sequentially
select and analyse the samples in the magazine. The instrument
can be configured to hold up to 3 or 6 magazines and each magazine
has a capacity of 88 samples. Thus a total of 528 samples can be
analysed fully automatically. The sequence of operation of the
sample changer can be varied by software control to suit the
specific requirements on any particular day or night. The
mechanism has proved extremely reliable throughout extended trials
running for 24 hours a day over several months and where faults
have occurred, loss of data is minimised by constant automatic data
saving. Computerised fault diagnosis cuts down time to a minimum
and self calibration to international or local mine standards
ensures that the instrument is correctly calibrated at all times.

The computer uses the 80386 microprocessor chip with 32 bit architecture and communicates with RS232, RS422 and IEEE488 interfaces. The 40MB hard disk allows other software packages such as a databases and word processing for more comprehensive reporting. The printer, sited alongside the computer on a mobile unit, prints out the gold and uranium levels showing their location. Further immediate transmission of confirmed data can then be made by the Chief Assayer via a modem or optical fibres to other interested parties such as the geostats department, mine manager, central computer or, in the case of a commercial laboratory, the client.

Laboratories with very high sample throughput can have several AZTECs interfaced to a local area networking (LAN) computer, which correlates AZTEC results with other information, such as sample identification, and prepares and transmits reports.

Maximum throughput of samples is of prime importance and every effort has been made to cut operation time. A choice of fixed precision with variable analysis time, or fixed time with variable precision, can be selected at will. Where gold levels are high, considerable improvement in sample throughput can be achieved by using the former mode. The facility to operate the computer in foreground/background mode complements the fast processing of the computer.

The ore sample is ground to less than 100 microns using a disc mill with as large a capacity as possible e.g. 2kg. Two types of sample container are available, to suit either 50g or 100g samples (the exact sample mass will depend on density).

Accessories such as the 'Topper', 'Tailer' and 'Vibrafil' are available to speed up this process.

No weighing of the sample is required as automatic density correction is incorporated throughout the density range from carbon to high density ore samples.

For an automatic analyser to be of use in a busy assay laboratory it is essential that a high degree of reliability is inherent in the design. From the early design stages this philosophy has been a major design criteria. Electronic circuits are modular using the Eurocard format. Computerised fault diagnosis is incorporated to cut down-time to a minimum. All instruments are put through extended soak testing before shipment. Any faults are displayed on the VDU and an audible alarm is initiated. AZTEC has been designed to work unattended through the day and night to maximise throughput. An extended calibration routine lasting two hours is implemented during the night to check zero calibration to within 0.05 g/t.

Economics

Taking the above concept into account the following figures can be expected.

As the AZTEC may be operated 24 hours per day with unattended operation the instrument may be loaded on a previous day and only Sundays and Bank holidays need be considered non productive. An allowance of five days for malfunction/servicing has been assumed. If the instrument has been down for a period, production can be restored by using the machine on a Sunday.

Sundays	52
Bank holidays	8
Malfunction/Servicing	5
	65 days

Working days in a year = 300 i.e. 82% availability.

To achieve a throughput of 1000 samples/day two AZTECs would be required. Taking all costs into account including premises, labour, servicing and sample preparation, the cost per sample for the first year is 19 pence. Cost of operation in the second year would be 32 pence and only the inflation rate need be applied to this figure for ongoing years.

If the capital cost is included, the cost per sample for the first year would be 59 pence and for the second year 72 pence. Again the inflation rate would be the only cost increase over a number of years. After five years, or the length of time over which the instrument was depreciated, the cost per sample would revert to the annual running costs only which would be 32 pence x correction for inflation rate.

The costs for the second year and onwards are increased due to contract servicing charges which do not apply in the first year as they are covered by warranty. The cost for power and labour represent 35% of the above running costs including the service contract. One technically competent person could look after a number of AZTECs with an allowance of say 20% of a manager's time to oversee and check results. It is considered that the type of person in charge of the instrument would not necessarily be a chemist but an instrument technician with a knowledge of computerised instrumentation and statistics.

Layout of Laboratory

AZTEC does not require a great deal of space to set up. Two or three rooms to house the sample preparation, AZTEC and services are all that are required.

The sample preparation room follows conventional lines which will house say three Labtechnic LM-1 disc mills or conventional ring mills to grind the sample. The ground ore is then poured into the sample container using a 'Vibrafil' vibrating funnel. The special sample container top is then inserted using the 'Topper' to align and insert the cap to a preset depth. The sample container is then placed in a numbered position in the magazine. A dumb terminal connected to the main instrument computer (or LAN computer, when several AZTECs are used) allows sample identification data to be entered for correlation with the sample when it has been analysed.

The magazine is then taken through to the AZTEC where it is loaded into the appropriate dock.

The room to house the AZTEC can have a minimum dimension of 6 x 4 metres. One wall should preferably be an outside wall to site the air conditioner and water chiller. The air conditioner should be set to keep the temperature at about 25°C. The X-ray tube is water cooled in a closed circuit by a chiller which will preferably be housed outside the laboratory.

A 50 litre dewar of liquid nitrogen is placed close to the small integral cryostat and filling is automatically initiated when the cryostat requires replenishing.

As there is absolutely no lead in air pollution, costly extraction units are not required.

Accuracy of Results

The preproduction instrument was installed in a large South African gold mine in October 1987. The results and functioning of the instrument have been beyond expectation in both the accuracy of results and the reliability of equipment. See Fig. 3.

The range of the equipment is 0-10 000 g/t. The accuracy is affected by a combination of random error, systematic errors, matrix influences and sample density.

(a)   Random errors are mainly due to counting statistics where the longer the analysis time the better. This must be a compromise between accuracy and throughput and general operating procedures. The examples shown in Table 1 are to 1 standard deviation for 100 and 300 second counting times, for the larger sample container (100g). Increasing the counting times by a factor of four will reduce the random error by a factor of two, i.e. there is a square-root relationship.

The smaller sample container (50g) will give an improvement of approximately 20%, due to lower attenuation in the sample. However, sampling error may be increased with the smaller sample, depending on the type of ore body and grinding technique.

Users with relatively homogenous ore bodies can use the smaller sample container to advantage. Similarly, the Model LM-1 disc grinding mill can be used to homogenise large samples to reduce sampling error and allow the smaller sample container to be used, irrespective of ore body characteristics, with consequent benefits in improved accuracy.

(b)   Systematic errors, in the instrument, include factors such as calibration accuracy, and electronic stability. The accuracy of sensitivity calibration is important and a combination of international and customer standards can be used to optimise it. An artificial gold standard of about 100 g/t is used to initially set up AZTEC. The accuracy of this calibration should be better than 1%.

The three factors affecting the electronics stability of the instrument are:

(1)   the sensitivity drift which is less than 0.2% between calibrations.

(2)   the accuracy of the zero calibration which at the 95% confidence level is 0.1 g/t.

(3)   the non-linearity of the calibration curve which is less than 0.5% up to 100 g/t and less than 2% up to 1000 g/t. A non-linearity correction can be provided.

(c)   Matrix effects are minimised because of the use of K X-rays. However, certain heavy elements will have an influence. For example:

(i)   1 kg/t of lead alters the zero by -0.3 g/t.

(ii)   1 kg/t of uranium (and associated lead in equilibrium) alters the zero by -0.3 g/t.

(iii)   1 kg/t of uranium (and associated lead in equilibrium) in a sample lowers the gold sensitivity by about 1.5%. In instruments with both gold and uranium counting channels, the uranium value is calculated first, and used to correct the gold value.

(iv)   1 g/t of mercury alters the zero by +0.4 g/t.
         Mercury interference is detected and advised.

(v)   10 g/t of tungsten alters the zero by +0.5 g/t.
         Tungsten interference is detected and advised.

(vi)   Samples containing high concentration (greater
         than 10%) of iron or heavier elements may need
         special matrix correction.

(d)   Sample density.

An important feature provided by AZTEC is the automatic
density correction for ground ore samples in the density
range 0.8 - 1.7 g cm^{-3}.   The zero reproducibility over
this density range is 0.1 g/t.   Sensitivity
reproducibility over the density range is 1%.   A similar
density correction is provided for gold-in-carbon samples
in the density range 0.3 - 0.8 g cm^{-3}.

OPTIONS

AZTEC has been designed specifically for gold but a very cost
effective option is the facility to simultaneously measure uranium
in the sample with no further sample preparation or analysis time.
The range of the instrument for uranium is 0 - 20 000 g/t, and the
random error primarily due to counting statistics is less than 10
g/t for ore concentrations up to 1000 g/t.

AZTEC can be mounted into a mobile laboratory which houses the
system and the grinding mill.   A surface mine can save
considerable sums of money by having this facility only minutes
away from the working face.   Both the AZTEC and grinding mill can
be powered from a single phase supply such as a portable generator
to make the laboratory autonomous.

Conclusion

The AZTEC has shown itself to be a very accurate and economic
alternative to fire assay for gold.   The automatic analysis of
samples at a high throughput rate makes the technique very
attractive for any laboratory with a requirement of 50 or more
analyses per day.

REFERENCES

Sharland, C.J., 1986, "AZTEC: A High Precision Automated Instrument
    for Gold Assaying". Proceedings SAIMM Johannesburg. Gold 100.
    International Conference on Gold Vol. 2: Extractive Metallurgy
    of Gold.

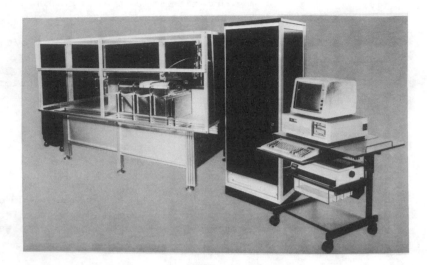

Fig. 1 - The AZTEC Gold Analyser

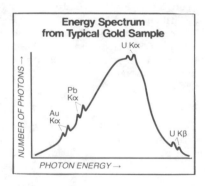

Fig. 2 - Energy Spectrum from a typical gold sample with gold present

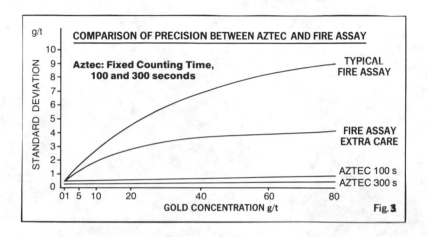

Fig. 3 - Comparison of Precision between AZTEC and Fire Assay

100s counting time	300s counting time
g/t	g/t
0 ± 0.5	0 ± 0.3
5 ± 0.5	5 ± 0.3
10 ± 0.6	10 ± 0.4
20 ± 0.6	20 ± 0.4
50 ± 0.7	50 ± 0.5
100 ± 0.8	100 ± 0.6
500 ± 1.5	500 ± 1.1
1000 ± 2.5	1000 ± 1.8

Table 1 - Errors to 1 standard deviation

2

Feasibility and Financing

Chapter 8

TECHNICAL AND ECONOMIC EVALUATIONS OF MINING PROJECTS

FOR BANK FINANCING

by Peter H. Grimley, Ph. D.

Robertson & Associates
Consulting Geologists & Mining Engineers
Toronto, Ontario, Canada

ABSTRACT

It is important to realize that lending institutions have a
very different risk-profile from that of an equity investor. For
bank financing, a feasibility study must address the concerns of
the lender and these can be summarized under the four 'P''s:
Project, Performance, Product and People.

Generally, feasibility studies focus on the measurable
project risks associated with a new venture, but frequently
insufficient emphasis is placed on key risk areas such as ore
grade and metallurgical recovery.

Before accepting full or even partial non-recourse risk,
financial institutions will demand evidence that the project was
constructed according to specifications and within budget and
that its Performance within agreed limits has been demonstrated.

Product risk should be discussed in detail in any feasibility
study. For industrial minerals, penetration of the market can be
critical to success and for commodities with volatile prices,
the sensitivity of the cash flow to a range of prices should be
carried out. Methods of reducing price risk such as long term
contracts or, in the case of gold, forward selling should be
carefully evaluated.

The People risk is rarely discussed in feasibility studies,
but can be a key element. Few projects turn out to be faithful
copies of the study and the ability of the operator to identify

and correct problem areas before they become serious can be crucial to financial success. Individuals or teams who have proven track records in project construction and operation will give considerable comfort to lenders.

INTRODUCTION

A lender might be found for virtually any risk provided the risk can be clearly defined. The higher the risk, however, the higher must be the expected reward. This may take the form of higher interest rates and/or the additions of an equity participation which has the effect of giving the lender a share in any up-side potential.

Lenders can be:

Acceptability of Risk	Entity
Low	– Commercial banks
	– Product or equipment loans from purchasers/suppliers
	– Institutional loans (e.g., insurance companies)
High	– Venture capitalists

This paper will address loans made by commercial banks, i.e., at the lower end of the risk/reward spectrum.

In very simple terms, there are two methods of bank financing: recourse and non-recourse (or project financing). There are variations on these two themes, but in the interests of brevity, I will take the two extremes.

In Recourse financing, the banks will look to the owner (or in banking terms, the sponsor) for their security so that if a mining project fails to generate enough cash flow to service the debt (that is to repay principal and interest) the total assets of the owners can be used to repay the debt. It is obvious that the bank will concentrate as much on the financial capacity of the owners (their credit-worthiness) rather than the project itself although the effect of potential failure of the project on the owner's overall financial health will be carefully evaluated.

By contrast, in Non-(or Limited-)Recourse financing, or as it is more commonly called, Project Financing, the banks will have only the cash flow from the project itself with which to service the debt. The banks will therefore assume all or part of the various risks associated with the project. These risks may include commodity price fluctuations, the risk that ore reserve estimations will not hold up under actual operations, capital and operating costs risks, political risks and so on. When a bank is structuring a project loan, it will seek ways to limit the risks either by introducing mechanisms to improve the quality of the cash flow or to have the owners support the project until at least some of the risks have been reduced. For example, the owners may be required to guarantee the loan during a completion period. During this period, the project will have to demonstrate that it can be built on time and on budget and that it can operate within specified limits of its previously projected performance before the funds provided by the lending institution can be considered non-recourse.

Whether considering recourse financing or project financing, a bank will be balancing its risk versus its reward. It is this risk reward scenario that is most frequently misunderstood by the mining community. Prospectors are forever complaining that banks will not finance their exploration ventures. The answer is very simple; the risks are too high for the reward. Even in an operating situation, the risk-reward ratio is significantly different for the equity-holder and a project lender. Consider for a moment the effect of just one variable, commodity price. If the price of the commodity doubles, the equity holder stands to gain much in terms of dividends and stock price. The lender will only receive an agreed interest rate. Should the price drop by 50%, the equity investor suffers and the bank may well have to write off a worthless asset. In other words, the lender has significant downside risk but, unlike the equity investor, has limited up-side potential.

The borrower in his presentation to banks must clearly identify and quantify wherever possible the risks associated with the project. If they are not identified or wherever any substantially important section can be challenged or is in question, the banks or their consultants will make their own assumptions and the risk may be pushed higher up the spectrum resulting in higher interest rates or a refusal of the loan.

You will note that I have used the word 'risk' a number of times. It has been my experience that the major risks involved in mining projects can be summarized under the four 'P''s — Project, Performance, Price and People. In preparing a document for bank financing, the lenders will wish to know how you propose to minimize the risks associated with each of these 'P''s.

'PROJECT'

Ore Reserves

Of the project risks, the most important risk element is the ore reserve. It has been my experience that, more often than not, reserves are given scant treatment in feasibility study documents presented for bank appraisal. It is not uncommon to find only one or two chapters devoted to reserves while the rest of a seven-volume feasibility study is devoted to detailed engineering. The quality of an ore-body and the accuracy of its assessment are key parameters in the evaluation of any loan proposal. Some of the factors that impact on reserve calculations are:

Regional Geology: Detailed descriptions of regional geology are seldom required. A brief description of the geological setting is all that is required unless it is felt necessary to demonstrate the similarity of the project to other producing properties in the region. A long-winded geological treatise may demonstrate the author's erudition but also demonstrates his lack of awareness of his reader's real concerns.

Local Geology: Description of project geology should not be simply a listing of rock-types but should focus on those aspects which appear to be significant from an ore-depositional or from an operational perspective. The reader should not be left guessing as to the extent to which a schistose hanging wall will affect mine dilution.

Sampling: It is important to state, in an appendix if necessary, how samples were taken, how the core was logged and sampled, how bulk material was sampled. If a bulk sample was taken how do the results compare with face, muck or drill samples from the same area? How did the different sampling methods compare and what was the rationale for choosing one method over another for reserve calculations? A brief commentary on assaying techniques should be made together with a statement of the internal and external checking procedures.

Cut-Off Grade: It is important to state the cut-off grades and whether development and stope preparation muck have been treated differently from stope ore in determining cut-offs. A comment as to what reserves would be using different cut-off assumptions is appropriate.

Cutting Procedures: If high grade assays are cut, is the procedure arbitrary? If not, give the rationale and a statement of un-cut reserves.

Reserve Categories: Standard reserve definitions should be used. I note here that banks will normally consider only Proven and Probable reserve categories in determining loan life. A statement as to the probable accuracy of reserve calculations, both tonnage and grade, should be given, otherwise the bank will assign an arbitrary range of values.

In assessing the suitability of a project for non-recourse financing, a lender will require that sufficient proven and probable reserves are established to cover the period of the outstanding loan plus a 'cushion' of an equal amount. In other words, if the loan life is anticipated to be three years at planned production rates, the lender will expect proven and probable reserves to be sufficient for at least six years.

Geostatistical Analysis: While great strides have been made in recent years in geostatistical analysis, cooperation with on-site geologists is still imperative. Geostatistical analysis without detailed, on-site in-put can be dangerous and misleading.

Mine and Mill Planning

It has been my experience that all too often there is a lack of coordination between an exploration department, responsible for the initial outlining of reserves and the planning departments responsible for both overall and detailed mine and mill design. This lack of coordination may lead to important physical parameters such as wall rock conditions, mineralogical changes, etc., not being recorded during the exploration phase. Early collaboration between departments can avoid later misunderstandings and improve immeasurably the final feasibility.

The lending institutions will wish to see that planning has been carried out in detail in the areas which will be mined during the early years of operation, i.e., the period when their loans are outstanding. This observation is not as trite as it sounds. I know of two projects where feasibility studies showed detailed planning had been carried out in one portion of the deposit while actual operations commenced in a less well known area. In one instance, the initial results were almost disastrous, in the other, the outcome was fortunately less severe.

Lenders will need to be assured that the mining method is appropriate, that there has been sufficient allowance for dilution and that estimated operating costs are reasonable. The seven-volume feasibility study referred to earlier is often composed of highly detailed descriptions of the construction of the mill and the infrastructure. Operating cost assumptions, like ore reserves are often given scant treatment and are relegated to one or, at the most, two chapters. Poorly estimated operating costs can have a very significant impact on cash flow. In my experience, mine operating costs are more poorly estimated than are mill or infrastructure costs, but whether the risk is technical (i.e., equipment selection, development advance, drilling and blasting costs) or social (i.e., labour rates, work stoppages, lack of skilled labour), it is important that the risks be clearly identified and quantified. It is useful to the lender to have for comparison actual operating data from similar deposits already in production and if there is a significant difference, this must be adequately explained.

The importance of accurate estimation of mill recovery is sometimes underestimated. As with head-grade and product price, an increase or decrease in mill recovery directly affects profitability.

Bulk sampling for mill testing is particularly relevant if the metallurgy is at all complex. It is also important that the bulk sample be representative of the material which will be mined in the early years, and is not taken only from the most accessible location.

'PERFORMANCE'

Lending institutions will attempt to lessen the risk associated with construction and operation by the use of a "completion test". A loan will be considered wholly or partially recourse (i.e., all the available assets of the borrower are pledged as security for the loan) until the project has been constructed essentially in accordance with a previously agreed plan and schedule. This is commonly known as "physical completion". Cost over-runs may, and generally do, require additional equity funds from the borrower, although there may also be an upper limit or ceiling on the total funds the borrower must guarantee to provide.

Lending institutions will usually require that the project demonstrate that it can operate for a period of time (usually one to six months) to within plus or minus 10% of the planned rate, recovery, and output and within 10% of the anticipated cost. This removes some of the uncertainty associated with operations, but in the case of projects with planned future expansions (on which ultimate profitability may depend), it does not remove the risk entirely. Some project loans also demand that profitability be established during this completion period thus demonstrating that product can be sold at least at a minimum price and giving at least a temporary assurance with regard to price risk.

Failure to achieve completion may mean that all or part of the security for the loan remains with the borrower and is not transferred to the project. The result is that loan remains recourse.

While the concept of completion is fairly straightfoward, in practice complications may arise. Take, for example, the case of a copper-lead-zinc project with head grade projected to be 1% Cu, 4% Pb and 12% Zn with estimated mill recoveries of 85% Cu, 75% Pb and 95% Zn. If actual production during the completion period averages head grades of 1.5% Cu, 2.5% Pb and 10.5% Zn with actual recoveries of 88% Cu, 73% Pb and 91% Zn, can the project be considered to have achieved completion? Assuming that projected and actual metal prices are the same, the actual net smelter revenue in the example above would be greater than projected net smelter revenue even though the actual lead and zinc grades and recoveries were lower than expected. Whether the lending institution would concede that completion had been achieved would be dependant upon an examination of the reasons for the discrepancy between anticipated and achieved results. If physical targets were not achieved but higher than planned revenue resulted from higher than anticipated commodity prices, I believe that few lenders would be willing to accept that completion had been achieved.

The stringency with which a lender will enforce the completion criteria will depend, to a large extent, on the financial robustness of a project. Most banks will be reasonable and flexible if there appears to be a sufficient 'cushion' safeguarding their loan. In exceptionally sound projects, completion may be almost a formality, but regrettably, such projects are few and far between.

'PRODUCT'

The ability to sell the full output of the product for an acceptable price is one of the key risks that the lender hopes will be covered. In industrial minerals, the quality of the product and the ability to penetrate the market will be of significant concern, whereas in base metals, assuming there are no major deleterious elements in the concentrate, price predic-

tions will play a major role. In both types of commodities, the
existence of long term contracts will give a degree of comfort
to the lender and even more so if there is some form of 'fixed
price' contract. Unfortunately, if markets change significantly,
there will be an attempt by either the buyer or seller to
re-negotiate even the firmest of contracts. Long term sales
agreements do not have the sanctity it was once believed they
had.

With precious metals, the situation is somewhat different.
Supply of new gold does not appear to have nearly the same
influence on price as does, say, a very significant increase in
supply of other raw materials. Opinion is mixed as to what
effect the actual and planned increases in gold production will
have on gold price over the next few years, but a goodly
proportion of analysts believe that whatever the effect, it will
probably be masked by other influences such as interest rates,
the relative strength of the U.S. dollar or the psychological
impact of major world events.

It is in the area of product price that the perception of
risk profile between a lender and an equity investor becomes
most apparent, and it is important that this be realized by the
borrower and that he is sensitive to the concerns of the lender.
As explained earlier, the lender shares a downside risk with the
equity investor but does not share in any upside potential. This
is why, in testing the sensitivity of the cash flow to price, a
lender will sometimes take what appears to be a very pessimistic
viewpoint.

While hedging of base metals does take place to a limited
extent, the hedging of gold production is now the rule rather
than the exception and as such tends to even the odds somewhat
between investor and lender. Hedging is generally welcomed by
the lending institutions. In addition, the use of gold loans has
the effect of transferring a price risk to an operational risk,
i.e., the ability of the project to produce the physical gold.

'PEOPLE'

Good explorers do not necessarily make good operators, and a
real concern of a lending institution may be the ability of the
borrower to attract and retain not only a competent workforce
but also a middle and upper management team capable of constructing and operating a mining facility. Canada is fortunate in
having a number of experienced engineering firms which can and
often do carry out the engineering, procurement and construction
management, but the borrower needs at least a small team of his
own to supervise the construction and this team should prefer-

ably include members who will later form part of the operating management. Attracting high calibre individuals to a new operation is never easy, but even if such individuals can be located, the time necessary to weld together a team is frequently and often seriously underestimated. The 'start-up' period is not only one of technical adjustment, it is also a period of management and worker adjustment.

Feasibility studies will frequently address the availability of man-power, wage rates, percentage turn-over and so on but will rarely address the equally important aspect of management. If the study is carried out by independent consultants, this is understandable. They feel, perhaps, that the subject of management should more rightly be dealt with by the owners or borrowers. In approaching a lending institution, borrowers should be prepared to discuss, in detail, their philosophy of management and identify key players.

In my experience, few projects turn out as planned in a feasibility study. With luck, and good management, pleasant surprises will offset the unpleasant ones, but the essence of good management is in anticipating potential problems and dealing with them promptly and effectively.

The issue of management, therefore, will always be important and assumes very significant proportions if the borrower does not have a proven track record of successful start ups and operations. Even having a track record, however, does not guarantee a smooth progression into full operation and most major mining companies have at least one project of which they are not particularly proud.

ECONOMIC EVALUATION

In a non- or limited-recourse loan, the lender will be looking to the cash flow from operations to service his loan. The way that he calculates his cash flow is little different from the way in which an investor or management would. It is, however, important to realize that the lender will be concentrating almost entirely on the period during which his loan is at risk plus a 'cushion' in case of unpleasant surprises. He is not particularly concerned with the return on investment or net present value of the cash flow, he will probably focus much more on the protection ratio, i.e., the multiple by which net cash flow in any given period exceeds the amount required to service the debt.

Whether or not a feasibility study includes an analysis of the sensitivity of the cash flow to changes in cost or revenue

parameters, the lender will usually carry out his own sensitivity analysis using his own assumptions. He will generally focus
on the key risk areas of grade, mill recovery and commodity
price since these have the greatest direct influence on cash
flow, but in complex loan transactions, other factors such as
capital cost over-runs may be important.

While the concept of non- or limited-recourse lending is
relatively simple, in practice, loan agreements can be very
complex and in some instances run to several volumes. Complexity
may be introduced, for example, by allowing the borrower to take
down part of the loan as non-recourse with progressively more
being transferred from a recourse portion to non-recourse when
certain milestones in the project's development are achieved.
The proportion of cash flow devoted to debt service may also be
determined by reference to the protection ratio. These, and
other refinements can, in fact, make cash flow calculations very
complex and the development of a computer model to reflect the
complexity a significant expense.

In general, non-recourse loans, unless they are very simple,
are unsuitable for small projects because legal and other outside expenses in drawing up the necessary documentation and in
monitoring progress can be disproportionally large. Smaller
projects are therefore more rationally financed by equity or by
straight bank loans if the borrower has sufficient other assets
to offer as collateral.

SUMMARY

The principal risks facing a lender of non-recourse funds can
be identified under the four 'P''s, Project, Performance,
Product and People. In preparing feasibility studies for the
purpose of bank financing, the different risk profiles of lender
and equity investor should be recognized and the study should
focus on principal risks faced by the lender. These risks can be
summarized under the four 'P''s —

Project

Are the ore reserves sufficiently well established and has
sufficient metallurgical test work been carried out? Are the
capital and operating cost estimates reasonable when compared
with similar projects?

Performance

Can the project be constructed within budget and on time?

Does actual operational data during the completion period compare favourably with the feasibility study projections?

Product

Can the price risk be mitigated by long term, fixed contracts, by hedging or by gold loans?

People

Has the borrower a good track record of construction, development and operation? Is a management team in place?

Small projects are less suited to non-recourse financing because of the complexity of arranging such a facility.

Chapter 9

AN OVERVIEW OF ORE RESERVE CERTIFICATION REQUIREMENTS
FOR PUBLIC STOCK OFFERINGS
by
David S. Bolin and Gary R. Peterson

Principal Geologist
Pincock, Allen & Holt, Inc.
12345 W. Alameda Parkway
Lakewood, Colorado 80228, USA.

Senior Mineral Economist
Pincock, Allen & Holt, Inc.
12345 W. Alameda Parkway
Lakewood, Colorado 80228, USA.

INTRODUCTION

In both the United States and Canada, a company making a public offering of securities must comply with regulations which have been promulgated by Federal and/or Provincial Legislatures. This paper focuses on the regulations pertaining to reserve reports required by the United States Securities and Exchange Commission (SEC).

SEC RESERVE FILINGS

Reserve reports made to the United States Securities and Exchange Commission are the only reserve reports which are directly regulated by the United States government. The requirements for SEC reserve reports are set forth in the filing forms as shown in Appendix A, from Item 17A of SEC Form S-18. These reporting requirements, which are intended to protect the investing public from inaccurate or misleading claims, are primarily oriented to S-1 or other filings for new stock issues but must also be followed for annual 10K reports which must be filed annually by U.S. public corporations as part of the Corporate Annual Report. These undertakings require independent audits of reserves, mine designs, plant facilities and a general economic evaluation of the planned or existing operation as part of the backup documents for filing.

An ore reserve is defined by the SEC as "That part of a mineral deposit which could be economically and legally extracted or produced at the time of the reserve determination". This definition is unchanged since it was last amended in 1985, but is currently being more rigorously enforced and supporting data must be more detailed and documented than was previously required.

Reserve Estimation Procedures

The reserve estimate must include a statement regarding the name of the person or firm responsible for estimating the reserve and the relationship of that person to the company filing the report. The report must also include a description of the method used to estimate the reserve. The SEC will accept most mining industry accepted methods from manual cross-section or polygon methods to sophisticated geostatistical methods. The reserve estimation method is typically examined by the SEC but the SEC does not attempt to re-estimate reserves. The method should be described in detail and should be done using industry accepted reserve estimation practices. A reconciliation against historical production is not required but is advisable, if available, to support the validity of the reserve estimation procedure.

Under currently accepted practice, it is important to clearly define the differences between "Reserves" and "Resources". A resource model and mineral inventory is usually generated either by manual or computerized methods, and is the resource base from which the minable reserve is developed by the application of the proper engineering and economic parameters. The term "Reserve" is not usually applied to the resource base, although the expression "Geologic Reserve" has been used in the past. The term "Ore" is used only in the context of the minable reserve.

The categorization of resources and reserves with respect to the level of confidence generally follows the terminology developed over the past century in North America, and formalized in 1980 by the U.S. Bureau of Mines and U.S. Geological Survey in U.S.G.S. Circular 831, "Principles of a Resource/Reserve Classification for Minerals". The categories developed include measured (proven), indicated (probable), and inferred (possible). The demonstrated resource category includes measured plus indicated tonnages, and the identified resource category includes the tonnage in the inferred category. The requirements for each category are clearly stated, although some degree of subjective judgement is allowed. The important factor here is that the method and justification of the categorization be clearly described. These categories are normally applied to the resource or mineral inventory base, and then carried into the minable reserve estimate. Only measured (proven) and indicated (probable) categories may be reported as reserves for financial evaluation purposes.

The categories of proven and probable ore require that continuity of ore must be demonstrated among the sampling points. Continuity may be demonstrated through geological evaluation by an experienced geologist based on subjective evaluation of the deposit and through comparison with other similar deposits.

Ore continuity may also be demonstrated by geostatistical methods although the SEC provides no guidelines regarding the requirements for geostatistical demonstration of continuity. In general, however, geostatistical evidence of continuity may be demonstrated by a variogram range which is greater than the sample spacing.

The SEC provides no quantitative guidelines to determine how accurately, or by what method, the size, shape, and mineral content of reserves must be estimated to be "well established" and classified as a proven reserve. The distinction between proven and probable is an area which requires considerable judgement to determine what level of accuracy is required for a particular deposit relative to the mining method, overall grade of the deposit, and the amount of risk which is acceptable for the venture. For example, the accuracy requirements for a small open-pit gold mine with minimal up-front capital are likely to be less demanding than the accuracy required for reserve estimation of a large operation with hundreds of millions of dollars of investment.

If the division into proven and probable reserves cannot be accurately made, it is acceptable to report a single category of proven plus probable reserves, which is generally referred to as "Demonstrated Resources or Reserves". Regardless of the method for determining the categories of proven and probable reserves, the company may not disclose estimates for material other than proven and probable ore reserves. PAH has found that the SEC discourages use of the terms "possible reserves" or "marginal reserves". Such mineralized material may be discussed in the resource section of the report under headings such as "Resource Potential" or "Other Mineralized Material".

Economic Data

More data is now required on the engineering and economic aspects of mineral projects, and that data will be checked thoroughly by the SEC technical staff. For example, a company must show the computations, costs, and metallurgical or processing recoveries used to derive the cutoff grade and must be able to demonstrate that all material above the cutoff grade may be economically produced. The company should also provide sufficient documentation to demonstrate that those costs and recoveries are fair and reasonable estimates of the costs and recoveries which will be achieved in the operation. Stated mining engineering factors such as mine recovery, dilution, pit slope or underground rock stability must be in accordance with normal engineering practice, or if different, justification for the use of the exceptional parameters must be provided.

The commodity prices used for cutoff grade computation must be no higher than current market prices. For example, recent audits performed by PAH have been performed at a gold price of $450 per ounce and prices between $350 and $450 per ounce are being accepted by the SEC for ore reserve reporting purposes. The only allowed deviation is

for an existing, operating mine where an earlier reserve statement may
be used without adjustment for short-term fluctuations in prices.

For a new mine, the company must also report the amount of capital
expenditures and must show that revenues are sufficient to cover
operating costs and to pay back the capital investment. In the case
of an operating mine, any additional capital expenditures to obtain
the stated production rate must be discussed. There is no requirement
at this time to show annual cash flows or to demonstrate an adequate
rate-of-return on the investment for SEC filings. Cash flows are,
however, required by the Canadian Securities Commissions, and these
should be prepared if a Canadian filing is contemplated in addition to
a U.S. filing. A sample cash flow statement, along with an
explanation of the cash flow line items appears in Appendix B.

Legal Rights

The company must also establish that it has legal rights to mine
and extract minerals from the property and must provide copies of
titles, leases, or other pertinent documents. Any necessary access
rights, water rights or waste disposal requirements must also be
addressed. The company also must disclose any significant permitting
or environmental requirements.

Mining Methods

The company must state whether the reserves will be mined by open
pit or underground methods and must state whether the ore reserve
statement is for in-place or recoverable material. If the statement
is for in-place material, the mining and metallurgical losses must be
stated. The ore reserve must be minable, and minability must be shown
through a mine plan. Dilution material must be included in the
estimate of tonnage and grade. Open pit mines must show an ultimate
pit including road access. Ultimate pits without roads, such as those
generated by floating cone algorithms, are usually not accepted unless
waste tonnage has been increased to approximate the additional
requirements for a haul road. Underground mines must show the
location of the shaft and access drifts, and a schematic of the mining
method.

SUMMARY

The trend in reserve estimation and reporting in the United States
has been to become considerably more restrictive with regard to
quantification of resources, the classification of resources and
reserves, and the demonstration of economic viability. Securities
agencies, banking and financial institutions and investors are
becoming increasingly sophisticated with regard to the evaluation of
various projects and companies. Most of the above groups are quite
selective regarding the acceptability of engineering auditors and
consultants. It has been our experience that any evaluation report

must incorporate sufficient detail to satisfy the SEC and adequately inform the average investor.

APPENDIX A

FORM S-18 (Partial)
U.S. SECURITIES AND EXCHANGE COMMISSION

Item 17A. Description of Property - Issuers Engaged or to be Engaged in Significant Mining Operations.

    a) Definitions: The following definitions apply to registrants engaged or to be engaged in significant mining operations:

        1) Reserve: That part of a mineral deposit which could be economically and legally extracted or produced at the time of the reserve determination. Note: Reserves are customarily stated in terms of "ore" when dealing with metalliferous minerals; when other materials such as coal, oil shale, tar sands, limestone, etc. are involved, an appropriate term such as "recoverable coal" may be substituted.

        2) Proven (Measured) Reserves: Reserves for which (a) quantity is computed from dimensions revealed in outcrops, trenches, workings, or drill holes; grade and/or quality are computed from the results of detailed sampling and (b) the sites for inspection, sampling and measurement are spaced so closely and the geologic character is so well-defined that size, shape, depth, and mineral content of reserves are well-established.

        3) Probable (Indicated) Reserves: Reserves for which quantity and grade and/or quality are computed from information similar to that used for proven (measured) reserves, but the sites for inspection, sampling, and measurement are further apart or are otherwise less adequately spaced. The degree of assurance, although lower than that for proven (measured) reserves, is high enough to assume continuity between points of observation.

        4) (i) Exploration Stage - includes all issuers engaged in the search for mineral deposits (reserves) which are not in either the development or production stage.

          (ii) Development Stage - includes all issuers engaged in the preparation of an established commercially mineable deposit (reserves) for its extraction which are not in the production stage.

(iii)   Production Stage - includes all issuers engaged in the exploitation of a mineral deposit (reserve). Instruction: Mining companies in the exploration stage should not refer to themselves as development stage companies in the financial statements, even though such companies should comply with FASB Statement No. 7, if applicable.

b) Mining Operations Disclosure- Furnish the following information as to each of the mines, plants and other significant properties owned or operated, or presently intended to be owned or operated, by the registrant:

1) The location of an means of access to the property.

2) A brief description of the title, claim, lease or option under which the registrant and its subsidiaries have or will have the right to hold or operate the property, indicating any conditions which the registrant must meet in oder to obtain or retain the property. If held by leases or options, the expiration dates of such leases or options should be stated. Appropriate maps may be used to portray the locations of significant properties.

3) A brief history of previous operations, including the names of previous operators, insofar as known.

4) a) A brief description of the present condition of the property, the work completed by the registrant on the property, the registrant's proposed program of exploration and development, and the current state of exploration and/or development of the property. Mines should be identified as either open-pit or underground. If the property is without known reserves and the proposed program is exploratory in nature, a statement to that effect shall be made.

    b) The age, details as to modernization and physical condition of the plant and equipment, including sub-surface improvements and equipment. Further, the total cost for each property and its associated plant and equipment should be stated. The source of power utilized with respect to each property should also be disclosed.

5) A brief description of the rock formations and mineralization of existing or potential economic significance on the property, including the identity of the principal metallic or other constituents insofar as known. If proven (measured) or probable (indicated)

reserves have been established, state (i) the estimated
tonnages and grades (or quality, where appropriate) of
such classes of reserves, and (ii) the name of the person
making the estimated and the nature of his relationship
to the registrant.

Instructions:

1. It should be stated whether the reserve estimate is
    of in-place material or of recoverable material. Any
    in-place estimate should be qualified to show the
    anticipated losses resulting from mining methods and
    beneficiation or preparation.

2. The summation of proven (measured) and probable
    (indicated) ore reserves is acceptable if the
    difference in degree of assurance between the two
    classes of reserves cannot be reliably defined.

3. Estimated other than proven (measured) or probable
    (indicated) reserves, and any estimated values of
    such reserves shall not be disclosed unless such
    information is required to be disclosed by foreign or
    state law; provided, however, that where such
    estimates previously have been provided to a person
    (or any of its affiliates) that is offering to
    acquire, merge, or consolidate with, the registrant
    or otherwise to acquire the registrant's securities,
    such estimates may be included.

6) If technical terms relating to geology, mining or
    related matters whose definitions cannot be readily found
    in conventional dictionaries (as opposed to technical
    dictionaries or glossaries) are used, an appropriate
    glossary should be included in the registration
    statement.

7) Detailed geologic maps and reports, feasibility studies
    and other highly technical data should not be included in
    the registration statement but should be, to the degree
    appropriate and necessary for the Commission's
    understanding of the registrant's presentation of
    business and property matters, furnished as supplemental
    information.

c) Supplemental Information:

1) If an estimate of proven (measured) or probable
    (indicated) reserves is set forth in the registration
    statement, furnish:

(i) Maps drawn to scale showing any mine workings and the outlines of reserve blocks involved together with the pertinent sample-assay thereon,

(ii) all pertinent drill data and related maps,

(iii) the calculations whereby the basic sample- assay or drill data were translated into the estimates made of the grade and tonnage of reserves in each block and in the complete reserve estimate.

Instructions: Maps and other drawings submitted to the staff should include:

1. A legend or explanation showing, by means of pattern or symbol, every pattern or symbol used on the map or drawing; the use of the symbols used by the U.S. Geological Survey is encouraged;

2. A graphical bar scale should be included; additional representations of scale such as "one inch equals one mile" may be utilized provided the original scale of the map has not been altered;

3. A north arrow on maps;

4. An index map showing where the property is situated in relationship to the state or province, etc., in which it was located;

5. A title of the map or drawing and the date on which it was drawn;

6. In the event interpretive data is submitted in conjunction with any map, the identity of the geologist or engineer that prepared such data;

7. Any drawing should be simple enough or of sufficiently large scale to clearly show all features on the drawing.

2) Furnish a complete copy of every material engineering, geological or metallurgical report concerning the registrant's property, including governmental reports, which are known and available to the registrant. Every such report should include the name of its author and the date of its preparation, if known to the registrant.

Any of the above-required reports as to which the staff has access need not be submitted. In this regard, issuers should consult with the staff prior to filing the

registration statement. Any reports not submitted should
be identified in a list furnished to the staff. This
list should also identify any known governmental reports
concerning the registrant's property.

3) Furnish copies of all documents such as title documents,
    operating permits and easements needed to support
    representations made in the registration statement.

APPENDIX B

ECONOMIC EVALUATION EXAMPLE

Economic analysis of the example gold mining project was performed
using conventional discounted cash flow methods used by the mining
industry. The following assumptions and parameters were used in the
economic analysis.

1) Stand-alone property basis.

2) Project financing on a 100 percent equity basis.

3) End-of-year cash flows for discounting at 10, 15, and 20
    percent rates.

4) Tax Reform Act of 1986 for Federal Tax Calculations.

5) Nevada Taxes.

6) A gold price of $450 per troy ounce and a silver price of
    $7.00 per troy ounce.

7) Capital and operating costs from respective study sections of
    the audit report.

8) Unescalated dollars with no escalation of costs or prices for
    inflation.

Table A-1 shows the basis for the calculation of each line item of
the cash flow statement and the cash flow statement for the example
project appears in Table A-2.

Table A-1
Basis for Calculation of Cash Flow Line Items

Cash Flow Line Item	Basis or Calculation
Total Ore Mined	Per mine production schedule.
Total Waste Mined	Per mine production schedule.
Total material Mined	The sum of ore plus waste mined.
Total Ore Processed Payable Gold, Troy Ounces	Per mine production schedule. Calculated as the product of the gold ore grade times the annual ore production times the mill and refining recoveries.
Payable Silver, Troy Ounces	Calculated as the product of the silver ore grade times the annual ore production times the mill and refining recoveries.
Gold Sales	Calculated as price of gold per troy ounce times payable gold.
Silver Sales	Calculated as price of silver per troy ounce times payable silver.
Total Metal Sales	Summation of gold and silver sales.
Refinery/Trans/Insurance	Based on refining charges, transport costs, and insurance costs.
Net Smelter Return	Total metal sales less refining, transportation, and insurance charges.
Royalty	Calculated per JV/Ownership agreements at 5% of net smelter return.
Gross Income From Mining	Net smelter return less royalty.
Operating Costs	Calculated as the mining, crushing, leaching, recovery, and G & A costs/ton of ore times annual ore production.

Table A-1 (Cont'd)

Cash Flow Line Item	Basis or Calculation
Depreciation	Calculated as depreciable assets as defined by Tax Reform Act of 1986. Units of production depreciation was used for depreciable assets.
Net Proceeds Tax	Based on the net revenue after deductions times the County mill levy rate.
Property Tax	Based on the assessment basis for Nevada of 35 percent and the County mill levy rate.
Net Before Depletion	Gross income from mining less the following: Operating Costs, Depreciation, Net Proceeds Tax, and Property Tax.
Depletion Allowance	The depletion allowance was determined as the larger of the cost or percentage depletion allowance. The percentage depletion allowance for both gold and silver is 15 percent of gross income from mining subject to a limit of 50 percent of net before depletion.
Income Before Loss Carry Forward	Net before Depletion less the Depletion Allowance.
Tax Loss Carry Forward	If the Income Before Loss Carry Forward is a negative value (net operating cost) for a given cash flow year, the net operating loss may be carried back or forward as a deduction prior to income taxes.
Taxable Income	Income Before Loss Carry Forward less the Tax Loss Carry Forward

Table A-1 (Cont'd)

Cash Flow Line Item	Basis or Calculation
Federal Income Tax	Calculated based on Tax Reform Act of 1986. Calculation requires comparison of normal tax liability and tax liability calculated as alternative minimum tax. Regular tax is calculated on a sliding scale with a maximum corporate rate of 34 percent. The alternative minimum taxable income (AMIT) calculation does not allow certain preference items such as depletion and accelerated depreciation. Federal income tax is the greater of normal tax liability or alternative minimum tax.
Net Income	Taxable Income less Federal Income Taxes (includes minimum tax on tax preference items).
Add-Backs of Depletion and Depreciation	The depreciation and depletion cash flow items as calculated above are considered non-cash add-backs to the net income in the annual cash flow.
Capital Investments	Based on the estimated project capital investments.
Working Capital	Working capital is the money necessary to operate the business on a day-to-day basis and can not be recovered as depreciation, depletion, or amortization. Working capital is typically calculated as a specified number of days for direct operating costs.
Working Capital Recapture	Recapture of the working capital at the end of the project.
Equipment Salvage	Sales value of equipment at end of mine life.

Table A-1 (Cont'd)

Cash Flow Line Item	Basis or Calculation
Income Loss Carry Forward	Income Loss Carry Forward is a non- cash add-back item similar to depreciation and depletion.
Annual Cash Flow	Net Income plus Depreciation, Depletion Allowance, Income Loss Carry Forward, working Capital Recapture, and Equipment Salvage less Capital Investments, and Working Capital.
Discounted Cash Flow Rate of Return (DCFROR)	Rate of return where the net present value of the annual cash flows equals zero.
Net Present Value (NPV)	The cumulative value of the annual cash flows discounted at specific rate of return.

Table A-2

EXAMPLE
EVALUATION OF NEVADA GOLD PROPERTY
CASH FLOW STATEMENT

```
CASH FLOW STATEMENT
*********************
          GOLD PRICE   = $450.00
          SILVER PRICE = $7.00
```

	YEAR 1	YEAR 2	YEAR 3	YEAR 4	Total
MATERIAL MINED AND PROCESSED (Tons):					
TOTAL ORE MINED	750,000	1,000,000	1,000,000	750,000	3,500,000
TOTAL WASTE MINED	1,125,000	750,000	750,000	562,500	3,187,500
TOTAL MATERIAL MINED	1,875,000	1,750,000	1,750,000	1,312,500	6,687,500
TOTAL ORE PROCESSED	657,534	1,000,000	1,000,000	842,466	3,500,000
PAYABLE GOLD (Troy Ounces)	15,623	23,760	23,760	20,017	83,160
PAYABLE SILVER (Troy Ounces)	490,685	746,250	746,250	628,690	2,611,875
GOLD SALES	$7,030,354	$10,692,000	$10,692,000	$9,007,646	$37,422,000
SILVER SALES	$3,434,793	$5,223,750	$5,223,750	$4,400,832	$18,283,125
TOTAL METAL SALES	$10,465,147	$15,915,750	$15,915,750	$13,408,478	$55,705,125
-REFINERY/TRANS/INSURANCE	347,342	372,000	372,000	340,658	1,252,000
=NET SMELTER RETURN	$10,417,804	$15,843,750	$15,843,750	$13,067,821	$55,453,125
-NSR ROYALTY AT 5%	520,890	792,188	792,188	667,391	2,772,656
=GROSS INCOME FROM MINING	$9,896,914	$15,051,563	$15,051,563	$12,680,430	$52,680,669
-OPERATING COSTS	6,596,403	8,300,000	8,300,000	6,641,097	29,837,500
-DEPRECIATION	1,582,778	2,407,143	2,407,143	2,027,936	8,425,000
-NET PROCEEDS TAX	944,000	113,900	113,900	999,687	373,687
-PROPERTY TAX	970,584	958,585	949,677	942,994	221,832
=NET BEFORE DEPLETION	1,401,148	4,171,973	4,180,843	3,848,524	13,822,550
-DEPLETION	600,574	2,085,948	2,090,421	1,902,064	4,879,028
=INCOME BEFORE LOSS CARRY FORWARD	800,574	2,085,948	2,090,421	1,966,440	8,943,423
-TAX LOSS CARRY FORWARD	800,574	0	0	0	90
=NET TAXABLE INCOME	0	2,085,948	2,090,421	1,966,440	8,943,423
-FEDERAL INCOME TAX	0	834,387	836,166	773,705	2,764,600
=NET INCOME	0	1,251,581	1,254,255	1,192,755	4,176,933
+DEPRECIATION	1,582,778	2,407,143	2,407,143	2,027,936	8,425,000
+DEPLETION	600,574	2,085,948	2,090,421	1,902,064	6,879,028
-CAPITAL INVESTMENTS	10,250,000	0	0	0	10,250,000
-WORKING CAPITAL	1,303,333	0	0	0	1,303,333
+WORKING CAPITAL RECAPTURE	0	0	0	1,303,333	1,303,333
+EQUIPMENT SALVAGE	0	0	0	1,825,000	1,825,000
+INCOME LOSS CARRY FORWARD	0	0	0	0	0
=ANNUAL CASH FLOW	($8,769,636)	$5,744,691	$5,751,817	$8,331,089	$11,057,960

```
ECONOMIC RESULTS:

          NPV AT 10%  = $6,786,953
          NPV AT 15%  = $5,263,281
          NPV AT 20%  = $4,027,630
          DCFROR      = 50.79%
```

Chapter 10

AN EMPIRICAL METHODOLOGY FOR THE VALUATION
OF RISKY GOLD CASHFLOWS

Robert T. McKnight, P.Eng.

Manager, Finance and Planning,
Getty Resources Limited
Vancouver, B.C.

Introduction

The minerals exploration business, like many
businesses, is faced with the problem of valuing future,
risky cashflows. These cashflows are derived from those
mineral deposits which have been successfully
commercialized. Many deposits, however, do not have the
quantities or grade of ore to be considered economic at
the projected metal prices using the traditional tools of
the Capital Asset Pricing Model and discounted cashflow
analyses. If one calculates the future revenues, costs
and cashflows of one these marginal properties, and
discounted it at the appropriate risk-adjusted discount
rate, it may have a negative present value. Yet such a
property may still have real, positive value in the
marketplace. Why would a firm pay money for a property
with an (apparent) minus present value? Mining
properties such as these are bought and sold in arms-
length transactions on a regular basis. Why is this so?

Part of the reason relates to the nature of ore reserve
information. To generate cashflows, the analyst must t
assume that the reserve information is known with
certainty when, in fact, reserves are not known with
certainty. Mineral reserves are a function of commodity
price, extraction costs and the level of (usually
drilling) information about the reserves. Stated, or
currently known, mineral inventories may vastly
understate the true state of nature. Some firms may
therefore be buying a property based on a completely

different (ie more optimistic) view of the geology and
mineral reserves.

Another reason why firms might buy "uneconomic"
properties relates to the value of the "options" which
the firm is buying when it commits to purchase a mineral
property. These options might include, for example, the
right to put the property into production if metal prices
go .up or close it down when prices fall. Do these
options have a value? The answer to this question forms
the basis of this paper.

Traditional Analysis Techniques

The traditional methods of valuing mineral
properties have changed little over the years. These
techniques consist basically of projecting expected risky
cashflows into the future and discounting them at a rate
considered appropriate in the particular instance.
Certain refinements, such as simulation (Monte Carlo)
methods and have been introduced but these, again, just
repeat the same basic DCF-type calculations many times
over in hopes of gaining new insights.

The main trouble with traditional DCF valuation
methods centers around their "static" nature. They take
no account of the stochastic nature of metals prices or
of the range of management's available responses to price
fluctuation. If one could think of the owner of a mining
operation as having the right to make decisions about the
mine in response to changing metal prices, then he has a
a series of "options" available to him. For instance,
the owner has the option of shutting the mine temporarily
during a period of low prices, or increasing production
or changing the grades of ore mined. All these options
have a definite value (and a cost) which the analyst may
find difficult to evaluate with his conventional tools
using DCF analysis.

Under certain circumstances, the value of these
"options" can be estimated by the use of certain
mathematical techniques and some computer "number
crunching". In a sense, the situation is analogous to
one who buys an option (a warrant) on a particular stock.
The market value of that warrant will always exceed the
net proceeds from an early exercise of the warrant, even
if the underlying stock price is less than the exercise
price of the warrant (ie. the warrant is "out-of-the-
money"). Similarly, the value of the rights conferred by
ownership of a resource should also be a positive number.
For instance, a gold property which is considered

uneconomic merely because current gold prices are below
the projected operating costs may still have considerable
market value. That is, a firm will pay money for the
right to explore and mine that property at some future
date. The value is that of the option to respond to
price movements.

     In the last few years, finance researchers,
recognizing the shortcomings of conventional DCF methods
(based on the Capital Asset Pricing Model), have applied
the principles of option pricing theory to the study of
cashflows which are dependent on stochastically-varying
metal prices. Some interesting new concepts have emerged
which may help analysts value mining cashflows.

Financial Model Methodology

     Previous efforts have been made to apply option
pricing theories to natural resource investments.[1] These
approaches were analytical in nature. That is, they
applied computerized numerical methods to solve the
system of equations which describe the particular
problem. The approach taken in this paper is empirical,
or intuitive, in nature, and allows one to examine a more
complex set of conditions than may be possible using an
analytical method. A simulation model with empirical
price and cost inputs was developed (Figure 1.) with
which to analyze the cashflow stream from a hypothetical
(but not unrealistic) gold property which is subjected to
randomly varying gold prices.

     This model was constructed using Lotus Symphony[2]
Version 1.2 on a AT-class microcomputer based on the
Intel 80286 chip. Symphony is a flexible spreadsheet
application for microcomputers, combining spreadsheet,
word processing, and graphics applications plus a
programming language. All of the graphs and some of the
tables in this document were prepared using Symphony.
Within the Symphony model, a program called a "macro" was

---

[1]   Michael J. Brennan and Eduardo Schwartz, Evaluating
      Natural Resource Investments, Journal of Business,
      1985, vol. 58, no. 2

      Scott Palm, Neil Pearson, James Read, Jr; Option
      Pricing: A New Approach to Mine Valuation, CIM
      Bulletin, May, 1986.

[2]   Symphony is a registered trademark of Lotus
      Development Corporation.

written to run the iterative calculations and store
intermediate results for analysis and graphing.

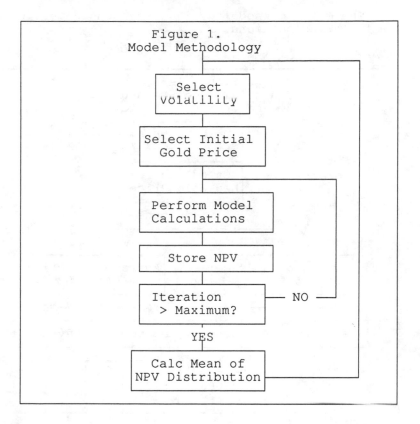

Figure 1.
Model Methodology

The main elements of the model encompass:

1. A hypothetical gold mine with its associated costs
   and gold output stream,

2. Randomly changing monthly gold price.

3. Mine operating policy.
   This dictates when the mine will open and close in
   response to gold price movements.

4. Cashflow analysis.
   This analysis which is described later, involves
   performing at least 100 iterations for each initial
   gold price, with each iteration projecting the
   random gold prices and mine status changes over a

100 month time period. The discounted cashflows
(DCF) of each iteration are stored and the mean DCF
of the distribution is calculated.

The model methodology is discussed in more detail in
the following sections.

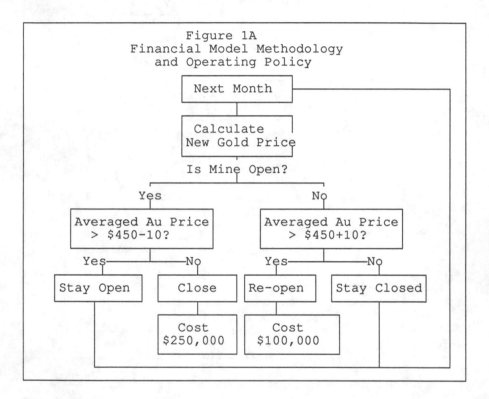

Figure 1A
Financial Model Methodology
and Operating Policy

The Gold Mine

The model employed contemplates a gold mine having
the following characteristics:

Gold Output Rate: 4,000 ounces per month

Initial State: Closed

Life of Mine: Variable depending on gold
                         price changes over a 100
                         month period. Maximum of

	100 months. minimum is zero months.
Operating costs:	$450 per ounce. The costs are assumed to be all variable costs. No fixed costs are assumed either on a short or long-term basis.
Closure Cost:	$250,000 per closure. There is assumed to be a one-time cost associated with each (temporary) closing of the mine. This may include severance payments, cleanup, etc).
Re-opening Cost:	$100,000 per re-opening.
Initial Gold Price:	Initial gold prices are varied in the analysis procedure from $350 to $550 per ounce. The prices then are allowed to vary randomly with a given volatility.

The logic employed in the model operating policy is shown graphically in Figure 1A.

Historical Gold Prices

Historical gold price data were obtained from a metal price database through I.P. Sharpe's database access facility. These data were downloaded to a diskette in ASCII[3] format and loaded into a Symphony file for analysis. The raw data and calculated price change data are contained the Appendices.

Gold prices have exhibited extreme variability since being released from the gold standard in the early 1970's. The analysis in this paper assumes that gold price changes are random in nature. Is this a reasonable assumption? A study of commodity prices in the 1800's lead to some important insights regarding the efficient market hypothesis. Gold, as a commodity, has a very

---

[3] For American Standard Code for Information Interchange

NORMAL DISTRIBUTION Figure 3

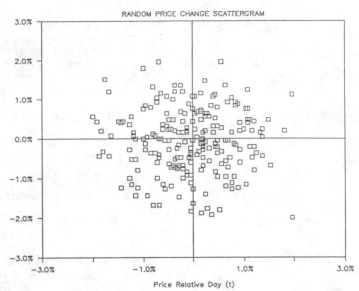

HISTORICAL GOLD PRICES Figure 4

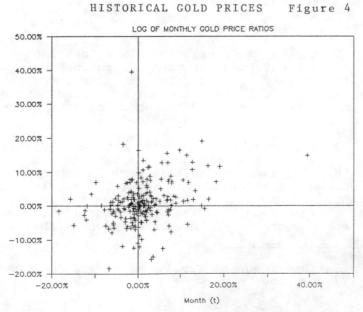

active market, with many buyers and sellers acting
independently based on their needs for the metal and
expectations as to future price movements. The current
spot market price for gold will therefore reflect the
expectation of the market as to its likely future value.
Future changes in gold prices from the current spot
market value should, therefore, reflect unexpected events
which, by definition, cannot be predicted. Thus gold
price changes should be random, in nature. Does the
historical record of gold prices demonstrate this
randomness?

Figure 3 shows a pattern of random daily price
changes generated from a normal distribution. The price
change on day (t) is plotted against the change on Day
(t+1)[4]. A random price change pattern is one which is
clustered evenly around the origin (0,0) and decreases in
density away from the origin. Now consider Figure 4.
This is the actual pattern of monthly gold price
movements from the gold price time series. The points
are clustered around (0,0) but there may be slightly more
points in the NE quadrant, although this may not be
statistically significant. Also, there are two outlyers
lying well away from any others. Erratic data points
such as these can be found in many data sets. If the
monthly gold price change data series are plotted on a
frequency graph, they look like Figure 5. Ideally, if
there is no cumulative drift in prices, a random pattern
would yield a normal distribution with a mean of zero[5].
The actual mean of this distribution is +1.1% with a
standard deviation of the sample distribution of 6.3%.
This may reflect the influence of inflation, or a
positive drift, on gold prices which would yield means
greater than zero.

With an historical time series, it is also possible
to calculate actual gold price volatility[6]. Figure 6.

---

[4]   Actually, the natural log of the ratio of
      $(price)_t/(price)_{t-1}$ is plotted.

[5]   For the natural log of the ratio of

      $(price)_t/(price)_{t-1}$

[6]   Volatility, as used herein, is defined as the
      standard deviation of a distribution composed of the
      natural logarithmn of the ratio of a price change

HISTORICAL GOLD PRICE   Figure 5.

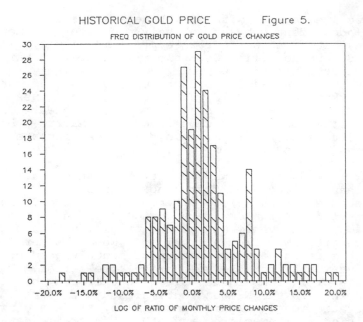

FREQ DISTRIBUTION OF GOLD PRICE CHANGES

HISTORICAL GOLD PRICE   Figure 6.

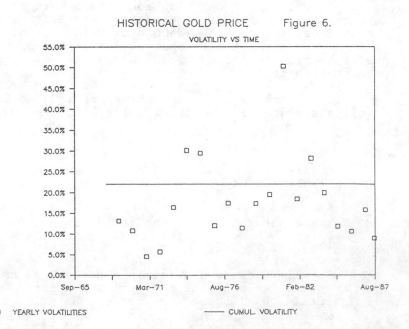

VOLATILITY VS TIME

shows actual computed gold price volatilities for each
year of the time series. The cumulative volatility (over
the entire time series) is calculated to be 21.9% per
year, but varies widely from less than 5% in 1970 under
the gold standard, to about 50% in the 1980-81 during the
very high gold prices in that period.

Random Walk Methodology

In the Symphony financial model, gold prices are
projected forward from an initial price using a "random
walk" method. The volatility of the random walk can be
varied to that desired for the particular run. From the
initial gold price setting, the random number generator
within Symphony was used to generate the next month's
gold price with the given degree of volatility.
Symphony's random number generator does not, however,
produce random numbers from the desired normal
distribution. It produces a uniform distribution of
random numbers in a range from zero to 1.0. (Figure 7.)

Figure 7.
Symphony Random Number Generator
(Uniform Distribution 0 to 1.0)

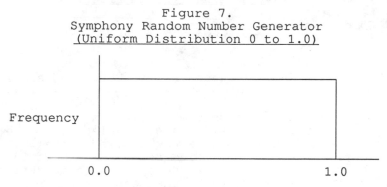

This distribution can be then altered to one with a mean
of zero as in Figure 8. A number randomly sampled from
this distribution was divided by the inverse of the
desired annual volatility.

from one period to the next.
ie. $\ln(Price_t/Price_{t-1})$

Figure 8.
Symphony Random Number Generator
Converted to a Distribution Symmetric about Zero

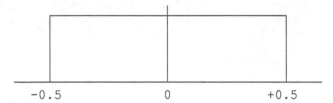

```
        -0.5              0              +0.5
```

This number is then added to one (1.0) and used to multiply the previous month's gold price to produce the current month price.

This calculation is summarized below:

Projection of Randomly
Varying Monthly Gold Price

Gold Price$_t$ = Gold Price$_{t-1}$ *
                {1+[(RAND-0.5)/(1/Annual
Volatility)]}

where RAND is random number between 0 and 1

Employing this procedure will yield a distribution with approximately the desired annual volatility. To test that this method produces the desired annual volatility, a test was conducted for 37 trials of 100 periods each. Figure 8A shows the resulting volatility distribution. As can be seen, the mean of the distribution in Figure 8A is 19.9% with a standard deviation of 0.9%, or very close to the desired result of 20%. This sampling method would produce the scattergram shown in Figure 9. This is less circular plot than that of the normal distribution but is still centered about the origin.

Mine Operating Policy

The mine operating policy dictates when the manager of the hypothetical gold mining operation will open and close the mine in response to gold price changes. The policy thus simulates, in a simple manner, the psychology of a business decision.

The operating policy employed in the financial model has the following rules:

1. The model (ie. mine manager) has no knowledge of future gold prices other than any information which might be available from a study of the gold futures market. The only other information available to the manager are those relating to his operating costs and past monthly gold prices.

2. At the end of each month, the manager can make a decision as to whether to:

   A. Either continue or cease mining, in the event that the mine was open in the previous month.

   RULE: if the average of the previous two (2) months gold price was more than $10/ounce <u>less</u> than the mine operating costs ($450 per ounce) then close the mine. That is, if the average gold price for months (t-1) and (t-2) is less than $440 per ounce then shut the mine (at a cost of $100,000), otherwise continue mining operations. A variation of this rule might be to consider the trend (sign of the first derivative) of the gold price series during the averaging period. That is, if prices are trending downward then modify the decision rule accordingly. Normal human reaction is to weight the most recent price more heavily than an older price. Another alternative (not examined) might be to vary the averaging period is reponse to the variation of gold price[7].

   B. Re-open the mine or continue the mine closure, in the event that the mine had been closed earlier.

   RULE: if the average of the previous two (2) months gold price was more than $10/ounce <u>more</u> than the mine operating costs ($450 per ounce) then open the mine. That is, if the average gold price for months (t-1) and (t-2) is more than $460 per ounce then re-open the mine (at a cost of $250,000), otherwise continue the shutdown. The same alternative decision rules could be examined here as well, including the effect of

---

[7] For instance, if price variance is low then perhaps a shorter averaging period is more appropriate than a longer one.

FREQUENCY DISTRIBUTION   Figure 8A.

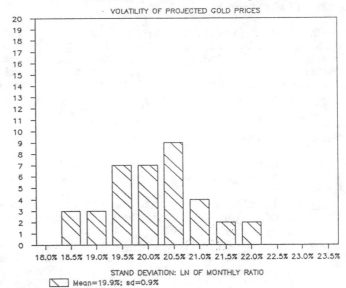

VOLATILITY OF PROJECTED GOLD PRICES

No. of Occurrences

STAND DEVIATION: LN OF MONTHLY RATIO

Mean=19.9%; sd=0.9%

RANDOM DAILY CHANGES   Figure 9.

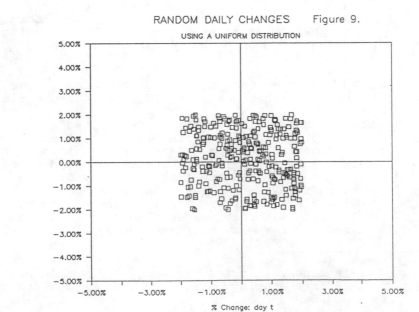

USING A UNIFORM DISTRIBUTION

% Change: day (t+1)

% Change: day t

changing the "hurdle increment", the increment
over breakeven costs before the manager will re-
open the mine.

This process of determining when to open and close
the mine is displayed graphically in Figure 10.   The
graph shows, on the upper line, the randomly varying
monthly gold price.   The lower hashured line shows the
changing status of the mine (mine open status is
indicated by a hashured line at the $450 level).   When
the manager of the mine (ie. the model) is deciding
whether to open or close the mine each month, there is a
financial decision being made.   For example, consider the
case where the mine is status is "CLOSED" but the current
gold prices have increased over the intervening time
period since closure so that they are now higher than the
mine operating costs.   The cost of opening must be
weighed against the liklihood of possible positive future
cashflows should gold prices continue to remain high.
The higher that gold prices go above mine operating costs
before a reopening decision, the lower the probability
that gold prices will decline below the breakeven cost
before a return is achieved on the reopening cost.   This
does not, however, consider the opportunity cost of
waiting to see if the gold price will reach the higher
hurdle price.   This   implies that there may be an optimal
hurdle price which is a function of re-opening cost and
price volatility, given a fixed output rate and cost
structure.   This may be an area of further study.

More complex operating policies are able to be
examined with the financial model.   For instance, the
financial model now allows the mine to be either open or
closed.   One refinement may be to allow increased or
decreased output levels on a more continuous basis.
Another might be to use decision criteria which looks at
not only the historical gold prices but also the
volatility of price in the recent months, or a change in
the averaging period of gold price.. No attempt was made
in this study to determine the existence of an optimal
operating policy.   Extensions to this study might examine
what optimalities, if any, may be present in an operating
policy or the use of gold futures .

Cashflow Analysis

From a given initial gold price, projection of
cashflows are made into the future for each simulation
(at least 100 iterations) and the resultant cashflows
discounted at the risk-free rate.   Each present value is
stored as part of a distribution of values (Figure 13)

STATUS OF MINE   Figure 10.

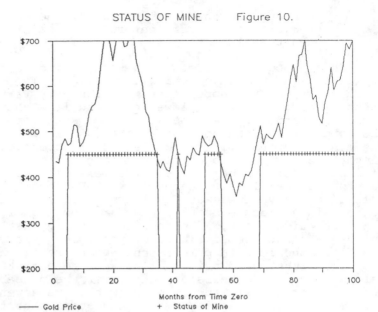

VALUE OF GOLD PROPERTY   Figure 10A.

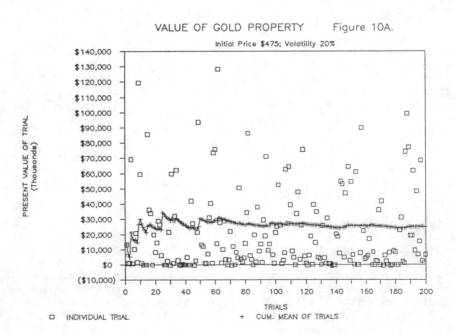

and a mean of the distribution is calculated and plotted on a graph (Figure 11)[8]. This procedure is then repeated for a different volatility and initial gold price. A risk-free (or near risk-free) rate is considered appropriate since the variability of the price component of cashflow, which is the major risk factor, is explicitly considered by the analysis method. A risk-free rate of 6.5% was employed for the analysis.

The results of the simulations are condensed into the following Table and are also presented in graphical form as Figure 11. The values of the cashflow streams for a given volatility increase monotonically with the increasing initial gold price. The effect of higher volatility is to increase the value of the mine. The data do not fall on a smooth curve, although with more iterations they would likely do so. Looking at the graphical representation of the data, it looks very much like a suite of European call option curves with an exercise price of $450 per ounce, equal to our mine operating cost (see Figure 12). The lower, straight line is the present value of the cashflow assuming no variability of gold prices. Thus, at a price of less than $450, equal to operating costs, present value is zero. At higher prices, the present value increases linearly. This is equivalent to the exercise value of call option before expiry as shown on Figure 12.

A typical frequency distribution of present values is shown in Figure 13. This displays a lognormal-type of distribution. The mean of this distribution is plotted on the graph (see Figure 11).

---

[8] Figure 10A shows the result of one iteration of 100 trials. The points denoted as square boxes are the present values of individual trials. Note that the mean (shown as the line of +'s) approaches a stable value as the cumulative number of trials increases. The resulting mean is plotted as a single point on Figure 11.

Value of Gold Property   Figure 11.

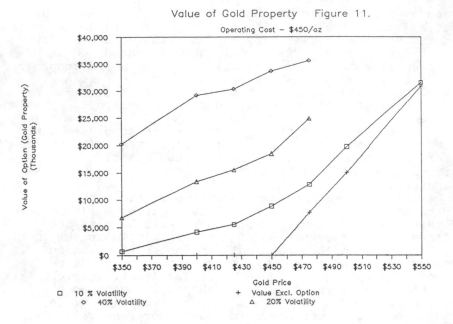

Operating Cost — $450/oz

Figure 12.

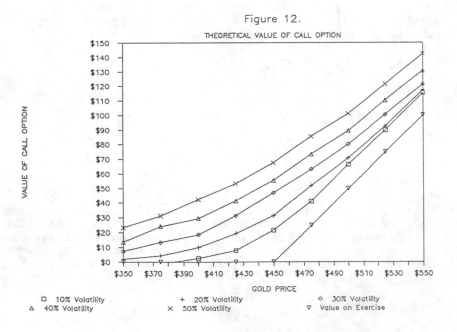

THEORETICAL VALUE OF CALL OPTION

Results of Financial Model
Means of NPV Distribution
($000's)

Initial Gold Price	Conventional DCF at 6.5%	VOLATILITY		
		10%/yr	20%/yr	40%/yr
$350	0	694	6,825	20,195
$400	0	4,188	13,403	29,217
$425	0	5,540	15,517	30,360
$450	0	8,867	18,450	33,646
$475	7,705	12,822	24,862	35,606
$500	15,000	19,698		
$550	30,821	31,482		

Conclusion

The results of this study strongly suggest that, under certain circumstances the value of a stream of risky cashflows which are dependent on the price of a randomly-varying commodity can, under certain conditions be compared with the value of a call option on the commodity with an exercise price equal to the mine operating cost. The study makes some simplifying assumptions with regard to operating costs. For instance, no on-going fixed costs are assumed during mine closure periods. Also, the procedure for projecting random gold prices is not normal in character. This may cause some distortion in the results. Further study and refinement of the financial model is contemplated to address these and other issues.

Chapter 11

GOLD BULLION FINANCING IN A POST-CRASH WORLD:
TRENDS, RESULTS & APPLICATIONS

Stephen B. Doppler

Doppler Mineral Research, Inc.
Evergreen, Colorado

INTRODUCTION

Following the world stock market debacle in October
1987, it would have been foolhardy to expect that
securing equity funding in 1988 would be easy. Indeed,
major equity financing's for gold mining companies have
until recently been few and far between in the weeks and
months following Black Monday. As a result, gold bullion
loans have taken on an increasingly important role among
producing firms in the industry.

During 1988, the role of bullion loans has become
increasingly complex and controversial. Some observers
herald the emergence of a new breed of bullion loans -
such as the 1,000,000 ounce loan to Newmont Mining, and
the 400,000+ ounce facilities drawn down by Bond
International Gold and Placer Pacific - as both
increasing the public's awareness of the technique, and
representing the dawn of a new age of gold financing.
But are these recent loan facilities significantly
different than those issued worldwide during the early to
mid 1980's?

Similarly, observers have voiced concerns regarding
the role that gold loans play in the world's bullion
markets, specifically in relation to their impact on
prices. Have the new generation of mega-loans depressed
current and future prices? Has the current surge of
bullion loans irrevocably changed the nature of the gold
market? Is there an upper limit on the number and size
of gold loans?

To answer these and other questions, the author draws upon his worldwide gold loan database. As of mid-1988, a total of 162 bullion loans (including offers of finance) have been identified, of which 114 fully documented transactions were compiled and analyzed for the purposes of this paper. The remaining 48 loans are either in the planning stage, have been established but not drawn down, or insufficient data precludes the inclusion of their parameters for analysis.

Background

Although gold loans have been utilized as a mining finance vehicle since 1981, it has only been in the last several years that bullion lending has emerged as a prominent method of financing development of gold mines.

Bullion loans were largely developed and refined in Australia to satisfy the needs of that country's re-emerging gold industry. To date, 73 of 162 gold loans involve Australian mining projects and/or Australian-based companies. The United States and Canada, despite sharing the distinction of having lent gold in mid-1981, trail Australia in absolute number of loans, with 41 and 36 bullion facilities respectively. A further 12 gold loans are known to have been issued for projects and/or companies elsewhere in the world.

Gold loan financing is generally viewed as being profitable for both the lenders - whose gold would otherwise be gathering dust in a vault - and the borrowers, who typically pay interest rates of 1.25 % to 3.50 % per annum - significantly lower than those charged on bank loans or bonds.

As a result, gold loans have become an increasingly popular, low-cost, financial instrument for raising funds in the post-Crash environment where the equity finance route is not as easy as it once was. During the first half of 1988, mining companies borrowed an estimated 3,246,723 ounces of gold ($US 1.38 billion at $US 425/troy ounce); estimates of total bullion borrowing for 1988 are in the vicinity of 5,000,000 ounces ($US 2.125 billion).

January 1988 is generally regarded as a turning point for gold loan financing, when Newmont Mining Corporation announced that it was finalizing negotiations for a six-year, one million ounce loan facility. The importance of this event, namely, in bringing this

financing technique to many people's attention for the
first time, should not be underestimated.

RECENT GOLD LOAN TRENDS & OBSERVATIONS

Since January 1985, an estimated 6,696,142 ounces of
gold bullion (equivalent to $US 2.845 billion at
$US 425.00/troy ounce) has been lent to mining companies
around the world. Of this total, 3,246,723 ounces
(48.5 %) have been lent during the first seven months of
1988, and a total of 3,616,161 ounces (54.0 %) since
Black Monday.

To evaluate this explosion of gold lending, data
from 114 documented bullion loans over the period June
1981 through July 1988 has been compiled into a series of
tables on the following pages in order to provide a more
specific analysis of worldwide bullion loan activity in
the period following Black Monday.

Frequency and Size of Gold Loans

From the data in Table 1, it is evident that a
tremendous upsurge in gold loan financing's has occurred
during the period immediately following Black Monday.
There is little doubt that many of these loans were
inspired by the combination of readily available funds,
comparatively low costs, and a greatly diminished
investor appetite for new gold shares in the wake of the
stock market crash.

The first half of 1988 has been especially active,
with the 27 documented gold loans surpassing the overall
total for the period 1981 through 1985. Of particular
note during the post-Crash period has been the number of
loan facilities greater than 100,000 ounces. On October
29, 1987, American Barrick Resources closed on a 263,713
ounce loan. The following quarter, seven firms entered
into gold loan agreements for facilities ranging between
120,000 and 1,000,000 ounces. Gold loans obtained during
the second quarter of 1988 (with the exception of a
400,000 ounce multi-option facility to Placer Pacific)
were significantly smaller.

Time will tell if the the 1987/88 surge in gold
lending activity has leveled off, or if the decrease in
the first part of the third quarter signifies a return to
a smaller, less active market. Certainly the decline in
gold prices to the $US 430.00/ounce range has reduced
enthusiasm for borrowing gold among those parties who

believe that gold is destined to rise above $US 500/ounce following the fall elections.

Table 1 Frequency and Estimated Size of Documented Gold
        Loan Financing's, June 1981 to July 1988.

YEAR	QUARTER	# LOANS	OUNCES BORROWED	CUMULATIVE TOTAL
1981		2	25,000	25,000
1982	-	1	12,000	37,000
1983	-	0	0	37,000
1984	-	6	201,554	238,554
1985	1	3	114,675	353,229
	2	1	22,000	375,229
	3	5	99,559	474,788
	4	8	<u>377,834</u>	852,622
			614,108	
1986	1	5	155,510	1,008,132
	2	9	109,304	1,117,436
	3	11	342,456	1,459,892
	4	2	<u>126,000</u>	1,585,892
			733,270	
1987	1	8	742,348	2,328,240
	2	9	464,535	2,792,775
	3	8	404,160	3,196,935
	4	6	<u>491,038</u>	3,687,973
			2,102,081	
1988	1	15	2,344,855	6,032,828
	2	12	693,907	6,726,735
	(July)	3	<u>207,961</u>	6,934,696
			3,246,723	

Length of Gold Loan Facilities

A commonly held notion by the gold mining industry is that bullion loans, at least until Newmont's million ounce, six year facility, tended to be of limited duration - generally 1.5 to 3.0 years.

As the data in Table 2 demonstrates, the average bullion loan facility since 1985 has a term of approximately four years. Contrary to popular opinion,

Newmont's benchmark loan is not a pioneer in terms signaling the willingness of lenders to establish longer term facilities.

To illustrate, Echo Bay Mines established two five-year loan facilities totaling 120,004 ounces with the Canadian Imperial Bank of Commerce during 1985 related to the acquisition of the Round Mountain and Sunnyside mines. Sumitomo Metals borrowed an estimated 5 Mt of gold from three metal trading firms in late 1985 as part of a seven year facility. Likewise, Hemlo Gold Mines drew down 380,000 ounces last year in an 8.5 year refinancing of the Golden Giant mine. Lastly, International Corona entered into a ten year, 106,000 ounce facility last October to refinance its share of capital costs for the David Bell mine.

It is erroneous to confuse the size of a bullion loan facility with the term of the loan. Lenders have very specific guidelines for structuring gold loans, and are generally reluctant to extend a loan beyond the mid life point of any producing property.

Table 2   Length of Gold Bullion Loans, Calculated from Date of First Drawdown to Final Settlement

YEAR	# LOANS	1	2	3	4	5	6	7	8	9	10	N/A
1981	2	2	–	–	–	–	–	–	–	–	--	--
1982	1	–	–	1	–	–	–	–	–	–	--	--
1983	0	–	–	–	–	–	–	–	–	–	--	--
1984	6	1	–	–	1	2	–	–	–	–	--	2
1985	17	–	2	1	2	5	–	1	–	–	--	6
1986	27	3	5	6	4	–	–	1	–	–	--	8
1987	31	2	2	2	3	1	3	–	–	1	1	16
1988*	30	–	3	1	5	3	2	2	–	–	--	14
TOTALS	114	8	12	11	15	11	5	4	0	1	1	46

* January 1 to July 30

Size & Frequency of Gold Bullion Loan Facilities

That there has been an explosion of bullion loans worldwide since Black Monday is without question. The 38 documented bullion loans issued between October 19, 1987 and July 31, 1988 averaged 95,162 troy ounces. To be fair, it must be noted that three loans - 1,000,000 ounces to Newmont Mining, 421,500 ounces to Bond International Gold, and 400,000 ounces to Placer Pacific

- account for a significant (50.4 %) of the total;
excluding these three loans from the analysis reduces the
average to 51,276 ounces per loan.

On the whole, gold loans issued during 1987 and 1988
have tended to be significantly larger (averaging over
82,000 ounces) than those over the 1984 to 1986 period
(30,000 to 35,000 ounces). There has been a sizable
number of large (> 150,000 ounces) loans over the past 18
months which tends to skewer the results, but in no way
diminishes the growth in absolute gold loan size
indicated in Table 3.

Table 3   Frequency of Bullion Loans by Estimated or
           Reported Size of Facility (Thousands of Troy
           Ounces).

OUNCES	1981	1982	1983	1984	1985	1986	1987	1988
< 10	–	–	–	3	3	14	8	2
10-25	2	1	–	1	6	4	4	8
25-50	–	–	–	–	4	5	3	7
50-100	–	–	–	2	3	2	6	3
100-200	–	–	–	–	1	2	3	6
200-500	–	–	–	–	–	–	3	2
> 500	–	–	–	–	–	–	–	1

However, it appears that the gradual growth in
frequency and absolute size of bullion loans during the
months leading up to Black Monday has accelerated during
the first half of 1988. It remains to be seen if the
drop off in loan activity during July 1988 represents a
short term slowing of this growth trend, or the start of
a more gradual decline in gold lending activity to pre-
Crash levels.

Uses of Gold Loans over Time

The most noticeable trend among published or stated
uses of gold loans is the increased utilization of these
facilities for debt refinancing or payback of existing
cash or bullion facilities. Of the 14 loans utilized for
this purpose, eight (57 %) have been secured since Black
Monday.

The rationale behind this re-financing activity
varies. American Barrick Resources borrowed 125,000
ounces in February 1988, in part to repay an earlier loan
(drawn at significantly lower gold prices) secured

against the Mercur mine. Similarly, NERCO borrowed
124,000 ounces in March 1988, a large portion of which
went to repay a 98,000 ounce loan taken out during
January 1987. Galactic Resources is reported to have
taken out a $US 13.75 million facility to repay its
existing US dollar loan at Summitville. Lastly, Newmont
Mining has borrowed 1,000,000 ounces to refinance debts
incurred during a corporate takeover defense.

Table 4   Published/Stated Uses of Gold Bullion Loan
            Facilities, June 1981 to July 1988.

	1981-84	1985	1986	1987	1988	TOTAL
Acquisition	–	5	3	1	1	10
Exploration	–	–	–	–	1	1
Working Capital	3	1	6	6	2	18
Mine Development	6	12	24	19	21	82
Debt Refinancing	–	1	1	4	8	14

        note: totals may not conform to other
                tabulations in this paper due to
                multiple declared intentions of
                borrowers.

    The future growth of bullion loans utilized for debt
refinancing will in all likelihood depend on each
borrower's future outlook for bullion prices. Should
sentiments favor a resurgence in gold prices during 1989
and beyond, it is very likely that firms who took out
longer term gold loans in the $US 300 to $US 350/ounce
range would become prime candidates for refinancing.

    The decline in gold loans utilized for funding
acquisitions is equally evident from Table 4.
Speculation for its demise includes the current industry
emphasis on amalgamations, and the comparative ease of
funding takeovers through equity or convertible debt
issues in the period preceding Black Monday.

    The use of gold loans to fund exploration programs
may prove to be a future growth area for the bullion
lenders. Cambior, Inc.'s June 1988 gold loan to Nova-
Cogesco Resources for exploration at the Sildor property
is the first definite example of this application,
although it is conceivable that previous loans
categorized as "working capital" facilities may have been
applied to this purpose.

Currency Conversion

To add flexibility to gold loans, a provision allowing the borrower to convert the outstanding loan balance between bullion and one or more currencies is often incorporated. For companies based in North America, this would normally allow for conversion between gold and US and/or Canadian dollars, although in principle, conversion to almost any could be arranged. Interest rates for converting gold loans are typically set in advance, generally at LIBOR plus a percentage. A minimum conversion fee is typically charged, and borrowers are sometimes limited as to the number of conversions allowed.

The ability to convert between bullion and currency provides the borrower with a significant advantage during times of volatile gold prices and interest rates.

To illustrate, consider the case of an Australian gold firm operating a producing property in Australia with a 10,000 ounce convertible debt facility to be repaid. Should the company forecast rising (falling) gold prices, the firm would switch its debt into currency (gold) while gold rises (falls), and then back into gold (currency) at a higher (lower) price. In the situation where the mining company expects the A$ to weaken (strengthen) against the US$, the optimal strategy would be to convert the debt to US$ (A$) while the A$ (US$) weakens, and then switch back into A$ (US$) at the stronger (weaker) exchange rate.

Impact of Newmont Mining Corp.'s Million Ounce Gold Loan

Newmont Mining Corporation's million ounce gold loan has been hailed as the financial vehicle which would enhance the popularity and credibility of gold-backed financing and further fuel the boom in North American and Australian gold production, and as signaling the growing importance of low-interest gold loans as the preferred method of financing among gold producers.

The prevailing opinion has been that lenders, at least until recently, tended to structure gold loans with terms of less than three years, and that these facilities were used primarily to fund working capital requirements. Newmont's five-year loan has been widely interpreted as signaling a new willingness to lend large amounts of gold (greater than 400,000 ounces) for longer terms.

The only unique feature of the Newmont loan is its absolute size. In terms of "rule of thumb" criteria utilized to evaluate bullion loans, this facility is well within the norm. Annual repayments of principal (200,000 ounces) equate to roughly 14.3 % of Newmont Gold's projected average production over the life of the loan, a level slightly below the 15 % to 25 % range common for gold loans the author has studied. Similarly, the 1,000,000 ounce facility represents just 6.7 % of Newmont Gold's published reserves (as of December 31, 1987), again, well within the range of similar gold loans.

STRUCTURAL CHANGES IN GOLD BULLION LOAN FACILITIES

Small mining companies have historically been at their most vulnerable stage as they make the transition from exploration to production. Often with limited fundraising abilities, these firms were often forced to sell off a major portion of their assets to gain financing. The advent of gold loans has made it possible for these small companies to gain access to readily available finance and avoid forced equity dilution of mining properties and shareholders.

The technique of lending gold, although first utilized in North America, evolved and matured in Australia during the mid 1980's for the development of low cost, bulk tonnage, low grade open pit gold mines. Having proved successful, gold loans are now widely used throughout the world for a variety of debt financing requirements.

The October 1987 stock market crash and the climate of falling interest rates in the first half of 1988 has prompted, if not accelerated, further developments and refinements in the structure of gold loans. While the basic financing mechanism remains essentially unchanged, new features have emerged to handle the needs of an increasingly sophisticated and demanding group of borrowers. In a nutshell, the key changes involve increased flexibility, something that most earlier gold loans lacked.

Gold loans of the early to mid-1980's were designed primarily to service the needs of junior mining companies who, more often than not, were trying to fast-track small gold properties into production. These companies typically possessed neither cash nor a proven track record, and in all likelihood, without gold loans, would probably have struggled to raise adequate financing

without serious dilution of property interests and
shareholders.

Companies in this situation required a simple, short
term solution, for which the early gold loans served
admirably, offering mine development funds for a one to
three year term with interest rates in the 3 % to 5 %
range during a time when rates on comparable cash
facilities were considerably higher.

As mining companies became more established,
borrowers found the "simple" gold loan to be increasingly
restrictive. The facilities were generally of limited
duration and forced producers to lock in the bulk of
their output at the prevailing gold price, surrendering
nearly all of the upside potential.

Among recent structural changes in gold loan
financing are:

Synthetic Gold Loans

This financing vehicle involves borrowing the
required funds in cash, and offsetting the currency loan
interest rates by the interest premiums earned on the
forward sale of gold bullion.

Assume that $US 10,000,000 is borrowed at 10 %, with
repayment scheduled over a four year period. The
borrower would then arrange sufficient future deliveries
of gold (with a value equal to $US 2,500,000 plus the
forward premium or contango) to offset the required
principle and interest repayments. Assuming a gold
forward sale premium of 6.5 %, the implied interest rate
of the $US 10 million "synthetic" gold loan is 3.5 %.

Gold Leases

Gold lease financing arrangements add the
depreciation benefits relating to ownership of plant
facilities and equipment to that of a gold loan. Overall
costs tend to be equal, or slightly less than that of
traditional gold loans.

Gold lease arrangements first require that a gold
loan be made to the lessor, who would sell the gold for
cash with which to purchase or construct the mine's
processing plant. The lessor, as owner, may claim
depreciation benefits, and is free to enter into a lease
arrangement whereby the mine operator has use of the

plant in return for a rental expressed in troy ounces of
gold. Rental receipts as such allow the lessor to
satisfy the delivery obligations of the gold loan.

Gold Hedging/ Forward Sale Arrangements

Gold forward sale facilities are in most respects
similar to a conventional gold loan. The primary
difference concerns the mechanics of gold delivery and
repayment; in lieu of a loan of gold bullion from the
metal trader to the borrower, the borrower contracts to
sell forward a portion of future production at prices
which are locked in up front, with payment made in
advance of actual delivery.

A large number of gold loan facilities in Australia
during the past few years have in fact been forward sale
agreements. The driving force behind the popularity of
this mechanism is the belief that gold delivered under
such a facility would be exempt from any future gold
production tax.

Floor Price - Price Protection - Hedging Programs

Lenders generally require that new gold producers
hedge that portion of annual gold production (generally a
sufficient quantity to cover operating costs) required to
ensure the mine's economic viability in the event of a
prolonged downturn in gold prices.

Floor price programs have been around for some time,
but today's products tend to be much larger and more
sophisticated. Mase-Westpac and Mocatta Metals have been
aggressively marketing these services of late.

In essence, floor price facilities involve a metal
trader guaranteeing a minimum price for a set amount of
production. In a participatory program, the borrower is
guaranteed a minimum floor price for a specific
percentage of annual production in return for granting
the bullion lender the option to purchase a percentage of
hedged production at the participation price. In
contrast, under a non-participatory program, the producer
pays a premium to lock in a floor price for specific
deliveries under which the mining company will receive
the higher of the floor price or the spot price.

Multi-Option Facilities

Over the last several years, as the gap between
market borrowing rates and gold loan rates diminished in
line with the general fall in interest rates, debt
facilities were developed that could be drawn down in
cash, gold, or a variety of currency and bullion-related
vehicles.

Should the recent drop in gold prices and/or rise in
interest rates continue, it is entirely likely that
producers may bypass these facilities in favor of the
older-style loans that lock in bullion prices.

Competitive Tender Facilities

Yet another variation has been the development of
facilities which allow the borrower to effectively put
the loan out to tender among bullion dealers at mutually-
agreed intervals (typically six months), so that the
borrower is assured of obtaining the best available
interest rates. Eastmet Resources Ltd. utilized this
approach for financing the development of its Youanmi
project in Australia, assembling a complete financing
package with a total fee (letter of credit and bullion)
of under 3.00 %.

IMPACT OF GOLD LOANS ON FUTURE BULLION PRICES & SUPPLY

In recent months, increasing attention and concern
has been focused upon the growing number and cumulative
size of gold loans, and the subsequent effect, if any, on
gold prices.

The decline in value of the world's equity markets
following the October stock market crash, combined with
the perception of a future economic slowdown, have had a
serious impact on the world's gold mining industry. The
reduction of investor confidence in the world's equity
markets, and the corresponding reluctance of most mining
companies to utilize them as a method for raising
finance, has resulted in gold loans becoming the primary
source of funding - at least in the near term.

Gold Loans & Future Gold Prices

Will the millions of ounces of gold outstanding
under existing bullion loan facilities and hedging
programs cause the market to find itself at some point in

the future to be short of supply, or burdened with excess
gold? Many investors, analysts and traders believe that
gold loans represent downward pressure on gold prices.

In response to those parties who have voiced
concerns that the flood of recent gold loans has caused
bullion to flood the market and depress spot prices,
there are others who maintain that the use of gold loans
is actually bullish for future prices.

If borrowers have committed to pay back gold loans
out of future production, then there should be less metal
available on the market in the months and years ahead.
This scenario would suggest a tightening of future
supplies, with resultant stronger gold prices.

One must bear in mind that gold loans are by nature
a form of forward selling. Bullion loans remove, or at
least reallocate, a portion of tomorrow's gold supply.
What is unclear, however, is the impact that gold loans
will have, if any, on future buyers. Similarly, the
impact of hedging mine output through the use of futures,
options and floor price programs has yet to be fully
assessed, and may have an equal or greater influence on
future prices.

The author supports the argument that the ultimate
effect of ever-increasing gold loans and total ounces
borrowed is to shift the flow of mine supply from the
future to the present. And, to the extent that future
production has been already sold, its negative influence
on price should already have been acted upon and
assimilated by the markets. As gold loans mature and are
repaid, the great majority of the borrowed metal will be
returned to the vaults of central bankers or to the
accounts of metal traders, thereby reducing future
supplies available to the market from current mine
production.

SUMMARY

October 19, 1987 will be long remembered as the day
a stunning 508 point drop on the New York Stock Exchange
set off a chain reaction of similar near-meltdowns of
stock markets around the globe. In the short term, as
equity funds became difficult, if not impossible to
raise, gold loans and their variations have taken on
renewed importance. Mining companies have increasingly
turned to alternative sources of funding as their
frustrations with the equity markets grew, looking to
gold loans, where available, to fill the gap. Since

Black Monday, the size and frequency of gold loans has
picked up dramatically, as mining companies find it
quicker, simpler and more cost-effective to raise funds
with bullion loans than through traditional equity
finance issues.

For many mining groups, gold loans have now become a
routine part of their financing strategy. Echo Bay Mines
reportedly plans to raise $C 125 million during 1988 from
bank loans convertible to gold. Galactic Resources has
recently secured gold loan facilities for the Ridgeway
and Summitville mines, and is reportedly seeking a
$US 100 million loan to develop the Far South East gold/
copper porphry deposit in the Philippines. American
Barrick has set up gold loan facilities totaling 399,713
ounces for its Gold Strike, Mercur and Renabie properties
since October 19.

A tremendous number of new gold mines will be
brought into production during the next several years.
However, not every eligible producer will require, or
desire, a gold-backed financing package.

Future gold loan activity levels are difficult to
project. Borrowers perceptions of future bullion prices,
interest fees and foreign exchange rates are likely to
continue to be a volatile and unpredictable force
governing gold loan activity throughout the world. The
author ventures to suggest that the rate of gold loans
will begin to trail off in the second half of 1988,
declining to the levels of 1986 and 1987, but
representing larger loan facilities.

Fear and investor disinterest have brought many of
the previously over-priced mining equity issues back to
more realistic levels. As this process continues, so
will it hasten the day on which investors view the market
as having stabilized, at which time equity financing will
begin to return to favor. Major equity issues planned
for 1988 by Echo Bay Mines, Bond International Gold and
BP Gold signify a willingness by major firms to test the
waters, and bode well for future equity financing - most
likely at the expense of gold-backed financing.

Although gold loans offer an attractive financing
route for established mining companies, the road is
studded with obstacles for emerging junior mining
companies. The variety of costs and charges associated
with documenting and implementing a gold loan means that,
in practice, these facilities are generally not an
economically viable alternative for firms or properties
requiring small facilities (less than 20,000 ounces).

Since October 19, only three gold loans of less than
10,000 ounces have transacted.

The outlook for the future appears to be one of
minimum loan sizes in the 20,000 to 25,000 ounce range.
Many lenders, as a matter of corporate policy, will
simply not consider loans of less than $US 10 or $US 15
million. Exceptions to this rule will exist, but are
likely to be confined to financing packages offered by
"high-risk" lenders willing to mediate their loan costs
and exposure through stock options, warrants and "equity
kickers."

On a cautionary note, a number of observers have
expressed misgivings about the recent upsurge in gold
loan activity, specifically concerning the involvement of
banks and metal traders who have little bullion market
experience. The potential for financial hardship caused
by this lack of experience should not be underestimated.
Extreme movements in bullion prices over the next several
years could spell disaster for a number of marginal
companies, banks and traders.

Chapter 12

GOLD PLANT START-UP PERFORMANCE - THE FINANCIAL IMPACT

Tostengard, G. R.

Performance Associates, Inc.
Danville, California

ABSTRACT

Gold plant performance during the start-up period, which extends until capacity production is sustained, has a major impact on the internal rate of return and the net present value of gold projects regardless of mining and processing configurations. The start-up period experiences substandard production, extra expenses, and sometimes reduced recovery. This results in lost profits, a portion of which are recouped at the end of the life of the mine. Poor start-up performance can put projects into financial jeopardy. The gold mining industry regularly experiences start-ups ranging from good to poor.

Problem areas, which affect start-up performance, include shortcomings in ore reserve sampling and testing, equipment design and installation, untrained plant operators, inability to cope with problems which arise, and organization of the maintenance program. An investment, equal to 0.5% to 1.0% of plant capital cost, in programs to prepare for start-up, in advance, is recommended. Investment in such a program will make a good start-up possible. Without it the best result which can normally be expected is a fair start-up.

Developers of new gold mines are urged to compare the cost of this investment with the anticipated increase in project net present value. Funds for this investment should be included in feasibility studies.

INTRODUCTION

For purposes of this paper, the start-up phase in a gold plant's life is defined as follows:

> Start-up begins with preoperational testing of equipment, after it has been installed. It includes functional testing of plant systems, introduction of feed, plant modification (if necessary) and operation of the plant until designed capacity is produced on a sustained basis.

We first became interested in start-up performance a number of years ago when Charles River Associates published a study, commissioned by the World Bank, which reviewed start-up performance of a selected sampling of mineral projects from around the world. The conclusion was that the vast majority of the plants experienced very serious start-up problems. It was the rare plant that achieved even 80% of designed production during the first year. Many were below 50%. Refer to Figure 1. Plants in developed countries, on average, fared no better than those in developing countries.

Figure 1

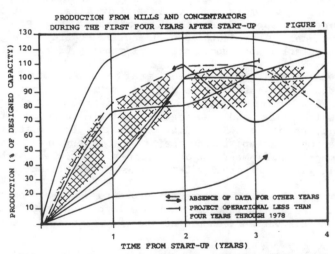

PRODUCTION FROM MILLS AND CONCENTRATORS
DURING THE FIRST FOUR YEARS AFTER START-UP   FIGURE 1

SOURCE: CHARLES RIVER ASSOCIATES. SHADED AREA INDICATES THE RANGE WITHIN WHICH THE MAJORITY OF PROJECTS FALL AND FOR WHICH INDIVIDUAL ARE NOT SHOWN. CURVES ARE REPRESENTATIVE OF 16 PROJECTS.

It became clear that our own experience with a very difficult start-up during the early 60's was not an isolated occurrence. Since that time we have observed numerous start-ups with varying degrees of success. We have been involved in the preparation for some very good start-ups and have been called in to help with start-ups which were experiencing difficulties. There were two major impressions which occurred to us:

o There were many problems which contributed to the poor start-ups, but a high proportion of the difficulty stemmed from inability of the operating and maintenance organizations to cope with the realities of a new plant and with the technical problems which arose.

o All of the production shortfalls must represent a staggering amount of money in lost profits.

We are therefore, very pleased to have this opportunity to do some testing of the financial impact of start-up performance using a number of hypothetical gold mine financial models.

THE FINANCIAL IMPACT

The financial consequences of substandard start-up performance stem from a number of factors, the most important of which are:

o Lost profits due to production shortfalls during the early months of the mine's life, which are, to some extent, recouped during the last months of the mine's life. The problems is that the recouped profits occur well out into the future, and due to the time value of money are worth much less than if they had been realized during the early months. In addition the fixed operating costs must be incurred a second time.

o Lost profits due to lower mineral recovery. These losses are normally not ever recouped over the life of the mine. Perhaps a future generation benefits from reworking the tailings.

o Substantial sums are spent making modifications to the plant. While some of the modifications are necessary, many are not. In the heat of a poor start-up, problem solving usually gets out of control. Many unnecessary modifications are made. Some modifications, in fact, create new problems which further inhibit production.

We have rather arbitrarily classified start-ups into three categories; poor, fair, and good as indicated by production as a percent of capacity during each of the months during the first year.

For this paper we have not considered start-up performance beyond the first year, though in some cases start-up extends beyond. Table 1 displays the production scenarios which we have assumed for each category. We have also calculated an average production percentage for the entire year.

Table 1

CATEGORY	START-UP PRODUCTION SCENARIOS BY MONTH												TOTAL
	1	2	3	4	5	6	7	8	9	10	11	12	
GOOD	50%	90%	100%	100%	100%	100%	100%	100%	100%	100%	100%	100%	95%
FAIR	50%	55%	60%	65%	75%	90%	100%	100%	100%	100%	100%	100%	83%
POOR	40%	45%	50%	55%	60%	65%	70%	75%	80%	85%	90%	95%	68%

While arguments can be made for scenarios which differ from those which we have chosen, the fact remains that actual start-ups cover the full range of these three categories and beyond.

To test the financial sensitivity, of a gold project, to start-up performance, we created five different mining and processing configurations which are described in Table 2.

Table 2

TYPE	RECOVERY	STRIP RATIO	OZ AU PER TON	CAPITAL COST MINE	CAPITAL COST PLANT	OPERATING COST/TON
O.P./HEAP LEACH 7,000,000 TPY	65%	2:1	0.045	$37,000,000	$21,000,000	$6.38
O.P./HEAP LEACH 360,000 TPY	65%	2:1	0.100	$6,100,000	$800,000	$16.00
U.G./MILL 180,000 TPY	95%		0.200	$8,000,000	$9,000,000	$62.50
O.P./MILL 5,400,000 TPY	92%	4:1	0.070	$43,000,000	$75,000,000	$15.00
O.P./MILL 300,000 TPY	95%	4:1	0.085	$5,700,000	$10,800,000	$24.25

It was not our intention to pattern these models after any specific mine. There are certainly mines of similar sizes which have been developed and operate for different costs. It was our intention to represent a reasonably realistic range which would provide a diversity as to size, type of operation, and financial return.

A financial model, which projects cash flow over the life of the mine, was constructed for each case. Net present value and internal rate of return were calculated for each. The following assumptions were held constant for each case:

o Price of gold at $400 per oz.

o Royalty at 4.5% NSR.

o Discount rate at 12%.

o State and local taxes equivalent to 8.5% of revenue.

o An allowance for working capital was included.

o U. S. Income taxes were calculated as though the mine were a stand alone tax paying entity.

o When 95% production, during the first year, was experienced, an allowance equal to 0.75% of plant capital cost was included as an investment in start-up preparation (e.g. preoperational testing, operator training, maintenance organization, etc.) based on our experience that only those plants which properly prepare for start-up will realize a good start-up.

o When first year production of less than 95% was experienced, an allowance for "start-up correction costs" was included. This allowance was estimated at 0 for 95% production level and 10% of plant capital costs for 50% production level. For production levels in between, the allowance represents a straight line interpolation.

o No debt financing was considered.

o All necessary equipment and facilities, including mining equipment, were assumed to be purchased.

o For each case, two models were prepared. One assumed a five year mine life and the other a ten year life.

In the interest of presenting a "conservative" evaluation of start-up performance impact, we did not project any substandard recoveries during start-up.

Table 3 illustrates, as one example, the format used in each of the financial models.

Each case was recalculated, at five percentage point intervals, for first year production levels ranging from 50% to 100%. Net present value and internal rate of return results of each of the ten year mines are presented in Table 4. Net present value curves as well as curves depicting the decline in net present value are presented in graphic form in Figures 2 through 11. Each of these graphs presents a curve for the 5 year mine as well as the 10 year mine.

Table 3

	20-Jul-88 STRTUPC2.WK1	OPEN PIT HEAP LEACH OPERATION PROJECTED CASH FLOW THOUSANDS OF US DOLLARS				RECOVERY OZ AU/TON ROYALTY NS	65% 0.045 4.5%	
	PRICE OF GOLD ANNUAL PRODUCTION	$400 7,000,000				DISCOUNT RATE START-UP FACTOR		12% 95%
				YEAR				
	0	1	2	3	4	5	6	TOTAL
CAPITAL COST								
MINE	$37,000							$37,000
PLANT	$21,000							$21,000
WORKING CAPITAL	$4,725							$4,725
OPERATION - PHYSICAL DATA								
AU OZ/TON		0.045	0.045	0.045	0.045	0.045	0.045	
AU RECOVERY		65%	65%	65%	65%	65%	65%	
PRODUCTION FACTOR		95%	100%	100%	100%	100%	0.05	
TONS ORE MINED AND PROCESSED PER YEAR		6,650,000	7,000,000	7,000,000	7,000,000	7,000,000	350,000	35,000,000
OZ AU PRODUCED		194,513	204,750	204,750	204,750	204,750	10,238	1,023,750
OPERATION - FINANCIAL DATA								
VARIABLE COST PER TON:								
MINING		$2.87	$2.87	$2.87	$2.87	$2.87	$2.87	
CRUSH, STACK, AGGLOM, RECOVERY		$3.00	$3.00	$3.00	$3.00	$3.00	$3.00	
VARIABLE COSTS PER YEAR		$39,036	$41,090	$41,090	$41,090	$41,090	$2,055	$205,450
FIXED COSTS PER YEAR		$8,799	$8,799	$8,799	$8,799	$8,799	$440	$44,435
ROYALTY NSR		$3,501	$3,686	$3,686	$3,686	$3,686	$184	$18,428
NON-INCOME TAXES		$6,613	$6,962	$6,962	$6,962	$6,962	$348	$34,808
START-UP CORRECTION COSTS		$0						$0
START-UP PREPARATION AND RECLAMATION	$158						$2,000	$2,158
TOTAL COST	$158	$57,949	$60,536	$60,536	$60,536	$60,536	$5,027	$305,277
TOTAL REVENUE	$0	$77,805	$81,900	$81,900	$81,900	$81,900	$4,095	$409,500
NET CASH INCOME	($158)	$19,856	$21,364	$21,364	$21,364	$21,364	($932)	$104,223
INCOME TAXES	$0	$1,403	$1,660	$1,660	$1,660	$1,660	($158)	$7,885
NET CASH FLOW	($62,883)	$18,452	$19,704	$19,704	$19,704	$19,704	($773)	$33,613
CUMMULATIVE NET CASH FLOW	($62,883)	($44,430)	($24,726)	($5,022)	$14,682	$34,386	$33,613	
DISCOUNTED CASH FLOW RATE OF RETURN								16.2%
NET PRESENT VALUE								$5,926

TABLE 4

SRTUPRSL.WK1	RESULTS OF FINANCIAL MODELS
20-Jul-88	NET PRESENT VALUE IN THOUSANDS AND INTERNAL RATE OF RETURN
	BASED ON TEN YEAR MINE LIFE

7MM TON HEAP LEACH				360M TON HEAP LEACH				180M TON UNDERGROUND			
1ST YEAR PRODUCTION	NPV	NPV DECAY	IRR	1ST YEAR PRODUCTION	NPV	NPV DECAY	IRR	1ST YEAR PRODUCTION	NPV	NPV DECAY	IRR
100%	$37,859		27.0%	100%	$6,065		32.4%	100%	$13,533		30.2%
95%	$37,088	$771	26.5%	95%	$5,969	$96	31.8%	95%	$13,324	$209	29.7%
90%	$36,306	$1,553	26.0%	90%	$5,890	$175	31.3%	90%	$13,110	$423	29.2%
85%	$35,383	$2,476	25.4%	85%	$5,788	$277	30.7%	85%	$12,836	$697	28.6%
80%	$34,460	$3,399	24.9%	80%	$5,686	$379	30.1%	80%	$12,561	$972	28.0%
75%	$33,536	$4,323	24.4%	75%	$5,584	$481	29.6%	75%	$12,287	$1,246	27.4%
70%	$32,613	$5,246	23.9%	70%	$5,482	$583	29.0%	70%	$12,012	$1,521	26.9%
65%	$31,690	$6,169	23.4%	65%	$5,380	$685	28.4%	65%	$11,738	$1,795	26.3%
60%	$30,767	$7,092	22.9%	60%	$5,278	$787	27.9%	60%	$11,463	$2,070	25.8%
55%	$29,844	$8,015	22.4%	55%	$5,176	$889	27.4%	55%	$11,189	$2,344	25.2%
50%	$28,921	$8,938	21.9%	50%	$5,074	$991	26.9%	50%	$10,914	$2,619	24.7%

5.4MM TON OPEN PIT				300M TON OPEN PIT							
1ST YEAR PRODUCTION	NPV	NPV DECAY	IRR	1ST YEAR PRODUCTION	NPV	NPV DECAY	IRR				
100%	$62,539		24.6%	100%	$11,653		28.7%				
95%	$61,240	$1,299	24.2%	95%	$11,472	$181	28.2%				
90%	$59,896	$2,643	23.8%	90%	$11,286	$367	27.8%				
85%	$58,051	$4,488	23.3%	85%	$11,027	$626	27.2%				
80%	$56,205	$6,334	22.7%	80%	$10,768	$885	26.6%				
75%	$54,360	$8,179	22.2%	75%	$10,509	$1,144	26.1%				
70%	$52,514	$10,025	21.7%	70%	$10,250	$1,403	25.5%				
65%	$50,668	$11,871	21.3%	65%	$9,991	$1,662	25.0%				
60%	$48,823	$13,716	20.8%	60%	$9,732	$1,921	24.5%				
55%	$46,977	$15,562	20.4%	55%	$9,474	$2,179	24.0%				
50%	$45,132	$17,407	19.9%	50%	$9,215	$2,438	23.5%				

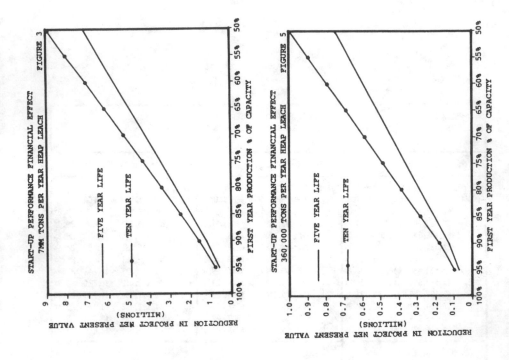

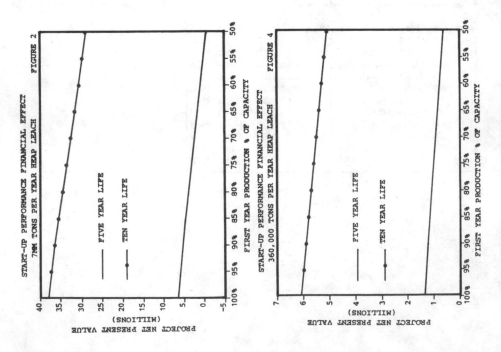

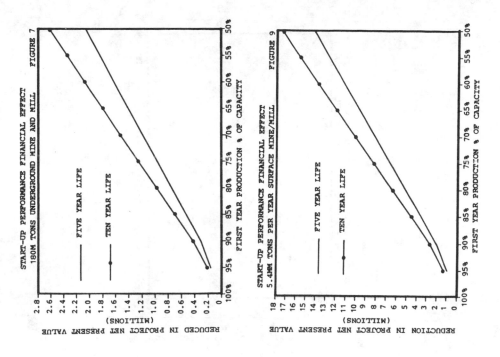

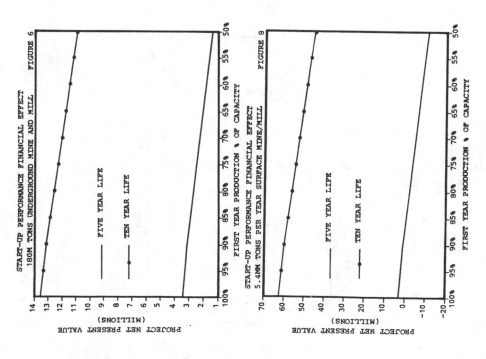

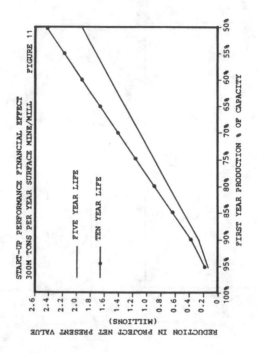

START-UP PERFORMANCE FINANCIAL EFFECT
300M TONS PER YEAR SURFACE MINE/MILL   FIGURE 11

FIVE YEAR LIFE

TEN YEAR LIFE

REDUCTION IN PROJECT NET PRESENT VALUE
(MILLIONS)

FIRST YEAR PRODUCTION % OF CAPACITY

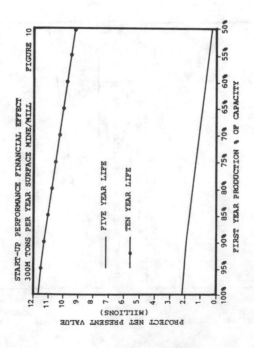

START-UP PERFORMANCE FINANCIAL EFFECT
300M TONS PER YEAR SURFACE MINE/MILL   FIGURE 10

FIVE YEAR LIFE

TEN YEAR LIFE

PROJECT NET PRESENT VALUE
(MILLIONS)

FIRST YEAR PRODUCTION % OF CAPACITY

We can draw certain conclusions from all of these data and curves:

o   Figures 2-11 tell us what we already knew; that a ten year reserve
    is more valuable than a five year reserve, assuming all other
    things are equal. They also demonstrate that the effect of start-
    up performance is very important to the worth of a project
    regardless of the mining and processing configuration. The penalty
    paid for having a poor start-up rather than a good one ranges from
    $600,000 to nearly $10,000,000 depending on the size of the
    project. These potential losses are too large to ignore. Had we
    included substandard recoveries during start-up, the indicated
    penalties would have been even larger.

o   They also point out that start-up performance is more critical to
    economic viability of projects which are more marginal to start
    with.

o   Even though start-up performance is more critical for the more
    marginal five year reserves, a longer life project pays a greater
    penalty in net present value than does the short life mine. This
    results from the fact that a longer life reserve must wait longer
    to recoup the lost start-up production at the end of the mine life.
    Therefore the recouped production is of less value.

In Figure 12 we have plotted the effect on the internal rate of
return for each of the projects. The loss in return rate is
surprisingly consistent and significant from one project to the next.

Another concern which should be on the minds of managers, is the
impact of start-up performance on the first year cash requirements.
To demonstrate this concern we made some pessimistic assumptions,
primarily the price of gold dropping to $340 per oz. We then
simulated a poor start-up, about 60% of capacity, and compared it with
a good start-up. The monthly cumulative cash generation, or deficit,
is displayed on Figure 13. Even a small "simple" operation cannot
afford to take a chance on a poor start-up. The first year start-up
performance could mean the difference between financial viability and
being forced to raise additional funds under inopportune
circumstances.

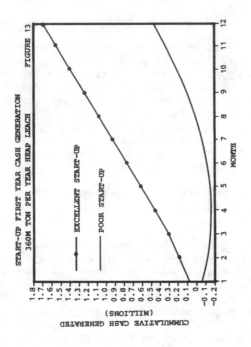

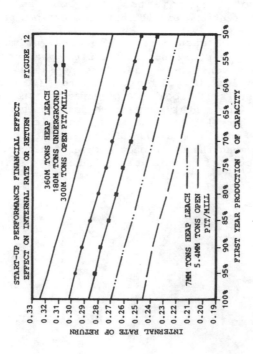

CAUSES OF POOR START-UPS

We have reviewed the financial impact of start-ups. Now we should consider the factors which can cause poor start-up performance and what can be done about them. The major causes include:

o Sampling and testing of the ore body was inadequate. The process does not work properly.

o Plant equipment design and/or installation is inappropriate. Equipment fails prematurely, lines are blocked, chutes plug, etc.

o Plant operators lack the knowledge and skills to properly operate the plant. They are unable to control the plant and respond properly to upset conditions. Their actions create rather than solve problems.

o Plant management is unable to cope properly with the many problems that do inherently arise during a start-up. Problems cannot be properly identified and solved.

o Plant maintenance is not executed properly from the very start. Again, equipment fails prematurely and down periods extend longer than they should.

The more of these causes which exist the more difficult is the start-up. We find that problems stemming from the various causes compound each other and worst of all confuse the issue. During a poor start-up it is almost impossible for the people on site, who are working with the problems day in and day out, to identify clearly the highest priority problems.

SOLUTIONS

Sampling and metallurgical testing

If errors have been committed during the sampling and testing of the ore body, the resulting operating problems cannot be avoided during start-up. Therefore, it behooves the mine developer to ensure that proper care has been exercised in the sampling and testing phase. However, if such problems do occur during the start-up, it is essential that they be identified as early as possible so that solutions can be devised.

Preoperational and Functional System Testing

The inevitable problems arising from equipment design or installation faults must be discovered early. Most of them can be detected and corrected before the introduction of feed. Therefore, we recommend a well organized preoperational testing program. Each unit of equipment is checked and tested to ensure that it has been installed properly, that it operates freely and that it responds properly to its controls. Lines are checked to ensure that they are not blocked and do not leak. After individual equipment units are checked, the various plant systems are run with water and then waste rock to ensure that they function properly.

If such a program is followed, the identification and solution of these problems is straight forward. If, on the other hand, they are not corrected prior to the introduction of feed, they will compound with other problems and may not be identified and corrected for many months. When they are finally discovered, the fix is performed under panic conditions. Costs which should have been the responsibility of the vendor or contractor, are borne by the owner. If they are discovered during the testing phase, responsibilities are clear.

Operator Training

An operator must have certain essential knowledge to be effective. He, or she, should understand the fundamentals of the process, the equipment (its purpose and the way in which it functions), and each control loop (what it controls, why, how, and where). He should be able to respond to each alarm (know what each alarm means, what is the likely fault, what are the possible causes, what are the consequences of an uncorrected fault, and how to correct the fault). He should be able to start the plant up and shut it down. He should be able to maintain process variables within set parameters. Most importantly, he should acquire these abilities prior to the introduction of feed.

This training can be accomplished through development of detailed plant operating manuals in sufficient time that the operators can be trained prior to the commencement of start-up.

It is a mistake to assume that this training is not necessary if "experienced operators" are hired. Since each plant is unique, an experienced operator will not have the necessary knowledge about the new plant. He may, in fact, add to the problems since the new plant may not operate with the same control logic as his last plant. He will assume that it is the same, unless instructed differently.

Problem Solving
<u>   </u>

Plant management and technical staff should receive training in how to solve problems. There are disciplined techniques and approaches which will help to isolate and solve problems. This sort of discipline is essential during periods when many problems are interacting on one another. This skill can be learned through formal training.

Organized Maintenance
<u>   </u>

Preventive maintenance and lubrication tasks need to be documented and organized into a schedule prior to introduction of feed. Spare parts must be cataloged and acquired. Maintenance personnel should have a systematic method of keeping track of work to be done, scheduling the work, and recording equipment maintenance work for future reference. Data accumulated during prepoperational and functional testing should be integrated into the maintenance records. The "honey moon" of trouble free equipment is almost nonexistent if the maintenance effort is not organized and if the maintenance personnel are not given the opportunity of becoming familiar with the specific equipment in the plant, prior to start-up.

COST

The preparation for start-up programs, discussed above, can be accomplished at a reasonable cost. Experience indicates that the cost will run between 0.5% and 1.0% of the capital cost of the plant. Very small heap leach operations should anticipate a minimum of $25,000.

The justification for the expenditure is the improvement in first year production. It has been our experience that a professional approach such as described in this paper is the difference between a fair start-up (83%) and a good start-up (95%). We recommend that you check your project to determine if the increase in net present value will more than offset the cost. You will very likely find that proper preparation for start-up is one of the best investments you can make. The necessary allowance should be included in the feasibility study, so that budget funds will be available when needed.

3

Legal and Ownership

Chapter 13

TITLE TO U.S. GOLD MINES:
AN INTRODUCTION FOR CANADIAN
JUNIOR MINING COMPANIES

by Richard Keith Corbin

Mining Lawyer
MOTHER LODE MINING LAW CENTER
Sacramento, California

ABSTRACT

Canadian junior mining companies whose gold exploration and exploitation programs take them across the border into the western United States should be prepared to deal with the myriad of title issues involved in acquiring unpatented gold mining claims.

The purpose of this paper is to serve as an introductory outline for Canadian junior mining companies to mining laws of the United States, as well as to relevant laws of some of the western gold mining states.

Topics will include the five w's (who, where, what, why and when) of Canadian junior mining companies' rights and responsibilities in acquiring interests in U.S. gold mining properties by claim staking, lease or purchase.

Historic terms of common parlance among U.S. miners, such as "claimjumping" are defined and discussed. Various legal doctrines, such as "pedis possessio" and "good faith" in the location of claims are covered as well.

Controversial subjects of contemporary interest are addressed in some detail, such as whether to locate a bulk-mineable disseminated gold deposit with placer claims or lode claims, and whether lodes may be staked over placers.

By thus heightening the awareness of Canadian junior mining companies, it is hoped that they can be properly prepared to avoid title disputes, and to win lawsuits when they are unavoidable.

The following important concerns are addressed:

1. Who may locate mining claims
2. Where may claims be located
3. What size claims may be located
4. How many claims may be located
5. Why locate as placer claims versus lode claims
6. When to locate lodes over placers
7. How to hold claims prior to discovery (pedis possessio)
8. What should be in a mining lease

1. WHO MAY LOCATE MINING CLAIMS

The General Mining Law which was enacted by the United States Congress in 1872 is basically unchanged to this day, and is commonly referred to as the 1872 Mining Law. Title 30 of the United States Code, entitled "Mining Lands and Mining," and specifically Chapter 2 "Mining Lands and Regulations in General," comprising Sections 22 through 54, inclusive, spells out the mining claimants' rights and responsibilities, and is often referred to as the "Miners' Bill of Rights."

As stated in the case of Shreve v. Copper Bell Mining Co., 11 Mont. 309, 332 (1891), 23 Pac. 315, Title 30 U.S. Code, § 22 and other sections of Chapter 2 are said to be the Chapter of the Miners' Rights upon the public domain.

Section 22 declares who may locate mining claims: all valuable mineral deposits in lands belonging to the United States shall be free and open to exploration, occupation and purchase by citizens of the United States, and those who have declared their intention to become such.

The U.S. Supreme Court interpreted this to mean that the opportunity is provided to "all citizens" to engage in mining. (O'Reilly v. Campbell, 116 U.S. 418 (1886))

Even minors who are citizens may locate mining claims. (Thompson v. Spray, 14 Pac. 182 (Cal. 1887); 43 C.F.R. 3832.1) Also parents of minors may locate claims on behalf of their children. (U.S. v. Haskins, 59 IBLA 1, 88 (1981); West v. U.S., 30 F.2d 739 (D.C. Cir. 1929))

Section 36 provides in pertinent part that: "two or more persons, or associations of persons, having contiguous claims of any size, although such claims may be less than ten acres each, may make joint entry thereof."

The rationale for allowing partnerships and associations to locate

mining claims is that mining requires large expenditures of capital,
and for this reason is most expediently conducted by an association of
individuals which allows its members to unite their capital.
(McKinley v. Wheeler, 130 U.S. 630, 633 (1889)) Such associations are
"little more than aggregations of individuals united for some legiti-
mate business ...." (McKinley v. Wheeler, supra, at p. 633)

According to the American Law of Mining, Second Edition, Chapter 31,
"Locators," "The Court's reasoning (in McKinley v. Wheeler, supra) has
never been contested, and it is universally assumed that partnerships
and other unincorporated associations may locate mining claims."
(Section 31.02[5], p. 31-8)

Furthermore, a corporation has been held by the U.S. Supreme Court
to be competent to locate mining claims, as long as all its members
are citizens. (McKinley v. Wheeler, 130 U.S. 630 (1889))

More recent court decisions have relaxed the citizenship requirement, to allow locations by aliens, within certain limitations. An
alien may own unpatented mining claims and protect his rights through
adverse proceedings, but is not qualified to obtain a patent. (Ginaca
v. Peterson, 262 Fed. 904 (1920)) If an alien should locate a claim,
his rights to the claim are not void but voidable, as he is subject to
losing his rights only by government action. (Manual v. Wulf, 152
U.S. 505 (1894)) For example, a locator who stakes a claim over a
prior locator who is not a United States citizen and has not declared
his intention to become such is not entitled to assert priority.
(Herrington v. Martinez, 45 F.Supp. 543 (D.C. Cal. 1945)) If a mining
claim is located by an alien and the alien subsequently declares his
intention to become a citizen and no adverse rights have been initiated, such declaration relates back to the date of location of acquisition of the alien's interest and validates the location. (Shea v.
Nilima, 133 Fed. 209 (1904))

If an alien conveys a claim to a citizen, the citizen has a title
good against all persons who had acquired no right before the conveyance. (North Noonday Mining Co. v. Orient Mining Co., 1 Fed. 522
(1880)) Furthermore, if a citizen and an alien jointly locate a claim
not exceeding the area allowed by one locator, the location is valid
as to the citizen and a conveyance from the two gives a valid title.
(Id.) There is also no restriction upon the location or acquisition
of mining claims by a corporation in which some or all of the shares
of stock are owned by persons who are not citizens of the United
States. (Alien Ownership of Shares by a Corporate Mining Locator,
M-36738 (July 16, 1968))

Accordingly, Canadian junior mining companies may now locate mining
claims in the United States, anywhere claims may be located.

2. WHERE MAY CLAIMS BE LOCATED

Title 30 U.S. Code, § 22 states in pertinent part that, "except as otherwise provided, all valuable mineral deposits in lands belonging to the United States ... shall be ... open to (location) ...."

Mining claims may be located in the states of Alabama, Arizona, Arkansas, California, Colorado, Florida, Idaho, Louisiana, Mississippi, Montana, Nebraska, Nevada, New Mexico, North Dakota, Oregon, South Dakota, Utah, Washington and Wyoming. (43 C.F.R. 3811.2-1(a))

However, there are a number of exceptions within these states where the public lands have been withdrawn from mineral entry. Withdrawal can take the form of power site withdrawal, railroad patents, patented mining claims, agricultural patents with reserved mineral rights, and pre-existing unpatented mining claims. In addition, large areas of the Western United States have been removed from mineral location through government reservations, government withdrawals, and national parks and recreation areas.

Moreover, there can be other constraints which limit the viability of a mining project, as much as direct legal restraints. In Nevada one of the extreme problems is lack of water. In California one of the particularly appalling concerns of miners trying to maintain mining claims is residential impact on historic mine sites. In El Dorado County, California, in 1986, residents passed a county ordinance called Measure A, which prohibited any new mining operations within approximately two miles of a residence.

In determining where to obtain a mine, the foregoing problems should be identified. The landman for the Canadian junior mining company should be particularly careful about locating claims over senior claims. Fortunately, there are a number of tools available to the landman to help avoid this and a host of other problems.

The principal resource is the State Office of the U.S. Department of the Interior, Bureau of Land Management (BLM). There, using the BLM Master Title Plats, one can identify private (fee) lands, acquired lands, patented mining claims, government reservations (such as Indian reservations and defense installations), and government withdrawals which preclude location of unpatented mining claims. The historic indices which accompany the Master Title Plats give a detailed chronology of all federal government transactions affecting each township.

In addition, the BLM geographic index allows the title examiner to determine the general location of unpatented mining claims. The federal recordation regulations, which went fully into effect on October 22, 1979, require every locator or owner of mining claims to file certificates (or notices) of location and proofs of annual labor with the appropriate BLM office. Failure to file the certificates and proofs of labor in a timely manner renders the claims absolutely void.

(43 C.F.R., § 3833.4. See also U.S. v. Locke, 105 S.Ct. 1785 (1985))
The location certificates and claim maps are used to compile the BLM
geographic index, which identifies all unpatented claims located within each quarter section of the public domain. Having identified a
potential conflict in the area of interest, the title examiner can
then review the relevant claim files which contain the location certificates, proofs of labor, transfers of interest, and all other documents and correspondence relating to the claims.

Another useful resource is the claim map recorded by the locator
in certain western states (Nevada, Arizona, Colorado, and New Mexico).
In Nevada the County Recorder maintains a "master claim map" which
depicts the location of all valid and abandoned unpatented mining
claims within the county, often using a BLM master title plat as the
base map.

A significant problem with the county master maps and the BLM geographic index is that they depend on the accuracy of information provided by the claim locator. If the locator erroneously describes the
location of his claims, the claims will appear in the wrong area of
the geographic index and the county claim maps. Since the true location of a claim is defined by its monuments on the ground and not by
the location notices or claim map filed by the locator (see 2 American
Law of Mining (2d ed.) § 33.04[9] and cases cited therein), it is
possible for an area to be completely occupied by valid pre-existing
claims even though the county and BLM records indicate that the ground
is open. The courts have usually upheld the rights of a senior locator whose documents are erroneous, particularly when the junior locator had actual knowledge of the claim's location in the field. (For
example, see Lombardo Turquoise Mining and Milling v. Hemanes, 430 F.
Supp. 429 (D. Nev. 1977))

Because of the possible inaccuracies of the geographic index and
county maps, a thorough land status check should include an examination of the ground for location monuments and other evidence of prior
locations. However, even this procedure has its limitations. In a
recent decision of the Nevada Supreme Court, a senior claim was upheld
even though there were no location monuments or corner monuments on
the ground and the location certificate and claim map described the
claim some two miles from its actual location. The Court found that
there was a "pattern of rock monuments" which was enough to establish
validity of the senior claim. (Kenney v. Greer, et al., 99 Nev. 40,
656 P.2d 857 (1983))

_ Recently a comprehensive checklist has been developed for mining
lawyers, landmen and exploration geologists to assist them in evaluating mining claim titles. (Maley, T.S., 1984)

3. WHAT SIZE CLAIMS MAY BE LOCATED

The standard placer claim is 20 acres, and is taken by legal sub-
division in aliquot parts when there has been a federal survey of the
area. Generally, when a placer claim is located in a surveyed area
only one claim post is necessary, a monument post. This is placed at
the point of discovery, and a notice of location is affixed. However,
in Nevada, the monument post must be "erected at any point along the
north boundary of the claim." (Nevada Revised Statute, § 517.090(1)
(a)) Moreover, under the 1872 General Mining Law, Title 30 U.S. Code,
§ 36, an association of individuals may locate a placer claim of up to
160 acres for eight persons or more.

Lode claims are usually located in rectangles ("The end lines of
each claim shall be parallel to each other") with a maximum of 300
feet on either side of the vein and 1,500 feet along the course of
the vein (Title 30 U.S. Code, § 23), or approximately 20.66 acres.
In addition to the discovery monument, there must be four corner monuments. In California the "center of each end line" for a lode claim
must also be monumented (California Public Resources Code, § 2302),
whereas in Nevada, "the center of each side line" of a lode claim must
also be monumented (N.R.S., § 517.030(1)). An exception to the lode
claim size as noted above, is where a lode claim is "stacked" over a
placer claim, in which case the lode claim is limited to 25 feet on
either side of the vein, but still up to 1,500 feet long. This has
been interpreted to mean 25 feet on either side of the vein not including the vein, so it is possible to have a claim somewhat wider than
50 feet. (Title 30 U.S. Code, §§ 23; 37)

According to the rules and regulations of the miners long before
patents were issued, miners were in the habit of consolidating and
joining claims, whether they consisted of one or more original locations, into one for convenience and economy in working it, this practice being permitted and approved by the Department, and under
Section 36 two or more persons, or an association of persons, having
contiguous claims of any size are permitted to make a joint entry
thereof. (St. Louis Smelting & Refining Co. v. Kemp, 104 U.S. 636
(Colo. 1882); see also Donnelly v. U.S., 228 U.S. 243, 265 (Cal.
1913); Peabody Gold Mining Co. v. Gold Hill Mining Co., 111 Fed. 817,
821 (Cal. 1901))

4. HOW MANY CLAIMS MAY BE LOCATED

The United States Supreme Court first held in St. Louis Smelting &
Refining Co. v. Kemp, 104 U.S. 636 (1882), that there is no limit on
the number of mining claims an individual, association, or corporation
may locate. Nevertheless, the government and rival claimants have
complained from time to time that one entity or another has "too much"
land. This "too much" land complaint was expressly overruled in Baker
v. U.S. 613 F.2d 224 (9th Cir. Ariz. 1980), wherein the Interior

Department took that position. The "too much" land complaint is no
more tenable for a rival claimant. Wrongdoing cannot be imputed
solely because an association of individuals locates a number of
claims. (U.S. v. California Midway Oil Co., 259 F.3d 343 (D.C. Cal.
1919), affirmed, 263 U.S. 682)

At the turn of the century, Congress had not yet found it proper
to put a limitation on the number of mining claims that an individual,
association or corporation may locate or acquire. (Last Chance Mining
Co. v. Bunker Hill, etc., Co., 131 Fed. 579, 583 (1904), cert. den.,
200 U.S. 617)

Although there have been a number of limitation proposals recently,
neither Congress nor the courts have yet limited the number of claims
a locator may locate.

5. WHY LOCATE AS PLACER CLAIMS VERSUS LODE CLAIMS

Recently there has been a lot of discussion on the subject of
whether to stake a bulk-mineable disseminated gold deposit as lode
claims or as placer claims. Technically, the definition of a lode is
"veins or lodes of quartz or other rock in place bearing gold ... or
other valuable deposits." (Title 30 U.S. Code, § 23) A placer is
defined negatively as including all forms of deposit, "excepting veins
of quartz, or other rock in place," which usually means sand or gravel. (Title 30 U.S. Code, § 35)

This problem was encountered with low grade and widely disseminated
deposits of uranium after World War II. Although such deposits may
occur in rock in place, if they do not have well-defined boundaries
separating them from country rock, they are placer deposits. (See,
e.g., Titanium Actynite Indus. v. McLennan, 272 F.2d 667 (10th Cir.
1959))

The Supreme Court in 1888 defined a placer deposit as "ground within defined boundaries which contains mineral in its earth, sand or
gravel; ground that includes valuable deposits not in place, that is,
not fixed in rock, but which are in a loose state, and may in most
cases be collected by washing or amalgamation without milling."
(United States v. Iron Silver Mining Co., 128 U.S. 673, 679 (1888))

Indeed, it is popular in the mining industry to say that a placer
deposit is a deposit of heavy minerals concentrated mechanically.
(Globe Mining Co. v. Anderson, 78 Wyo. 17, 318 P.2d 373, 377, fn. 4
(1957))

Nevertheless, these definitions are not broad enough to encompass
many widely disseminated and low grade placer deposits which are
found in rock in place but do not have distinct boundaries on both
sides of the mineralized area. The only safe definition, one which

is critical to the validity of a location, is that a placer deposit
is any mineral deposit which is not a vein or lode.

Generally speaking, deposits of broken, loose, or scattered material are to be located by placer claims. (Layman v. Ellis, 52 L.D. 714
(1929))

A placer claim may, however, include ground containing lodes or
veins, or ground having no value for placer mining. (Hogan & Idaho
Placer Mining Claim, 34 L.D. 42, review denied, 34 L.D. 178 (1905);
Adams v. Quijada, 25 L.D. 24, review denied, 25 L.D. 285 (1897))

There are a number of other criteria which may be considered in
making the determination of whether to stake placer or lode claims, as
summarized in 1 American Law of Mining (2d ed.), § 32.02[4][a]:

> "Six factors have been delineated as relevant to a determin-
> ation of whether a particular mining claim should be located
> as a lode or placer: (1) whether the mining methods used to
> extract the minerals are traditionally hardrock operations
> or traditionally placer methods; (2) whether the deposit it-
> self is best characterized as 'hardrock' or sand or gravel;
> (3) whether the surrounding rock is sufficiently different
> from the character of the deposit; (4) whether a boundary
> can be identified between the mineralized deposit and the
> surrounding deposit; (5) whether similar claims in the area
> have been located as lodes or placers; and (6) the depth and
> surface extent of the mineralization."

In a particular area, whether similar claims have been located as
lode or placer is a practical guideline.

A good example is the Awakening Mining District on the western
slope of the Slumbering Hills, northwest of Winnemucca, Nevada. On
the pediment, a placer mining association called the Dry Lake Placer
Association originally staked a number of 160-acre placer claims.
Later on, Amax Exploration, Inc., came in and blanket-staked several
thousand acres in the area which is now known as the Sleeper Mine.

Just south of Amax' holdings, Lacana, a Canadian company, staked
174 claims, 168 of which were placer and six were lode. Just south
of that, Sundesert Mining Company, a placer mining association, staked
a large block of 80-acre placer claims.

Just to the west of Amax, on the lakebed, another placer association staked 160-acre placer claims for miles north and south along the
range line. These claims were in turn leased to another Canadian
company, Freegold Recovery, Inc.

In summary, in the immediate area of the Sleeper, five corporations
or placer associations all staked placers. Amax commenced drilling

and later staked lode claims over its own placer claims.

Staking placer claims in this manner in the Nevada desert appears to comply with the better-reasoned case law on the subject, as the areas of bulk-mineable disseminated gold deposits are typically covered with softer alluvial fan material under which there are scattered gold particles, although there may be isolated bonanza lodes as well.

In this regard, in Leadville Mining Co. v. Fitzgerald (Fed. Case No. 8, 158 (1879)), it was held that a vein lying on fixed immovable rock, but covered by alluvial material is not a lode.

Moreover, decisions of the United States Supreme Court indicate that placer claims should be located on "softer" material. In Reynolds v. Iron Silver Mining Co., 116 U.S. 687 (1886), the Court said:

> "Placer mines, though said by statute to include all of the deposits of mineral matter, are those in which this mineral is generally found in the softer material which covers the earth's surface, and not among the rocks beneath."

Somewhat later in Cole v. Ralph, 252 U.S. 286, 295-296 (1920), the Supreme Court further said:

> "... and to sustain a placer location it must be of some other form of valuable mineral deposit (citations omitted), one such being scattered particles of gold found in the softer covering of the earth."

"Good faith" of the locator is usually the deciding factor in determining whether a claim is properly located as a lode or placer.

In Bowen v. Sil-Flo Corporation, 451 P.2d 626 (Ct.App.Ariz. 1969), the Court determined that good faith locations based on local customs should be protected. The Court said:

> "If our decision needs reenforcement, we find it in the fact that at the time this lode claim was initiated it was a custom in this mining district to locate similar ore under the lode statute. We agree with the following:
>
> 'A locator who makes a good faith location in reliance upon past practice should be protected regardless of the court's view as to what might have been the most appropriate form of location.' 1 American Law of Mining § 5.20, at 764."

6. WHEN TO STAKE LODES OVER PLACERS

In the wake of the recent Nevada Federal District Court decision in Amax v. Mosher, et al. (Case No. R-85-162, D.Nev. 1987), if and when a locator may stake lode claims over placer claims, and related issues, have been the hot topic at gold mining seminars in the United States, and the subject of a number of authors' papers. (E.g., Harris, R.W., Esq.*, 1988; Fisk, T., Esq., 1988)

There is statutory authority for locating lode claims over placer claims, commonly called "stacking." If appropriate, the preferred method is to stake placer claims first, and then lode claims. Normally lode claims are 1,500 feet by 600 feet. However, under Title 30 U.S. Code, § 37, the maximum size of the stacked lodes are 25 feet on either side of the vein, or 50 feet total.

The controversy which is in the forefront of attention in the gold mining industry is when a junior locator can locate lode claims over a senior locator's placer claims. Historically this depends on the facts of the situation. If the senior locator consents, then it is permissible. However, without the senior locator's specific consent, it is generally not permissible. The United States Supreme Court has ruled on this question in what has become generally known as the "consent doctrine."

The consent doctrine was created by the U.S. Supreme Court to protect placer claimants against "occupation by strangers seeking to discover [unknown] veins beneath the surface." (Clipper Mining Co. v. Eli Mining & Land Co., 194 U.S. 220 (1904)) Therein the Court ruled that a person going onto a valid placer location to prospect for unknown lodes and veins against the will of the placer owner is a trespasser and acquires no rights in any lodes or veins he may discover during such trespass. (Id., 194 U.S. at pp. 230-231)

In Duffield v. San Francisco Chemical Co., 205 F. 485 (9th Cir. 1913), the Court considered a case wherein a prospector went on unoccupied placer claims owned by another and made lode locations. Since the junior locator entered the placer claims without permission, the Court ruled in favor of the owner of the placer claims.

---

*Richard W. Harris, Esq., Reno, NV, drafted the Articles of General Partnership Creating the Dry Lake Placer Association on behalf of Mr. Mosher and his associates. He represented them for a brief period after Amax located the Sleeper claims, but before Amax filed the subject lawsuit. In a letter prepared on behalf of the association members, addressed to the agent who located Amax' Sleeper claims, he charged that his clients' pre-existing 160-acre placer claims had been "jumped."

The Duffield Court stated the following:

"... But the important facts shown by the record are that
the appellants went upon the land when it was vacant and
unoccupied, made a discovery, and performed all the necessary acts to perfect lode locations thereon. This they
had no right to do if there were valid placer claims covering the same ground."

In a later federal appellate case the court upheld a jury instruction, stating as follows:

"No person other than the owner of the placer claim has
the right to enter upon the same for the purpose of discovering [an unknown] vein or lode and locating the same,
and one who attempts to do so without the owner's consent,
or without his knowledge, is a trespasser and can acquire
no rights to such lode claim ...." (Campbell v. McIntyre,
295 Fed. 45, 46 (9th Cir. 1924))

However, there are a few notable exceptions to the consent doctrine.
A lode location may overlap a placer claim, or for that matter another
lode or even a patented claim to complete the required rectangular
configuration for a lode, so long as the point of discovery is on
ground not previously appropriated. Also, small fractions can be
located as lode claims, but not as placer claims.

There is another major exception to the consent doctrine, wherein a
claim lapses due to failure of the senior locator to perform assessment work, and record a proof of labor with the county and with the
BLM. The U.S. Supreme Court ruled in U.S. v. Locke, 105 S.Ct. 1785
(1985), that failure to file a proof of labor with the BLM by December
30th will result in forfeiture of the claim. The only exception is a
proof of labor postmarked by December 30, which is allowed 20 days to
reach the appropriate BLM office.

If there is no placer deposit, and the ground is obviously not
placer, then a lode locator can overstake a senior placer claimant.
However, this can be quite risky, as the court may decide the junior
locator is wrong, and accordingly, a "claimjumper." This can happen
whether it is a lode over placer situation, placer over lode, lode
over lode, or placer over placer. Courts in California and Nevada
have usually held in favor of the senior locator in these cases, even
where the senior locator's paperwork is defective in some respects.

Typically, California courts have looked to the doctrine of "good
faith," and have dealt strictly with claimjumpers, equating their
locations with "naked trespass." In the case of Brown v. Murphy, 36
C.A.2d 171, 181 (1939), the Court held:

"Good faith confronts any subsequent locator who enters
upon the actual possession of a senior locator's land
for the purpose of initiating a claim to the same ground,
although the senior locator be invalid, and when such
entry is in bad faith, such intrusion constitutes a naked
trespass."

Early on a Nevada appellate court, in the case of Nelson v. Smith,
42 Nev. 302, 176 Pac. 261, 265, stated that filing a location within
the staked boundaries of another claim located prior thereto, so that
the new location covers the workings of a prior locator, is what in
mining circles is known as "claimjumping." Since then the Nevada
Supreme Court decision in Claybaugh v. Gancarz, 81 Nev. 64, 398 P.2d
695 (1965), held that the bad faith of a junior locator precludes him
from challenging the title of a good faith senior locator.

More recently, in the case of Lombardo Turquoise Milling & Mining
v. Hamenes, 430 F.Supp. 429 (1977), the Federal District Court for the
District of Nevada cited Claybaugh as the principle "touchstone" in
such cases, and went on to state:

"[E]quitable principles will be applicable in a contest
between a claimjumper and a senior locator claiming under
a defective location, and the demonstrated bad faith of a
junior locator will preclude recognition of his claim to
the property." (Lombardo, supra, 430 F.Supp. at p. 440)

The hazards of a junior locator staking lode claims over a senior
locator's placer claims are clearly demonstrated in a Nevada State
court decision assessing punitive damages of $250,000.00 against the
junior locator. (Industrial & Petroleum, Inc. v. Victor Industries (6th
Judicial District Court for Pershing County, Nevada, Case No. 5498))
Although the Nevada Supreme Court reduced the award to $100,000.00, it
upheld the lower court decision, noting the evidence clearly demonstrated the senior locator had placer location monuments in place as
well as 50 recently-dug discovery trenches. (Unpublished Order,
Appeal No. 15594)

The most recent guide to lode over placer conflicts is the federal
district court decision in Amax v. Mosher, et al. (Case No. R-85-162,
D.Nev. 1987). Following discovery of its Sleeper ore body in 1982,
Amax Exploration Company blanket-staked lode claims to the north and
south, over placer claims previously located by the Dry Lake Placer
Association.

Amax sued for quiet title against the placer mining association,
asking for $30,000.00 damages against each association member, and
alleged their claims were not procedurally proper. The association
members countersued Amax for $50,000,000.00 for "claimjumping."

Prior to trial the association filed a motion for a preliminary

injunction to prevent Amax from mining on the contested claims, and
the Court entered a mutual restraining order, stating, <u>inter alia</u>:

> "Plaintiff and Defendants are each restrained and enjoined
> from mining or removing commercial or substantial quanti-
> ties of minerals or material from the claims in conflict,
> including claims 4, 25 and 28 North (sic) Lake and Dry Lake
> Placers or from substantially altering, covering, burying,
> or otherwise disturbing the claims in conflict. (<u>Prelimi-</u>
> <u>nary Injunction Order</u>, page 3, lines 24-28, Sept. 20, 1985)

On the eve of trial, Amax reached an out-of-court settlement with
the association members on the majority of the disputed claims. However, the precise terms of the settlement, because of a stipulation
therein, remain confidential and disclosed only to the parties, their
counsel and the Court.**

The trial on the balance of the claims in the small claim block
identified as the GR claims resulted essentially in a draw. In reaching its decision the Court noted that both parties' claims were "procedurally proper." Nevertheless, the Court held that neither party
had a valid discovery, finding that Amax' efforts demonstrated only "a
promising potential for exploration" that fell short of a legal discovery on each Amax claim, and that the "haphazard operation" of the
defendants, coupled with questionable assay techniques, did not constitute a discovery of mineral sufficient to satisfy the "prudent man"
test.

After finding that neither party had satisfied the requirements of
the doctrine of pedis possessio, the Court then concluded that "the
land remains in the public domain open to peaceable exploration by the
parties or by any other citizens."

The <u>Amax</u> case suggests that the doctrine of pedis possessio may be
given a more restrictive interpretation than in the past, at least in
some federal courts.

7. <u>HOW TO HOLD CLAIMS PRIOR TO DISCOVERY (PEDIS POSSESSIO)</u>

Technically, under the 1872 Mining Law, discovery must precede
location. However, when discussing bulk-mineable disseminated gold
deposits which are low grade, as a practical matter, many companies
locate large blocks of claims prior to discovery. The doctrine of

---

**Counsel: Sherman & Howard, Denver, CO, and Hill Cassas de Lipkau and
Erwin, Reno, NV, for Amax; Richard Keith Corbin, Mother Lode Mining
Law Center, Sacramento, CA, and Cartwright, Sucherman & Slobodin, Inc.,
San Francisco, CA, for Mosher, et al. Court: Honorable Bruce R.
Thompson, Judge, United States District Court, District of Nevada.

pedis possessio may protect a company after location and prior to
discovery if it is diligently pursuing discovery. The doctrine of
pedis possessio was stated by the U.S. Supreme Court in Cole v. Ralph,
252 U.S. 286, 295-296 (1920):

> "In advance of discovery an explorer in actual occupation
> and diligently searching for mineral is treated as a lic-
> ensee or tenant at will, and no right can be initiated or
> acquired through a forcible, fraudulent, or clandestine
> intrusion upon his possession. But if his occupancy be
> relaxed, or be merely incidental to something other than
> a diligent search for mineral, and another enters peace-
> ably, and not fraudulently or clandestinely, and makes a
> mineral discovery and location, the location so made is
> valid and must be respected accordingly."

In other words, the doctrine of pedis possessio prohibits encroach-
ment by other locators in the area of exploration or development if
three criteria are met:

1. There must be actual physical occupancy of the ground.
2. The locator must be engaged in diligent bona fide work directed
toward making a discovery of valuable minerals.
3. The locator must exclude competing mineral locators from his
claim. (But note that the locator must allow other users of the public domain, such as hunters, fishermen, and fossil hunters, to have
access to the claim so long as they are not in danger.) (Multiple Use
Act of 1955, 30 U.S. Code, § 612(b); U.S. v. Curtis-Nevada Mines, Inc.,
611 F.2d 1277 (9th Cir. 1980))

The extent of prediscovery protection remains unclear. Some courts
have held that the right extends to the entire block of claims, while
others limit the right to those portions of the claim group actually
in possession of the exploration crew. In the case of MacGuire v.
Sturgis, 347 F.Supp. 580 (D.Wyo. 1971), the Wyoming Federal District
Court adopted an expansive view of pedis possessio and stated various
factors for asserting possession to an entire group of claims:

> "... Plaintiff is presently entitled to the exclusive
> possession on a group or area basis where, as here, the
> following exists or was done for his benefit: (a) the
> geology of the area is similar and the size of the area
> claimed is reasonable; (b) the discovery (validation)
> work referred to in [the Wyoming statutes] is completed;
> (c) an overall work program is in effect for the area to
> be claimed; (d) such work program is being diligently
> pursued, i.e., a significant number of exploratory holes
> has been systematically drilled; and (e) the nature of
> the mineral claimed and the cost of development would
> make it economically impracticable to develop the mineral
> if the locator is awarded only those claims in which he

is actually present and currently working. Plaintiff is
entitled to the future exclusive possession thereof so
long as he, or his successors in title, remain in possession thereof, working diligently towards a discovery."

However, the Supreme Court of Arizona rejected this expansive view
of prediscovery protection in the case of Geomet Exploration v. Lucky
Mc Uranium Corp., 124 Ariz. 55, 601 P.2d 1339 (1970), cert. dismissed,
448 U.S. 917 (1980). There Geomet had overstaked and drilled a portion of Lucky Mc's 200 claim block without challenge. The Court upheld the "claimjumper" by ruling that pedis possessio protection
extended only to those areas under actual occupation:

"If the first possessor should relax its occupancy or
cease working toward discovery, and another enters peace-
ably, openly, and diligently searches for mineral, the
first party forfeits the right to exclusive possession
under the requirements of pedis possessio. We hold that
pedis possessio protects only those claims actually occupied (provided also that work toward discovery is in progress) and does not extend to contiguous unoccupied
claims on a group or area basis."

In order to maximize the degree of protection and minimize the risk
of loss, it has been advised that a claim holder should undertake the
following measures:

1. The claim boundaries should be well marked and maintained.
This may require considerable effort and expense if a mining company
has acquired an older claim group in which the location and corner
monuments are misplaced or missing altogether.
2. Everyone involved with the project, from the senior geologist
to the part-time sampler, should be aware of the claim boundaries in
relation to various topographic features. Every member of the project
should be directed to advise the project geologist or other responsible person of any unusual activity within the claim group.
3. The claim group should be patrolled on a regular basis in order
to detect intruders at the earliest possible stage. The company
should also review the BLM geographic index and county claim maps
every quarter to determine whether the claim maps have jumped "on
paper."
4. The company should develop a procedure for challenging rival
locators. Any person who appears to be sampling or prospecting on the
claims should be politely confronted and asked to leave the property.
The geologist or other company employee should take the name, address,
and license number of any intruder and note the encounter in his field
book.
5. If the company discovers any kind of mechanized activity on its
claim group (e.g., bulldozing or drilling), it must take immediate
action to restrain the rival locator before he makes a discovery. The
company should contact its legal department or private law firm and

request that a restraining order be obtained as soon as possible.
Unfortunately, such a legal action will usually be time-consuming and
expensive, but the lesson of Geomet is clear: protect your claims or
lose them.

   It is inadvisable to use force against a claimjumper since the law
considers a human life to be paramount to property rights. Some form
of physical restraint, such as blocking a drill road, might be permissible in certain circumstances (but do not block a public right of
access). (Harris, R.W., Esq., 1988)

   8. WHAT SHOULD BE IN A MINING LEASE

   Unpatented gold mining claims may be transferred by verbal conveyance accompanied by a change of possession of the premises, if there
is no statute providing otherwise. (Doe v. Waterloo Mining Co., 70
Fed. 455 (1895)) Claims may also be sold (St. Louis Smelting and
Refining Co. v. Kemp, 104 U.S. 636 (1882)), and they may even be sold
in part (Little Pittsburgh Consol. Mining Co. v. Amie Mining Co., 17
Fed. 57 (1855)), and they may be leased.

   Probably no area of natural resources law has generated more disputes, litigation (and legal fees, for that matter), judicial decisions and scholarly analysis than the determination of the mineral
rights, mining rights and surface use rights granted or reserved under
deeds, leases, purchase agreements and other instruments regarding
mineral rights. Hardly a year passes without the publication of an
article dealing with these issues. These issues are particularly
common in cases of severed estates and multiple mineral and surface
uses. (Erwin, T.P., Esq., 1988)

   A prime example of the draftsman's need to consider all of these
issues is the case Christensen v. Chromalloy American Corp., 99 Nev.
34, 656 P.2d 844 (1983).

   In Christensen, the Nevada Supreme Court held that a mineral reservation clause was ambiguous because it was silent as to the method of
mining operations the original grantor contemplated. The Court noted
the reservation did not clearly establish that the parties to the original reservation intended to allow open pit or strip mining of the
"other minerals" referred to in the reservation. The Nevada Supreme
Court remanded the action with instructions that extrinsic evidence be
admitted regarding the intent of the original parties to the reservation. (Id., p. 40)

   The Christensen case indicates the pitfalls of which the draftsman
must be wary when drafting any grant, reservation or lease of mineral
and mining rights. Though it appears there exists no clause which
will satisfy every court, among the items which should be included
are: (a) a specific description of the minerals sought; (b) a specific
description of minerals known to exist or which possibly exist; (c)

provision for "all other" minerals; (d) other interests of possible use or benefit such as water and geothermal resources; (e) authorization to use all mining methods including open pit, and all extractive and processing methods which may be discovered or developed <u>after</u> execution of the contract; (f) all other surface and incidental uses deemed, in the acquiring party's discretion, useful in its mining and processing.

In addition, a Canadian junior mining company seeking to acquire mining claims by lease should insist on provisions for the following: (a) a purchase option; (b) "areas of interest" or "boundary protection"; (c) lesser interests – intended to reduce the lessee's minimum payment and production royalty payment obligations if the lessor's title is less than originally contemplated; (d) cross-mining, processing and stockpiling rights; (e) Commingling of ores extracted from parcels having different owners; (f) unitizations of several ore producing parcels as a single mine; (g) establishing boundaries and boundary ore agreements with adjoining property owners; (h) amendment, relocation and patent of unpatented mining claims, including the right to locate lodes over placers; (i) retention of "nonmineral" rights after lease termination; (j) negation or limitation of implied covenants; (k) an adequate primary term; (l) payment method and credits, and liability for taxes; (m) title representation and warranties; (n) encumbrance, assignment and sublease of mining lease; (o) risk sharing in the event of <u>force majeure</u>; (p) intellectual property considerations to assure the lessee has full authorization to use all protectable processes and technologies.

<u>CONCLUSION</u>

Before a Canadian junior mining company seeks to acquire an interest in a U.S. gold mining claim, it should first seek advice from a competent U.S. mining lawyer, as well as from its counsel in Canada.

Based upon the authority of such cases as <u>Estate of Waring</u>, 47 N.J. 367, 221 Atl.2d 193, 197 (1966), a Canadian junior mining company should be able to choose any competent mining lawyer in any state, to represent it regardless of where the gold mining claim is located, as "Multistate relationships are a common part of today's society and are to be dealt with in commonsense fashion."

In this regard, the American Bar Association Committee on Professional Ethics has succinctly concluded: "[i]t is a matter of law, not of ethics, as to where an individual may practice law." (A.B.A. Comm. on Prof. Ethics, <u>Opinion</u> No. 316 (1967))

Accordingly, any competent mining lawyer may provide the necessary legal assistance in negotiating and drafting gold mining claim acquisition documents. Moreover, under multi-state <u>pro hoc vice</u> admission provisions effective in most state and federal courts, all competent

mining lawyers may represent their clients in mining litigation in
courts outside their respective licensing states. For example, in
the Amax case, supra, filed in Nevada, the two parties were represented by four different lawfirms from three different states. Yet only
one of the lawfirms maintains its practice in Nevada.

REFERENCES

American Law of Mining, Matthew Bender & Co., Inc., Martz, C.O.,
    ed., R.M.M.L.F., Boulder, CO (1st ed. 1983, 5 vols.).

American Law of Mining, Matthew Bender & Co., Inc., Outerbridge, C.,
    ed., R.M.M.L.F., Boulder, CO (2d ed. 1984, 6 vols.).

Erwin, T.P., Esq., 1988, "Considerations in Drafting Gold Leases and
    Development Agreements," Gold Mine Financing, Mineral Law Series,
    R.M.M.L.F., Denver, CO., Paper No. 8, 53 pages.

Fisk, T., Esq., 1988, "Pedis Possessio After Amax v. Mosher," 34th
    Annual Rocky Mountain Mineral Law Institute, R.M.M.L.F., Seattle,
    WA.

Harris, R.W., Esq., 1988, "Location and Maintenance of Large Claim
    Groups," Bulk Mineable Precious Metal Deposits of the Western
    United States, Schafer, R.W., Cooper, J.J., Vikre, P.G., eds.,
    G.S.N., Reno, NV, pp. 681-689.

Harris, R.W., Esq., 1988, "Location of Lode Claims Over Placer
    Claims," 34th Annual Rocky Mountain Mineral Law Institute,
    R.M.M.L.F., Seattle, WA, 62 pages.

Lindley, C.H., 2 Lindley on Mines, § 425(B), p. 1005 (3d ed.).

Maley, T.S., 1984, "Mining Claim Title Examination," Mineral Title
    Examination, Mineral Land Pubs., Boise, ID, pp. 345-353.

Chapter 14

LETTERS OF INTENT IN MINING TRANSACTIONS

by David R. Reid and Lawrence W. Talbot

BULL, HOUSSER & TUPPER
Barristers and Solicitors
Vancouver, British Columbia

INTRODUCTION

In the mining industry, timing in land acquisitions can be of the essence. In order to secure promising lands, deals are often negotiated quickly and evidenced by letters of intent containing what the parties feel are the essential terms of the deal. In some cases, a company is willing to acquire land only if it can lock up a number of promising properties, and so a "non-binding" letter of intent is used to formalize the essential terms, while leaving the company without legal liability. Later, the letters of intent will be sent to a lawyer to prepare a formal acquisition agreement. Many times, however, especially in a land acquisition rush, another company will attempt to better the deal prior to the execution of that formal agreement, and the first company will be forced to rely on the initial letter of intent, only to find out to its chagrin that it has no deal.

In other cases, the company holding the desired property is another mining company, and the negotiation of an option or joint venture is the only way the first company can acquire an interest in the desired land. In order to make its offer more attractive, it may have to reveal confidential or proprietary information about the property or the surrounding lands. Here, too, letters of intent are often used to formalize the essential points of the agreement. Once again, however, prior to the execution of a formal joint venture agreement, the transaction may collapse and the letter of intent prove to be only that.

In this paper, the authors examine the dangers associated with the use of letters of intent and the specific provisions courts look for to find a binding contract, and consider how to increase the

likelihood that a letter of intent will withstand the attempts of a
party to avoid the deal. The authors also consider the problems
associated with the revelation of information during the course of
negotiations and review the use of confidentiality provisions in
letters of intent, with a view to ensuring that such information
cannot later be used to defeat the revealing company's objectives or
interfere with its land acquisition plans.

CHARACTERISTICS AND OBJECTIVES OF THE LETTER OF INTENT

Black's Law Dictionary (5th ed.) at p. 814 describes a letter of
intent as a document "customarily employed to reduce to writing a
preliminary understanding of parties who intend to enter into
contract." Such a description, while doubtless accurate in many
instances, may be misleading in others simply because the objectives
of those who sign their names to that which are commonly called
letters of intent vary greatly.

A letter of intent may indeed represent a preliminary understanding of parties to a negotiation who are not yet willing to
commit themselves. By contrast, it may represent an agreement on all
essential points, reached after negotiation, and intended by the
parties to constitute a binding contract, with or without stipulation
for the subsequent preparation of a formal document comprising its
terms.

Alternatively, in a complicated negotiation, it may reflect the
parties' desire to set down that which they have agreed upon to date
and, those points being regarded as settled, continue negotiating
with respect to the remaining issues.

Again, the party drafting the letter of intent may have a more
complex intention: it may seek to bind the other party to an agreement while not committing itself, or may wish the document to be non-
binding but determine to cast it in broad terms for fear that an
explicit statement to that effect might weaken the confidence of the
other party in its good faith. Such covert objectives may lead to
deliberate use of ambiguous language in the letter of intent.

In general, however, the letter of intent is by nature an ambiguous document. The parties, gratified at the success of their negotiations, fail to anticipate and provide for the problems that may
ensue. The letter of intent, generally prepared without resort to
legal advice, is typically brief and informal, leaving much unstated.

From those characteristics, valued by businesspersons, arises the
particular hazard associated with the use of the letter of intent,
namely that some crisis in the relationship of the parties will focus
attention on the letter of intent, which, subjected to judicial
scrutiny, may prove not to fulfil the intentions of either or both of
the parties to it. The problem appears to have its source in the

failure of the parties precisely to express the results they hope to
achieve by the execution of a letter of intent.

THE CONTRACTUAL ENFORCEABILITY OF THE LETTER OF INTENT

If the worst should happen, and one party be forced to assert the
contractually binding nature of the letter of intent in court,
against the denials of the other party, the central issue will be
whether, at the time the letter of intent was signed, the parties
intended to enter into a binding and enforceable contract. The
extent to which a letter of intent evidences an intent to create
contractual relations is a question of fact in every case.

There are surprisingly few cases on letters of intent in mining or
other transactions, considering the extent to which they are used,
but the authorities do provide some guide to the means by which a
party to a deal evidenced by a letter of intent might increase the
likelihood that a court will interpret the letter of intent as a
contract, and not merely an agreement to agree.

The leading Canadian authority is Calvan Consolidated Oil & Gas
Company Limited v. Manning, [1959] S.C.R. 253 (S.C.C.). The defendant made an offer in writing to the plaintiff to exchange an
interest in an oil and gas development permit he held for an interest
in a similar permit held by the plaintiff. The offer was accepted
unconditionally. The letter authorized the plaintiff to deal with
its permit on behalf of both parties as it saw fit, and provided that
if the plaintiff wished to develop the land instead of farming it out
or selling it, an operating agreement, the disputed clauses of which
could be arbitrated, was to be drawn up. The contents of the letter
were to be reduced to a formal agreement the terms of which, if
disputed, were likewise to be settled by arbitration. The plaintiff
entered into a farmout agreement with a third party. The defendant
refused to ratify the agreement or to sign the formal agreement
contemplated in the letter.

The two substantial questions before the court were, first,
whether because of vagueness or uncertainty in the terms, there was
no enforceable contract, and second, whether the offer letter and its
acceptance constituted an immediately binding contract even though
there was provision for a formal agreement to follow. The former is
an issue, distinct from the subject of letters of intent, with which
we need not here concern ourselves.

As to the latter, Judson J., delivering the judgment of the court,
said at pp. 260-261:

"My opinion is that the parties were bound immediately on the
execution of the informal agreement, that the acceptance was
unconditional and that all that was necessary to be done by the
parties or possibly by the aribtrator was to embody the precise

terms, and no more, of the informal agreement in a formal
agreement. This is not a case of acceptance qualified by such
expressed conditions as 'subject to the preparation and
approval of a formal contract', 'subject to contract', or
'subject to the preparation of a formal contract, its execution
by the parties and approval by their solicitors'. Here we have
an unqualified acceptance with a formal contract to follow.
Whether the parties intend to hold themselves bound until the
execution of a formal agreement is a question of construction
and I have no doubt in this case. The principle is well stated
by Parker J. in Hatzfeldt-Wildenburg v. Alexander, [[1912] 1
Ch. 284 at 288-9] in these terms:

> "It appears to be well settled by the authorities that if
> the documents or letters relied on as constituting a
> contract contemplate the execution of a further contract
> between the parties, it is a question of construction
> whether the execution of the further contract is a
> condition or term of the bargain, or whether it is a mere
> expression of the desire of the parties as to the manner
> in which the transaction already agreed to will in fact
> go through. In the former case there is no enforceable
> contract either because the condition is unfulfilled or
> because the law does not recognise a contract to enter
> into a contract. In the latter case there is a binding
> contract and the reference to the more formal document
> may be ignored.'"

Judson J. then added at p. 261, in an obiter dictum, that he was,
moreover, "fully satisfied that the parties thought they were bound
until very close to the institution of [the] action". Obviously
impressed by the finding that there had been substantial performance
on both sides, he noted at the same page that:

> "neither party felt the necessity of a formal agreement when
> they were dealing in a very serious way with the subject-
> matter of their contract and there was no difficulty."

In view of these remarks, we think we may take it as read that if
the parties to a letter of intent abide by its terms for some time
before the binding effect of the document is challenged, an inference
may be drawn that they intended to be bound by it.

However, in a recent Alberta case, Angus Leitch & Associates
Limited v. Legrand Industries Limited (1983), 31 Alta. L.R. (2d) 158
(Alta. C.A.), it was made clear that the mere fact of the parties
having acted on the letter of intent will not necessarily determine
the outcome. In that case, the parties had signed a letter of intent
pursuant to which they were to combine their resources in a new company involved in process engineering for the oil and gas industry.
The preamble to the letter of intent clearly contemplated the execution of a further agreement "according to, but not necessarily

limited to, the following general principles, terms and conditions".
The company was formed and commenced operations, but no formal contract comprising the terms of the letter was ever executed. When the
defendant ceased working with the new company, the plaintiff sued for
breach of contract.

The trial judge, emphasizing the subsequent performance by the
parties, and relying in part on the authority of Calvan Consolidated
Oil, referred to above, concluded that there was a contract.

The Court of Appeal allowed the defendant's appeal, holding at p.
166 that "the intentions expressed in the letter of intent did not
amount to an enforceable agreement", but merely an agreement to
agree. The decision in Calvan was distinguished on the basis that
here, unlike in that case, fundamental terms in the letter were left
to be settled in the future, with no mechanism provided for their
resolution.

In another Alberta case, Alta-West Group Investments Ltd. v. Femco
Financial Corporation Ltd. (1984), 34 Alta. L.R. (2d) 5 (Alta. Q.B.),
an agreement to act as joint venture partners that contemplated the
parties subsequently entering into a formal agreement, but otherwise
contained all essential terms, was held to be enforceable. Having
regard to the parties' intentions, as reflected in the letter of
agreement and in the parties' conduct during the negotiations and in
abiding by the terms of the letter for some four months, the requirement for formal documentation was held not to be a condition
precedent to a binding contract.

Moreover, it was the court's view that the otherwise valid agreement did not fail because the parties had failed to agree on the
terms and details of a clause where those details were unimportant to
the parties.

Block Bros. Realty Ltd. v. Occidental Hotel Ltd., [1971] 3 W.W.R.
51 (B.C.C.A.), in which the owner of certain properties accepted a
written offer to sell them which contained, among the provisions for
payment, the words "Balance by Mortgage and or Agreement for Sale,"
was another case in which the reasoning from Calvan was applied. The
judgment of the court, delivered by Bull J.A., was that the use of
those words clearly anticipated a further meeting of the minds of the
parties in order to determine an essential term which had not yet
been agreed upon, as distinct from a mere expression of desire as to
the manner in which the contract was to be carried out. The contract
was, therefore, unenforceable.

There are a number of other authorities to similar effect but, as
they do not break new ground, we are of the opinion that they do not
merit discussion here.

It is possible to extract a number of general principles from the
authorities considered above and, based upon them, to formulate

certain recommendations with a view to ensuring that a letter of
intent will have the effect in law that was contemplated by the
parties to it at the time of its making.

First, it is apparent from the cases that the courts will to a
great extent defer to the language used by the parties to express
their intentions with respect to the enforceability of a letter of
intent. It is, therefore, advisable to state the parties' intentions
in the clearest possible language, particularly where the parties
intend that the existence of any agreement is conditional on the
execution of formal documentation. To that end, it is recommended
that the parties enlist the aid of lawyers in drafting the letter.

Second, the courts attribute great significance to the completeness or lack of completeness of the purported agreement. The Alta-
West case demonstrates the principle that so long as the essential
terms of the agreement are set out in the memorandum, it is not
objectionable that the agreement is not worked out in meticulous
detail.

Similarly, if not all of the essential terms are present in the
letter, but a mechanism is agreed upon as the means of resolving
matters not yet decided upon by the parties, that will suffice. If,
then, the parties desire their letter of intent to bind them, they
would be well advised to ensure that all essential terms are provided
for, either initially or by some means agreed upon at the outset.

The conduct of the parties during and after the negotiations will
be viewed as an important indication of their intention. In particular, the authorities demonstrate that a willingness to abide by the
terms of the letter of intent, especially for a significant length of
time, will in most instances provide the basis for an inference that
the parties intended to be bound by those terms. Accordingly, the
parties should take care that they do not demonstrate by their
conduct an intention inconsistent with that expressed in the letter
of intent.

Our final recommendation is that, wherever possible, parties negotiating joint ventures or other commercial agreements should avoid
the use of the letter of intent. The extent to which a letter of
intent evidences an intention to contract is a question of fact in
every case. We have outlined the principal factors to which the
courts look to determine the enforceability of such a document, but
it is not possible to say with confidence to which factor a court
will accord the greatest weight in a particular instance. This uncertainty of result is indubitably much better avoided in commercial
transactions.

THE LETTER OF INTENT AND THE MISUSE OF CONFIDENTIAL INFORMATION

The subject of the use and misuse of confidential information
arises in the context of letters of intent because it frequently

happens in the course of negotiations toward a joint venture or other commercial arrangement that one or more of the parties must reveal to the other or others confidential information about the subject of the negotiations. Should the relationship between the parties, evidenced by a letter of intent, subsequently break down, the party in receipt of the confidential information may attempt to use the information given to it in confidence for its own benefit, without the consent of the revealing party and, indeed, so as to defeat the latter's objectives or interfere with its plans.

The letter of intent signed by the parties might contain provisions intended to ensure that confidential information disclosed in the course of negotiations is not misappropriated, but should the letter of intent prove not to be binding upon the parties, the confidentiality provisions will prove to be scant protection for the revealing party.

Fortunately for the party revealing the confidential information, in circumstances such as those outlined above, the courts have formulated and applied the broad equitable principle that information received in confidence should not be used to the detriment of the person who provided it. An equitable obligation of confidence may arise from the mere receipt of information in the knowledge that it was confidential.

The leading case in this area is <u>Seager</u> v. <u>Copydex Ltd.</u>, [1967] 1 W.L.R. 923 (C.A.). The plaintiff had invented and patented a carpet grip. He negotiated with the defendants regarding the marketing of the grip and, during these negotiations, the plaintiff talked about some improvements which he believed would lead to a better grip. The negotiations broke down and no business relationship was established between the parties. Subsequently, the defendants made a carpet grip which did not infringe the plaintiff's patent, but did include the improvements he had suggested.

The court, comprising Lord Denning M.R. and Salmon and Winn LL.J., made two interesting findings of fact. First, the judges found that the defendants realized that the information regarding the carpet grip was given to them in confidence. Second, they found that the defendants believed that the new grip was a result of their own ideas but that they had unconsciously made use of confidential information given to them by the plaintiff.

Each of the three judges gave separate reasons, concurring in the result. All three found that there had been an unconscious breach of confidence by the defendants. The leading judgment was that of Lord Denning. At page 931, he said:

"The law on this subject does not depend on any implied contract. It depends on the broad principle of equity that he who has received information in confidence shall not take unfair advantage of it. He must not make use of it to the

prejudice of him who gave it without obtaining his consent.
The principle is clear enough when the whole of the information
is private. The difficulty arises when the information is in
part public and in part private.

...

When the information is mixed, being partly public and partly
private, then the recipient must take special care to use only
the material which is in the public domain. He should go to
the public source and get it or, at any rate, not be in a
better position than if he had gone to the public source. He
should not get a start over others by using the information
which he received in confidence. At any rate, he should not
get a start without paying for it."

Stephenson v. Baliy Motors Ltd., [1978] 5 W.W.R. 645 (B.C.S.C.)
was a case in which the plaintiff inventor and the defendant producer
had entered into negotiations with the common object of forming a
joint venture for the manufacture and distribution of a gas-saving
device. When the deal collapsed, the defendant made use of information received from the plaintiff for his own benefit. Hutcheon J.,
holding the defendant in breach of its confidential relationship with
the plaintiff, referred to the Seager case and applied the words of
Megarry J. in Coco v. A.N. Clark (Engineers) Ltd., [1969] R.P.C. 41
at 48:

"It seems to me that if the circumstances are such that any
reasonable man standing in the shoes of the recipient of the
information would have realized that upon reasonable grounds
the information was being given to him in confidence, then this
should suffice to impose upon him the equitable obligation of
confidence. In particular, where information of commercial or
industrial value is given on a business-like basis and with
some avowed common object in mind, such as a joint venture or
the manufacture of articles by one party for the other, I would
regard the recipient as carrying a heavy burden if he seeks to
repel a contention that he was bound by an obligation of
confidence."

A case still more directly to the point with respect to transactions in the mining industry is International Corona Resources Ltd.
v. Lac Minerals Ltd. (1987) 62 O.R. (2d) 1 (Ont. C.A.). The headnote
accurately summarizes the facts:

"The plaintiff company owned mining rights on certain land, on
which it was drilling exploratory holes. The defendant company
approached the plaintiff company with a view to a possible
partnership or a joint venture, and the plaintiff revealed the
results of the drilling, from which it was clear that an
adjacent property was likely to include mineral-bearing
deposits. The plaintiff attempted to acquire the mining rights

in the adjacent property but the defendant put in a successful
competing bid, and developed the mine on its own account.

In an action by the plaintiff the trial judge found an industry
practice that imposed an obligation on parties seriously
negotiating a joint venture not to act to the detriment of each
other. He found that the information given to the defendant
was confidential, and revealed only for the purpose of a
possible joint venture. He found the defendant in breach of
its duty of confidence, and also of a fiduciary duty, in
acquiring the adjacent property, and concluded that the
defendant held the property on trust for the plaintiff."

On appeal by the defendant to the Ontario Court of Appeal, the
court, unanimously upholding the conclusions of the trial judge, made
a number of important statements which we think bear repeating at
some length.

Discussing the claim based upon breach of confidence, the court
stated at p. 22 that the elements of such a claim are (1) that the
information in question was confidential, (2) that the information
was communicated in circumstances in which an obligation of confidence exists, and (3) the unauthorized use of the information by the
defendants to the detriment of the plaintiff. It was the court's
view that the fact that the information was transmitted to the defendant with the mutual understanding that the parties were working
towards a joint venture or some other business arrangement giving
rise to an obligation of confidence was sufficient to satisfy the
second requirement.

The central issue with respect to the claim based upon breach of
confidence was whether the information furnished to the defendant was
confidential, in view of the fact that some of the information had
been public for a number of years. The court described the confidential information and concluded at p. 38 that:

"The fact that parts of the information in the same or
differing forms...were in the public domain does not detract
from the fact that the total package furnished by [the
plaintiff] to [the defendant] was something that was not in the
public domain and was confidential."

With respect to the submissions of the defendant that it owed no
fiduciary duty to the plaintiff not to act to the detriment of the
latter, the court said at p. 44:

"We agree that the law of fiduciary relations does not
ordinarily apply to parties who are involved in arm's length
commercial transactions. Nevertheless, it appears to be clear
that the law of fiduciary relations does apply in certain
circumstances to persons dealing at arm's length in commercial
transactions.

It becomes a question of fact in each case whether the
relationship of the parties, the one to the other, is such as
to create a fiduciary relationship. The circumstances which
give rise to such a relationship have not been fully defined
nor are they forever closed."

The court approved the express finding of the trial judge that
there is in the mining industry a practice that imposes an obligation
on parties who are seriously negotiating a joint venture in circumstances such as those found to have existed in that case not to act
to the detriment of each other, and continued at pp. 49-50:

"In order to establish a fiduciary relationship between two
persons, it is not necessary that they enter into a contract or
agreement. We are of the opinion that the trial judge was
correct in finding that a fiduciary relationship came into
existence between [the plaintiff] and [the defendant] as soon
as they entered into serious negotiations with respect to a
joint venture between them even though they never reached a
final agreement with respect to a joint venture. It may very
well be that a fiduciary relationship will not arise in all
cases where persons are negotiating with respect to an intended
partnership or joint venture. The practice established by the
evidence that intending joint venturers are under an obligation
not to act to the detriment of each other related only to the
mining industry and to a situation such as existed in the case
at bar.

It was a case of negotiations between a junior mining company
(Corona) whose primary activities were those of locating,
staking and evaluating mining claims and a senior mining
company (LAC) whose activities included all of the above
together with the practice and experience of bringing into
production and operating gold mining properties. It was a case
of the senior company seeking out the junior company in order
to obtain information with respect to mining claims already
owned by the junior company and to discuss the joint business
venture. Having regard to the practice found to exist in the
industry with respect to the obligation not to act to the
detriment of each other, particularly with respect to
confidential information disclosed, it was to be expected that
Corona would divulge confidential information to LAC during the
course of their negotiations. In those circumstances, it is
only just and proper that the court find that there exists a
fiduciary relationship with its attendant responsibilities of
dealing fairly including, but not limited to, the obligation
not to benefit at the expense of the other from information
received by one from the other."

The court summarized its conclusion with respect to the claim of
breach of fiduciary duty at p. 53:

"Unlike negotiations with respect to an ordinary commercial
transaction where there is not a common object but each party
seeks to obtain an advantage without concern as to the effect
of such advantage on the other party, the parties to these
proceedings were negotiating with respect to a common object
with an established practice in the industry that neither one
would act to the detriment of the other during the course of
such negotiations with respect to the proposed subject-matter
of those negotiations.

Having regard to all of those facts, we are satisfied that the
trial judge was correct in holding that there existed a
fiduciary relationship between the parties notwithstanding the
fact that a partnership or joint venture agreement was never
consummated. In so holding, no broad addition was made to the
law of fiduciary relationships. The holding merely recognizes
... a usage in the mining industry which is consistent with
business morality and with encouraging and enabling joint
development of the natural resources of the country."

The remedy granted the plaintiff by the trial judge was that of
restitution of the property by way of a constructive trust together
with an accounting of profits realized as a result of the breach of
fiduciary duty and breach of confidence by the defendant.

The Court of Appeal considered the two claims separately and
concluded that the appropriate remedy in each case was that of a
constructive trust. At p. 67, the court noted that:

"We have already said that if the value of the [property in
question], minus the cost of improvements expended on it by
LAC, is the proper measure of the plaintiff's claim, then it is
just and appropriate that the remedy take the form of a
transfer of the property on payment of the cost rather than the
form of a judgment for damages. To repeat, having regard to
the uncertainties and variables involved in placing a monetary
value on the property (the extent of the ore reserves and the
future fluctuations in the price of gold, the rate of
inflation, the rate of exchange between United States and
Canadian dollars, and a host of other factors) it is far more
just to grant a restitutionary proprietary remedy than damages."

The court's reasoning on this point has obvious relevance in mining
cases generally.

It is to be noted that the court in this case expressly refrained
from stipulating the proper legal approach to deal with a breach of
confidence or good faith after negotiations have ceased with respect
to the formation of a joint venture or partnership or after the

termination of a joint venture or partnership formed as a result of such negotiations. Presumably those situations, too, will be covered by the equitable principles discussed above.

4

Mining

Chapter 15

MINING THROUGH OLD UNDERGROUND STOPES AT MASCOT GOLD

by D.S. Cavers, C.O. Brawner and A. Bellamy

Senior Geological Engineer
Hardy BBT Limited

President
C.O. Brawner Engineering Ltd.

Mine Superintendent
Mascot Gold Mines, Nickel Plate Mine

ABSTRACT

Pit slope designs at Nickel Plate Mine have used relatively steep interim and final wall slopes of 60 degrees overall. The rock is unusually strong and tough with most discontinuity sets being steeply dipping. An integrated program of monitoring and drainage has been used. Stopes of large extent which are generally unsupported occur on the property. For the most part, no deterioration of the stability of the stopes has occurred over the last 30 or more years. Mining methods through these areas have used a combination of fills placed via VCR drop raises and crater blasting of the backs. One of the major challenges in mining through the stope areas is to maintain slope stability conditions and to reduce dilution of the higher value ore zones occurring immediately around the stope areas. The paper discusses geology, slope stability and mining methods at Nickel Plate Mine.

INTRODUCTION

The Nickel Plate Mine is a gold bearing skarn deposit located in the Southern Interior of British Columbia above the town of Hedley at an elevation of 5000 - 6000 feet. The mine is approximately 300 kilometres east of Vancouver. The history of the mine below is based in part on Millis (1983).

The Nickel Plate Mountain ore deposits were initially discovered in 1898. The Nickel Plate, Sunnyside, Bulldog, Horsefly, and Copperfield

203

claims were staked and by 1904 the Daly Reduction Mill had been erected and was in operation.

In 1909, the Nickel Plate Mine, under the name of Hedley Gold Mining, changed ownership from the Daly Estate to the U.S. Steel Corporation. In the same year, the Great Northern Railway linked Hedley to Oroville, Washington, and Princeton, B.C. The mine ran continuously, with one closure in 1920, until December 1930, because the then price of $20.60/oz was too low to sustain mining. Production at that time was 200 TPD.

The Kelowna Exploration Company Limited (Kelowna) acquired the Hedley Gold Mining Co. in 1932 and developed sufficient underground reserves to commence mining again in 1935 at 325 TPD. From this period until its closure in 1955, a total of six separate underground ore zones were mined. These included the main Nickel Plate zone, four Sunnyside zones, and the Bulldog zone. The total published production for the Kelowna and Mascot mines up to this period was 1,556,749 oz (4.4×10^7 g) gold, 188,139 oz (5.3×10^6 g) silver, and 4,077,305 lbs (1.85×10^6 kg) copper.

In 1908, the Mascot Fraction Claim was staked by other interests on the west slope of Nickel Plate Mountain. A deal could not be struck with the owners of the Nickel Plate Mine and the Hedley Mascot Mine was formed. This mine went into production in 1936, producing over 250,000 oz (7×10^6 g) of gold from only one fractional claim which formed part of the main Nickel Plate ore zones.

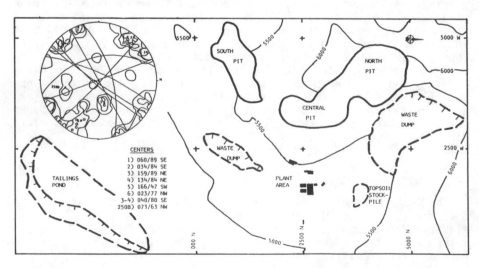

Figure 1. Schematic site plan. Stereonet is a lower hemisphere equal angle (Wulff) net. Contours are in 2% increments.

The Mascot-Nickel Plate Mine today encompasses the Kelowna Exploration and the Hedley Mascot properties. Earlier mining was predominantly from underground stopes which were generally unsupported due to
the very competent ground. Figure 1 shows the overall layout of the
three main open pit areas in the present mine. There are three main
pit areas:

a) The Bulldog zone is the southernmost pit. The planned floor
   elevation of the pit is 5100 feet which will be accomplished using
   two pushbacks. There are no major stopes in the Bulldog area.
b) The Central Pit area comprises the North and South Central Pits with
   planned floor elevations of 5420 and 5410 feet, respectively The
   South Central Pit, in which mining has recently been completed, was
   excavated through the Sunnyside 3 Stope area. The North Central Pit
   area will penetrate the 450 Stope area.
c) The Nickel Plate pit is the northernmost pit area and will be the
   major production area within a few years. The largest stopes occur
   in this area.

GEOLOGY - ORE RESERVES

The following discussion is based in part on Simpson and Ray (1986),
McGonigle, Tremaine and Johnson (1948) and input from Nickel Plate
geologists as noted in the acknowledgements. The Nickel Plate Mountain
ore deposits are one of the largest metasedimentary gold bearing skarns
known in the world. The Nickel Plate deposit lies within the upper
Hedley Sequence where a large skarn zone developed peripherally to a
stock of diorite and gabbro. The Hedley Sequence consists of thinly
bedded siltstone, black argillites and minor wackes interbedded with
impure limestone beds from 1 to 10 m thickness. The bedding generally
dips 30 degrees to the west and is interrupted locally by small scale
folds. The Nickel Plate skarns form a large west-dipping bowl centred
around a stock of hornblende diorite, quartz diorite and gabbro, which
in turn is part of a series of stocks termed the Hedley Intrusives.

The rock is unusually strong and tough. Measured rock strengths in
endoskarn and skarn units range from 16,000 to 65,000 psi. Some weaker
and more brittle limestones and intrusives occur locally.

Sills and dykes of diorite porphyry comprise about 40% of the
skarned interval and commonly show skarn alteration. The remainder of
the interval consists of varying amounts of iron rich pyroxene, garnet
and calcite. Gold is associated with arsenopyrite and pyrrhotite and
to a lesser extent chalcopyrite, pyritye and sphalerite. Gold occurs
in native form as minute blebs (about 25 microns across) along the
cracks and grain boundaries of the sulphides. Gold bearing zones often
follow bedding and are usually localized by small scale folds and
diorite porphyry sill/dyke contacts.

ROCK MECHANICS FOR OPEN PIT DESIGN

Structural geology

Prior to mine startup, an unusually large amount of structural geology data was available from mapping carried out during mining of the original stopes. Limited surface mapping was available from contemporary work. In addition, two oriented core holes were drilled and oriented using a clay orientor (Call et al, 1982).

Structural geology data from the different sources was all reasonably consistent, although small scale and local variations occurred. Figure 1 shows a typical summary stereonet for the Central Pit area. The main discontinuities which are present include joints and faults. Predominant joint sets are typically steeply dipping and, in many cases, are subparallel to the major trend of the walls. Continuity is typically greater than 10 m, although some smaller curved features occur locally. Bedding joints are less frequent but joints approximately parallel to bedding are important on some parts of the east wall of the Central Pit area where they are subparallel to the District Footwall Fault System.

Due to the prevailing discontinuity patterns, the predominant potential failure mode is toppling. In addition, wedges 1 to 2 benches high may occur in some areas on "random" discontinuities that do not fall into well defined sets. Local block ravelling may occur on some of the curved discontinuity surfaces.

It is noteworthy that the underground stopes show little evidence of failure, although they have been abandoned for at least 30 to 35 years. Except for the square set stope area, the stopes are unsupported and spans between pillars may, in some instances, be 30 to 60 m or more. Vertical openings range up to 45 m in the case of the 450N stope. The general lack of failure is due to the extremely competent rock, the generally steeply dipping discontinuities and to the shallow cover.

Blasting

In order to develop a blasting program unique to the conditions of the Nickel Plate Mine, three key factors had to be considered. These factors are grade control, wall control, and fragmentation. Each factor in itself has a major role in the successful, efficient development of the three Nickel Plate ore zones.

In a high waste-to-ore ratio mine as is the case for the Nickel Plate, grade control is essential to maximize ore production at minimum dilution. The maintenance of safe, stable pit walls requires controlled blasting, which is an inherent part of the Nickel Plate blasting program. Ore release is contingent upon the efficient removal of waste rock. Ultimately, good fragmentation governs the efficient loading of the broken muck.

<u>Production Blasting</u>.   The Nickel Plate skarn formations are both variable in hardness and toughness.   To deal with the different rock types, different production pattern sizes have been developed based on the drillability and fragmentation characteristics of the rock.   Waste pattern sizes, subgrade drilling depth and powder factors are listed below for the different rock types.

TABLE 1 - WASTE BLAST PATTERNS AT NICKEL PLATE

ROCK TYPE	BURDEN x SPACING Feet (m)	SUBGRADE Feet (m)	P.F. Lbs./S.Ton (kg/metric ton)
Gnt - Px. Skarn	12 x 14 (3.6 x 4.3)	2.5 (0.8)	0.77 (0.34)
Calc. Px. Skarn	13 x 15 (4.0 x 4.6)	3.0 (0.9)	0.69 (0.31)
Porphyry Sill	14 x 16 (4.39 x 4.9)	3.0 (0.9)	0.60 (0.27)
Limestone Skarn	15 x 17 (4.6 x 5.2)	3.5 (1.1)	0.55 (0.25)

In order to reduce subsequent pit wall damage, sequential blasting is incorporated in order to reduce the explosive charge per delay to tolerable levels.   A typical waste initiation tie-in is shown in Figure 2.

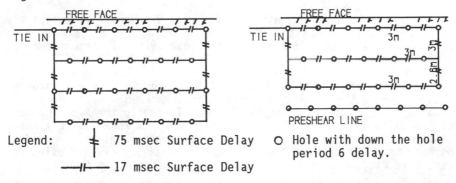

Legend: ⸺ 75 msec Surface Delay   O Hole with down the hole period 6 delay.

⸺ 17 msec Surface Delay

Figure 2. Tie in for production blast in waste.

Figure 3. Tie in for blast in ore.

Typical bench heights are 20 ft (6 m), but during pit development, waste areas are drilled to 40 ft (12 m) bench heights.   Blast hole diameters are 6.5 in (16.5 cm).

All surface delays are double ended. Trunk lines are primacord.
Down hole delays enable the surface to be ignited first to prevent
cutoffs and thus misfires. Emulsion type bulk explosives are used with
12 oz primers. Sequencing can be varied depending on how it is
intended to cast the muck. The powder factors have been designed to
give optimal fragmentation relative to rock type. Ongoing experimenta-
tion is used to maintain blasting results at close to optimum under
constantly changing conditions.

Control Blasting. Preshear blasting is generally used at the Nickel
Plate Mine to form the final bench face surface. All holes are 6.5
inch diameter. The preshear holes are drilled at the design bench
crest line to exact bench elevation at 40 ft (12 m) depth which
represents a double bench height.

Preshearing two benches simultaneously has been found to improve
stability conditions of the face due to elimination of the stepout area
which would otherwise be required at 20 ft (6 m) depth (Cavers, 1987).
This stepout area has been found to be a source of rock fall, to
require additional scaling and can project rock fall farther out into
the pit. Elimination of two of the three stepouts required over the
quadruple bench interval between safety berms also potentially
increases the width of the safety berm.

The final production blast in a bench sequence normally consists of
three rows. The back row in front of the preshear row is the buffer
row and is shot last. Sequential blasting is used. A typical blast
pattern for ore is shown on Figure 3.

Groundwater

Groundwater pressures at Nickel Plate Mine are generally relatively
small. Precipitation is low and, in many areas, the old stopes have
provided bottom drainage for many of the near vertical discontinuities
which are present.

Pneumatic piezometers are installed on an ongoing basis to provide
data concerning the presence of permanent or sporadic water pressures
which may develop in the discontinuities. Typically three piezometers
are placed in a single 160 ft (50 m) hole drilled from a safety berm.
Deeper holes are not efficient since the piezometers at depth would be
too far behind the pit wall. The piezometer holes are drilled using the
blast hole rig, which has been found to be fast and efficient.

To date, appreciable water pressures have only been observed in part
of the South Central Pit area where there were no underlying stopes or
drifts. In this case, maximum water heads of some 60 ft (20 m) were
observed in local areas of the pit during the wet season.

Horizontal drains inclined upward at 5 to 10 degrees are used to
drain water from the discontinuities. Drilling targets are established
on the basis of geological observations and water pressure measure-

ments. These drains occasionally tap water "pockets" but more frequently drip occasionally. In spite of the small volumes of water released, it should be noted that high water pressures can be generated on near vertical discontinuities by small volumes of water. The most important function of the drains is to reduce peak water pressures during heavy precipitation events or spring melt.

Slope Angle Recommendations

To date, the slope angles used at Nickel Plate Mine have been as follows:

- slope angle: 60 degrees overall
- interbench and intersafety berm heights: 20 ft and 80 ft (6 and 24 m).
- intersafety berm slope: 78 to 82 degrees.

Important considerations in using these angles include the following:

a) The rock is extremely competent, strong and tough.
b) Most of the discontinuities are relatively steep and would require some initiating event, such as high water pressures to precipitate failure of the overall wall.
c) More than one pushback is planned for many areas of the pit which allows operating experience to be obtained at the steep slope angles before the final slope is excavated. In areas of favourable conditions, mining at flatter angles would reduce cash flow and would not give a good indication of the conditions which would prevail for steeper walls. For example, the problem of drill stepout is not often realized for flatter walls.
d) Monitoring and drains are an integral part of the mine plan.
e) Unusually good structural data was available from the old mine plans.

INFLUENCE OF OLD STOPES ON SLOPE STABILITY

Three general configurations of old stopes are present at Nickel Plate Mine:

a) Relatively narrow stopes, such as the Sunnyside 3 stope which was approximately 60 ft (20 m) wide. Several narrower stopes also occur.
b) High, wide stopes. The 450N stope is the best example with heights and widths of approximately 120 to 150 ft (35 to 45 m).
c) Tabular stopes, such as the Nickel Plate stopes which may have widths of several hundred feet, spans between pillars of 100 to 300 ft (30 to 90 m) and variable heights. Some of the tabular stopes have additional tabular stopes underlying them.

The presence of the old stopes influences slope stability, stability of the pit floor and ore dilution. Wall conditions can be influenced by the presence of filled or unfilled stopes. Boundary element analysis of some of the larger stopes suggests that they are only stable as long as the roof arch is maintained in both directions. To date, detailed measurements of deformation above old stopes have not been made. Visual observations suggest that for the smaller stopes intersected to date, the deformation above the stope has been relatively small. In view of the potential problems with ore dilution, safety on the pit floor, and possible wall stability problems, detailed plans to deal with each significant stope intersected to date have been developed. It is stressed that much additional experience remains to be gained, particularly with mining methods through the large tabular Nickel Plate stopes.

CONSIDERATIONS FOR MINING THROUGH STOPE AREAS

Modelling Stopes

Underground plans and two directions of sections showing detailed mine development and sequencing dating back to the initial days of mining enabled the construction of plexiglass models for the Sunnyside 450N and Nickel Plate Glory Hole stopes. The models typically require only a few man days to construct.

Three dimensional CAD drafting equipment running on a VAX computer was used to develop a three dimensional (3D) model of the Sunnyside 3 stopes (see Photos 1 and 2). It was found that a large amount of memory was required and constructing a 3D surface around the perimeter of the sections was complex and only approximate, particularly in areas of overhangs or complex topography. Interpretation of the sections without the 3D surface construction proved to be difficult due to their complex shapes. The computer program used for the model was produced by Intergraph and was originally designed for modelling more regular shapes. Recent 3D modelling developments may offer increased possibilities with respect to easier modelling on less expensive computer equipment.

To date, the plexiglass model is considered the most cost effective. A major disadvantage is that it is not possible to incorporate detailed ore reserves and other data in the mine's computer data base into the model.

Dilution

Dilution of ore reserves around the old stopes forms an important constraint on the methods used for mining around the old stopes. In the underground stopes, pillars left behind contain high grade ore. Exact location of the pillars is critical to their recovery. Similarly, ore reserves in the back can be subject to high dilution for some mining methods.

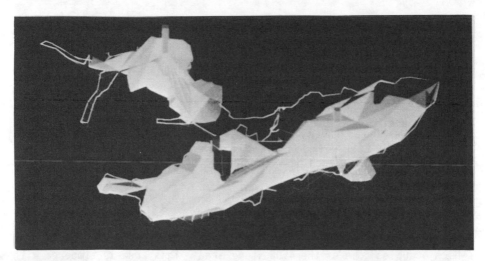

Photo 1. Sunnyside 3 computer generated stope model showing isometric view of outside of stope with back of stope in place in some areas. Existing pillars are dark vertical shapes (not trimmed to back line). Raises are shown as rectangular solids. Some cross sections are visible.

Photo 2. As Photo 1, but back of stope has been removed to show fills and existing pillars. Glory Hole fill is at right. "Cutouts" in conical fills are where fill hits back. Modelling for two subsequent stopes has used plexiglass sheets.

Safety

A major concern in working above old stopes is that the working
bench floor does not suddenly drop or cave. Parameters that should be
considered to decide what is a safe floor thickness before backfilling
or blast caving a stope are faulting and slips, roof span, stope height
and rock type.

After a stope has been backfilled, selected production blast holes
are drilled deeper to search for possible voids. If small voids are
found, another drop raise is not necessarily needed, however, the area
may be marked off to keep heavy equipment out until mining has advanced
through the stopes.

Square Set Area

As discussed above, most of the stopes at Nickel Plate Mine are
unsupported. One exception is the square set area or Red Mud Stope
which occupies some of the 450S stope area. This area is timbered with
square sets using 12 in (30 cm) timbers on a 6 ft (1.8 m) cubic
pattern. Drilling showed that some caving had occurred onto the top of
the square set area, and the square set area itself appeared to be
decked. It was not possible to achieve usable drop points for raises.
Therefore, the area was blasted in, using heavy charges. Walls were
blasted in first, followed by the back in-blasted down into the area
using sequenced charges. Excavation occurred from the edges towards
the centre.

Large Stopes vs Flat Lying Stopes vs Multiple Stopes

The most frequently used method to date at Nickel Plate Mine for
dealing with the old stopes has been filling via drop raises. VCR
blasting of the walls and back has also been used. The optimum stope
for drop raise filling has a sloping floor and back and/or large
vertical extent in order to allow significant spreading of fill from a
single drop raise point. Flat lying stopes or stopes with low vertical
extent require relatively large number of VCR drop raise points in
order to achieve a reasonable percentage filling.

Multiple tabular stopes in which one comparatively flat-lying
tabular stope overlies another are considered some of the most
difficult to mine. Procedures for mining through these areas are still
being developed at Nickel Plate Mine.

FILLING STOPES

A variety of methods were considered:

a) Filling with muck or low grade ore dropped through vertical crater
   retreat (VCR) raises or pushed into existing glory holes.
b) Pneumatic stowing of waste or low grade material into the stopes.

c) VCR blasting of the backs to fill the stopes with bulked material.

In addition, various slope geometries to reduce stress concentrations are presently under consideration for mining through the large tabular stopes.

Filling of the old stopes via drop raises or some method of stowing offers the best chance to reduce dilution. Operational experience shows that some dilution still occurs since it is impossible to completely fill the old stopes in all areas. In addition, it has frequently proven difficult for shovel operators and mine geologists to identify the transition between blasted high grade muck from the stope back and the lower grade ore placed as fill.

To date, pneumatic stowing has been viewed as being too expensive, particularly considering the high abrasiveness of the rock which would produce high equipment wear.

Drop Raises

Drop raises have formed one of the main methods for placing large volumes of fill into the underground stopes. The drop raises used at Nickel Plate Mine are 10 x 10 ft (3 x 3 m) and have been up to approximately 100 ft deep. The raises were blasted using VCR methods, using the drilling pattern shown on Figure 4. The holes were drilled using the blast hole rig with 6 1/2 in (16.5 cm) holes. It has been found possible to keep holes in line to depths of at least 100 ft (30 m).

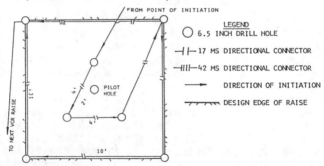

Figure 4. Drop raise blast layout.

The angle of repose of fill placed through the raises is variable. Measurements underground of fills placed using fills dozed into glory holes gave angles in the order of 38 degrees. This was also the angle for an upwardly unconfined fill placed through the drop raise points for the Sunnyside 3 opening. Where the fill approaches the back, however, the additional confinement may give an effective angle of repose of 45 degrees or more due to the fill tending to choke.

VCR Blasting

VCR blasting of the backs and/or walls of stopes relies on choking the opening with bulked material from the sides of the opening and back. VCR blasting has been routinely used on drifts and smaller stopes. In addition, the method is used where voids exist above fills in the larger stopes. VCR blasting suffers from a major disadvantage in that dilution is significantly higher and the distribution of the ore after blasting is much more difficult to predict.

Monitoring

Monitoring is considered to be an integral part of the overall program. Monitoring specifically related to stopes includes the following:

a) Checking of the distribution of stopes, back elevations relative to the old mine plans and fill configuration by deepening selected production blast holes.
b) Monitoring of the volume of fill placed through drop raises versus the volume which it should theoretically be possible to place. This helps to show areas in which premature choking of the fill may have occurred, raising the possibility that there are voids within the workings.

PRELIMINARY EXPERIENCE

To date, stability of pit slopes and backs of intersected underground openings has been good. Significant operational experience has been gained in blast design for production, wall control, stope filling and fill raises. Stope filling experience has been close to design. The following paragraphs briefly summarize experience to date.

During mining through the Sunnyside 3 stopes, the stability of the wall and the stope backs remained good. Photo 3 shows the north wall of the Central Pit which intersected the openings. There was no visual evidence on the pit floor that the opening existed below. There was no movement of the pit floor during mining through the filled Sunnyside 3 filled stopes. Both large and small scale structural instability have not been significant problems with the design and blasting practices used.

Photo 3.   North side of Central Pit showing ends of Sunnyside 3 Stope
           (left side of Photos 1 and 2) intersecting final wall. Note
           edges of fills present in stope. Safety berms are at 80 ft.
           intervals.

Production and wall control blasts proved to be an important part in
achieving safe, stable walls at the designed 60 degree slopes. VCR
drop raises have been used for the Sunnyside 3 stopes and are planned
for the 450N and 450S Stopes. The unreinforced raises were placed
quickly with no problems over a short period of time. They stood up
well during filling. VCR crater blasting of the back and walls has
been used for the Sunnyside 4, Bulldog and Red Mud Stope areas. There
were no significant mining problems in these areas; however, appreciable unavoidable dilution did occur.

The schematic outline of fill distribution in the Sunnyside 3 stope is shown in Figure 5. The stope locations were close to original design. The volumes of fills placed through the different drop raises were similar to those predicted. One additional raise was placed in order to place additional fill in the bottom end of the stope. The total volume of fill placed via the drop raises was 6700 cubic yards. From a production point of view, dilution occurred since it was not possible to completely fill the opening in all areas, and in many areas it was difficult to tell where the blasted ore/waste rock interface occurred.

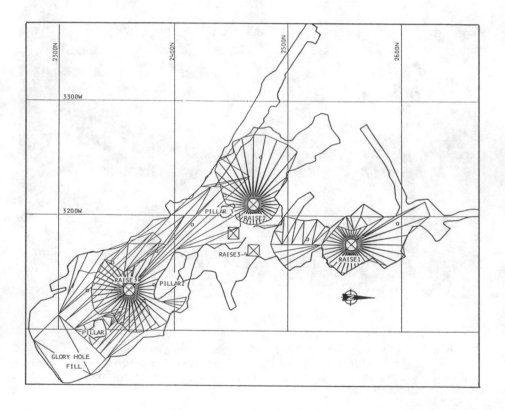

Figure 5. Plan of Sunnyside 3 Stope in South Central Pit showing distribution of raise fills. Generated from 3D computer model shown in Photos 1 and 2.

Major fills are planned via drop raises into the 450N and 450S Stopes (see Figure 6). These have not yet been placed but will have a total volume approximately ten times that of the Sunnyside 3 opening.

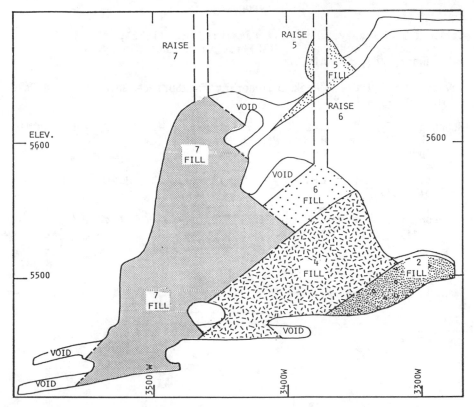

Figure 6. Composite longitudinal section showing 450N Stope fills in part of stope. Elevations and distances in feet.

ACKNOWLEDGEMENTS

Permission from Corona Corportion to publish this paper is gratefully acknowleged. Messrs. Dave Pow, Bill Wilkinson and Peter Thomas have contributed appreciably to the work discussed in this paper. Mr. George Headley assisted with design of fills for the 450 Stope and provided much needed help with the final production of this paper. Wilma Hartmann typed the manuscript.

REFERENCES

Call, R.D., Savely, J.P. and Pakalnis, R. (1982), "A Simple Core Orientation Technique." Stability in Surface Mining, Vol. 3, C.O. Brawner, ed., SME-AIME.

Cavers, D.S. (1987), "Rock Engineering for Surface Gold Mines." Gold Mining 87, C.O. Brawner, ed., SME-AIME.

McGonigle, F.A. Tremaine, C.W.S., Johnson, E.W., 1948, "Notes on the Nickel Plate and Hedley Mascot Mines, Hedley, B.C." - CIM Field Trip.

Millis, S.W., 1983, "A History of the Nickel Plate, Hedley Mill". (Not Published).

Simpson, P.G. and Ray, G.E., 1986, "Geology of the Nickel Plate Gold Deposit" - CIM, Oct. 1986.

Chapter 16

GROUNDWATER EVALUATION AND CONTROL
FOR GOLD MINING PROJECTS

by Adrian Brown

Adrian Brown Consultants,Inc
Denver, Colorado

ABSTRACT

Groundwater inflow and impacts have become major aspects in the
permitting and operation of many gold mines. First, the hydrothermal
or placer genesis of many of the orebodies being investigated or
produced today almost guarantees that mines will be "wet", involving
groundwater control as part of operations. Second, the style of ore
deposition also often involves alteration of the host rock, resulting
in relatively weak, low permeability materials that require extensive
groundwater pressure control for stable mining. Third, the location
of many mines in arid areas guarantees that impacts on water resources
will be matters of key focus during permitting and operation.
However, groundwater control and protection is a technically difficult
area and often involves managerial decisions and activities that do
not fit comfortably into traditional mining. In an attempt to
overcome some of these difficulties, this paper provides a guide to
mine managers and hydrologists as to appropriate strategies for
designing, setting up, operating, and evaluating groundwater programs
at gold mines that get the job done and are cost effective.

INTRODUCTION

The objective of gold mining is generally to obtain the maximum
production of gold at the minimum cost. However the achieving of this
aim is increasingly threatened by the impact of groundwater issues in
the development, production, and reclamation phases of many gold
mines.

It is almost inevitable that groundwater will be an issue at a gold
mine. The ore genesis of most deposits is via water, in one way or

another, essentially guaranteeing that the mine will be wet and will require some form of dewatering or groundwater control. Further, the emplacement of at least hydrothermal deposits is likely to lead to the alteration of the host rock, creating ground condition problems that are exacerbated by the presence of groundwater pressure. Finally, water is generally needed for processing the ore; this water is often obtained from the groundwater regime.

In addition to these engineering issues there are a range of groundwater impact issues. The extraction of groundwater for dewatering, ground control, or water supply purposes leads to a depression of the groundwater table in the vicinity of the mine, which has the potential to impact the availability of groundwater for other users. In addition, the excavation of a pit and the disposal of waste rock creates the possibility of acid mine drainage being generated on the site. Finally, the deposition of process tailings, often containing at least traces of metals, cyanide or other process additives, creates the potential for a threat to water quality near the project. The permitting process requires that the design of the facility and the reclamation system accommodate these possibilities, and to do so they need considerable information about the groundwater system.

The managers of gold mines have to make a continuing series of decisions based on predictions of what will happen as a result of groundwater flow and solute transport. These decisions relate to design, permitting, operations, safety, reclamation, and impact. All of them rely on a prediction of what will happen to the groundwater system. Accordingly, managers are principally interested in the predictive side of groundwater hydrology. Unfortunately, these predictions are hard to make.

Mining geohydrology is a complex area of technical activity. It deals with the behavior of groundwater in natural systems, and after perturbations to that system resulting from mine activities. The water with which it deals can only rarely be seen, and the conduits through which it flows are difficult and expensive to investigate. The study of groundwater flow and transport developed mainly for water supply purposes. In general, the locations where such supplies are sought can be characterized by relatively extensive, homogeneous, predictable circumstances. In most gold mining situations, the conditions are the opposite: great local and regional disturbance ("ground preparation") which leads to considerable complexity in geology and geohydrology. As a result, there tends to be a high level of uncertainty reported for the results of gold mine hydrology studies, specialists in the field are renowned for disagreeing with each other, and the predictions of behavior of such groundwater systems frequently turn out to be sufficiently incorrect that they lead to the wrong managerial decision. Accordingly, it is understandable that there is a considerable degree of skepticism about the reliability of results of gold mining groundwater investigations and a reluctance to base important decisions upon them.

The author of this paper does not believe that this situation is inevitable. Experience in mining geohydrology allows mines to avoid the mistakes of the past, and to obtain usable, reliable results from appropriately planned groundwater investigations. This paper sets out and illustrates a series of strategic concepts which, if put into place by management early in development of a gold property, and if followed by the technical staff, will produce results that will allow management to make decisions with confidence.

CONCEPTS FOR MINING GEOHYDROLOGY

Concept #1: Groundwater Hydrology is Part of the Hydrologic Cycle

Mining geohydrology systems are located within the overall hydrologic cycle. In that cycle, any water that will enter the mine must have come from somewhere. There are two principal sources: infiltration from precipitation or surface water, or from storage within the subsurface materials. As water is essentially incompressible, it follows that the water flow in a system must also be going somewhere; particularly prior to development, it is often possible to utilize this fact to better understand the overall groundwater flow system.

In particular, there is a considerable amount known about the way that water enters the deep groundwater flow system. Water reaches the surface by either precipitation or overland movement in rivers, streams, or other conduits. Once on the surface, the water either runs off, evaporates, enters the shallow groundwater system and is transpired by plants, is stored in the shallow groundwater system, or enters the deep groundwater system. Extensive study by the author and others (see for example the discussion in Freeze and Cherry, 1979, p. 211 et seq.) has provided a considerable amount of insight into the source of recharge to groundwater systems. It has been found in a wide range of studies that the average deep percolation to the groundwater table is a proportion of the average annual precipitation. For the purposes of most mine geohydrology studies, the deep groundwater contribution can be estimated by relating it to the incident precipitation and the surface conditions at the site, using the proportions presented in Table 1.

TABLE 1. Approximate Infiltration as a Percentage of Precipitation

Evaporation	Low (0-0.5 m/y)	Medium (0.5-1.5 m/y)	High (1.5+ m/y)
High (2+ m/y)	5%	15%	25%
Medium(0.5-2 m/y)	10%	20%	30%
Low (0-0.5 m/y)	15%	25%	40%

The significance of infiltration is that it tends to limit the total amount of water that is available for flow to a system, once the stored water that is in the site area has been depleted. In high permeability locations this limitation may be critical in evaluations of inflow and impact. As an example, a complex analysis of a mine in high permeability materials in the Rocky Mountains produced a predicted mine inflow of 6,000 liters per second. Review of this value indicated that the possible catchment area for the inflowing water to enter the groundwater system was only about 500 hectares. The total precipitation in the area is about 1.5 meters per year, so that the total precipitation to the catchment averages only 250 liters per second. This suggested that the analysis was incorrect, and that the long term inflow to the mine would probably be in the order of 100 liters per second.

Concept #2: The Foundation of Mine Geohydrology is Geology

The geology of the mine site and its environs forms the framework within which groundwater flow and solute transport associated with the mine will take place. The geology of the unsaturated and saturated portions of the potential flow paths are both important, at both the local and the regional scales. The mine-site geology is established in some detail as a result of the ore exploration program; in addition some information about the flow of water in the system may be available by evaluation of the ore distribution.

The geological description of the materials that are encountered in the site drilling program allows a reasonably close approximation to be made as to the important hydrological parameters to be made. To allow this approximation with respect to the granular materials, it is important to have information as to the detailed grainsizes encountered, as the permeability of the granular materials is dependent on the finest 10% of the material. With respect to the rock materials, some information about the degree of fracturing of the rock elements is needed. This may require special efforts, as rotary and hammer drilling do not provide this information directly. Note also that it is important to have information on the nature of the geological materials above and below the water table, as both are of importance in the analyses to be performed.

The geohydrological parameters relating to the materials encountered in gold mining have been summarized from extensive experience and published sources in Table 2. In using these values, care should be taken to note that the values are in general order of magnitude estimates, and if precise values can be shown to be needed they should be obtained in the field.

TABLE 2. Typical Geohydrological Parameters(1) for Mining Studies

Material Type	Hydraulic Conductivity m/s*1E09	Effective Porosity m3/m3	Specific Yield m3/m3	Specific Storage m3/m3/m*1E06
SOIL (2)				
Clay (3)(4)	1	0.2	0.001	100
Silt (3)	100	0.2	0.01	10
Sand (3)	10,000	0.2	0.1	1
Gravel (3)	1,000,000	0.2	0.2	1
ROCK (5)				
Slightly fractured	1	0.0005	0.0001	1
Normally fractured	100	0.005	0.001	1
Highly fractured	1,000	0.05	0.01	1
Karstified	1,000,000	0.1	0.1	1

Notes:
(1) The parameters are as follows:
    Hydraulic conductivity..measure of the resistance to fluid flow
    Effective porosity......porosity of pathways that transport fluid
    Specific yield.........porosity which is drainable under gravity
    Specific storage........water volume released from a unit volume
                            of material due to unit reduction of head

(2) "Soil" here means a material in which flow is mainly within the
    bulk of the material, rather than through heterogeneities.

(3) The description here refers not to the mean size, but to the sieve
    size which passes 10% of the material.

(4) Unfissured clay - if clay is fissured, treat as "slightly frac-
    tured rock".

(5) "Rock" here means a material where flow is primarily through
    fractures and voids, rather than through the intact material
    itself.

Concept #3: Flow of Groundwater is Subject to the Laws of Hydraulics

    The flow of groundwater is required to take place under the same
principles as the flow of water in any other system. There appears to
be a tendency to think that groundwater behaves somewhat unpredictab-
ly. Experience suggests the opposite: because water is essentially
incompressible, the flow and head conditions that exist in groundwater
systems are predictable within the accuracy needed for gold mining
purposes.

    The fundamental relationship that is used to describe groundwater
flow is known as Darcy's Law, which relates flow to head gradient, and

flow tube area. The constant of proportionality in the relationship
is known as the coefficient of hydraulic conductivity. Darcy's Law is
at the base of almost all flow relationships in mining groundwater
flow analysis, and flow relationships for other situations can be
derived from it, as shown in Table 3.

TABLE 3. Flow and Velocity Relationships (Steady State)

ITEM	PLANAR	RADIAL	SPHERICAL
EXACT:			
Flow to inner opening	$\dfrac{kHA}{(R-r_0)}$	$\dfrac{2\pi kHL}{\ln(R/r_0)}$	$\dfrac{4\pi kH}{(1/r_0-1/R)}$
Particle velocity	$\dfrac{kH}{n(R-r_0)}$	$\dfrac{kH}{r\,n\,\ln(R/r_0)}$	$\dfrac{kH}{r^2 n(1/r_0-1/R)}$
APPROXIMATE (R much greater than r_0)			
Flow to inner opening	$\dfrac{kHA}{R}$	kHL	$4\pi kH$
Particle velocity	$\dfrac{kH}{nR}$	$\dfrac{kH}{2\pi r n}$	$\dfrac{kH r_0}{r^2 n}$

where: k = hydraulic conductivity H = head drop at inner opening
        A = area of flow R = distance where flow starts
        r_0 = inner opening distance L = length of flow cylinder
        r = general radial distance n = effective porosity

The significance of this table is that it contains almost all the
steady-state flow equations that are needed for interpretation of
tests and for the computation of the inflow and solute transport
associated with gold mining.

Concept #4: Mining Groundwater Flows and Heads are Mainly Steady State

In general, transients in groundwater flow associated with gold
mines die out relatively rapidly when compared with a typical mine
life. Accordingly, steady-state conditions tend to be arrived at
relatively early in the mine life, after the initial perturbations to
the groundwater system caused by mining have died out. The reason for
this is that the water stored in the ground is generally a relatively
small volume when compared with the amount of water that infiltrates
to the mining area from the surface over the time periods of significance in mining projects.

The significance of this concept is that it simplifies both the testing and analysis of groundwater systems in mining, as steady state parameters are easier to test for and steady-state analyses are easier to perform. The validity of this simplification is relatively easy to check. The volume of water that is in storage can be calculated by multiplying the volume of the "drawdown cone" and multiplying it by the specific yield (in the case of water table lowering), or by the specific storage multiplied by the aquifer thickness (in the case of confined flow). A typical computation for radial flow to a pit in British Columbia produced the following results:

Quantity of water in storage in cone = 112,000 cubic meters
Flow of water to pit = 300,000 cubic meters/year
Time to remove storage if no other source = 0.4 years

This is short compared with the project timespan, so it is expected that the groundwater system will approach steady state quickly.

Concept #5: Mining Solute Transport Conditions are Usually Transient

The size of most mining project sites is relatively large, and the velocity of movement of solutes through the subsurface materials of those sites is in general slow. As a result, most computations of solute transport conditions at mines (for example movement of fluids from a tailings pond) need to consider transient transport conditions (concentrations, mass flux) based on steady state hydraulic conditions (flow, head).

As an example, the velocity of movement of water in saturated volcanic bedrock beneath a proposed tailings impoundment in South Carolina was computed using the information presented in Table 2, and the velocity equation for planar flow presented in Table 3. The computed mean groundwater velocity was 3 meters per year. Even for a 30 year project life, this result indicates that a solute plume would only move 100 meters, a short distance compared to project area size. Thus in this case it is appropriate to consider the development of the area affected by any chemically modified flow for periods considerably greater than the life of the project.

The most critical parameter in this evaluation is the effective porosity, particularly in flow through rock materials. The effective porosity exhibited by a fractured rock system can be highly variable, being as much as 10% and as low as 0.01%. In addition effective porosity is difficult and expensive to evaluate, and available precision of the testing is low. This sets a very distinct limit to the precision of evaluations of the rate of movement of solutes through groundwater systems, which cannot be remedied by sophisticated and accurate evaluation of the other aspects of the process.

Concept #6: Evaluations In Mining Geohydrology are Uncertain

"Are you sure?". For a mining hydrologist, this is perhaps the
most difficult question to respond to in respect of a prediction about
the behavior of a future mine. It is impossible to be absolutely
"sure" of the result of evaluations that use considerable simplifications of the system. (Note that this kind of uncertainty also occurs
in many other predictive aspects of mining, for example ore reserve
and price predictions.) One never knows exactly what the flow will be
to a mine, nor exactly what the head impact will be, nor exactly what
will happen if there is leakage from a tailings impoundment. There
are inherent limitations in the understanding of the groundwater flow
system, the parameters that describe the groundwater behavior in that
system, and the parameters that describe the behavior of chemical
species in that system that all militate against absolute accuracy.
Based on experience in mining geohydrology studies over the last two
decades, it appears that the limits to predictive accuracy in groundwater studies are as presented in Table 4.

TABLE 4. Best Accuracy Obtainable in Mining Groundwater Evaluations

PERFORMANCE	-----UNCERTAINTY FACTOR (1)-----	
MEASURE	SOIL SYSTEMS	ROCK SYSTEMS
Flow of water	1.3	1.4
Head changes	1.2	1.2
Velocity of water	1.5	1.7
Concentration	1.7	2.0

Note:To establish the best obtainable range of uncertainty
associated with each performance measure, multiply and
divide the value obtained by the factor in the table. The
bounds obtained approximate the 95% confidence limits.

While it is not possible to have a high degree of confidence in
even the best analysis, it is often possible to have a high degree of
confidence that estimates of these matters are sufficiently accurate
that decisions made by management based upon them have a low probability of being in error.

An example of this concept occurred recently in the investigation
of solute transport of cyanide from a possible leak in a lined pond.
In this case, there was very little information directly relating to
the question. An analysis was performed using the information that
existed to evaluate the concentration that would be expected to reach
the accessible environment; this level was substantially lower than
applicable standards. The question which had to be answered was
whether the uncertainty was so great as to render the result unreliable. To ascertain this, a statistical analysis was performed of the
uncertainty of the analysis: it was large. However, even with this
large uncertainty, there was better than a 95% probability that the

regulatory standard would be met, and the agencies concerned issued the permit on that basis, saving a very expensive additional investigation. In this case there was large technical uncertainty, but very small decision-making uncertainty.

Concept #7: Simple Evaluations are More Convincing than Complex Ones

Simple analyses are those that are performed using the standard closed form equations of the type presented in Table 3 (there are a corresponding set of formulae for transient conditions). The advantages of using these formulations are that they are easy to use, the results are easily checkable, and the assumptions and simplifications used in the evaluation are clear and stated. They also have the significant benefit of being inexpensive to "run".

The advent of the computer has injected a variable into mining hydrology which did not previously exist: the ability to conduct analyses with very considerable apparent precision and complexity. There has been a period of near infatuation with numerical tools, as they appear to overcome some of the difficulties of not having enough data for analysis. Today it is unusual for any mining groundwater proposal not to offer a major modeling exercise. Such exercises provide an apparently precise result, and precision makes managerial decision-making easier. However if the precision is not real, then the decision-maker is probably going to be misled.

The reasons that modeling is not always superior to simple methods are the mirror image of the reasons why simple methods are good: they are difficult to use and check, there may be built in assumptions that are not clear, and they are often expensive, both in computer time and person time. There are also some more subtle difficulties. First, many computer codes provide constraints to analysis because of their design; codes may not be able to include such features as infiltration, detailed evaluation near a mine, or conditions in three dimensions. Second, computer models are very demanding of data. Each mesh point and/or element in the analysis domain requires input of geological, hydrological, and state variables to allow analysis. It is a small and often-made jump to find an analyst demanding data "because it is needed in the model", without asking whether the decision-maker's answer might be affected by the data. Finally, sensitivity and probabilistic analyses are time consuming and expensive to do on a numerical model, so the uncertainty aspects of the problem are rarely thoroughly addressed using this approach.

There are certainly circumstances where numerical modeling is needed and useful. These include situations where decisions can only be made using detailed evaluations, or where the geology, infiltration, water withdrawal, and project complexity are so great as to make simple analyses unreliable. However the author has rarely found a situation where computer modeling provides a significantly different or better result from the simple approaches available using Darcy's

Law and some common sense, and conversely has found frequent situations where modeling has proven to be significantly incorrect. For
this reason, it is a rule in the author's organization that modeling
can only be used once an answer to the problem has at least been
estimated by other means.

An example which helped lead the author to this concept occurred in
the evaluation of inflow to a small underground mine in Quebec. The
decision that was required was whether to continue with the mine.
Early drilling and testing indicated that the orebody was very permeable, and the inflow might be large enough to make dewatering the mine
a prohibitive cost element. To resolve the matter, a very extensive
geohydrology data collection program had been undertaken at considerable expense to "provide input for the model". The model had
been run, and the conclusion was that more information was needed
before a critical prediction of inflow could be made, and the fate of
the mine decided. A simple analysis of the inflow was made using
steady-state equations, and it was concluded with acceptable confidence that the inflow would be low enough to allow development.
Subsequent development of the mine indicated that the estimate of
inflow made in this way was almost exactly correct.

Concept #8: Prototypes are the Best Tests of Geohydrology Systems

As stated above, prediction of the effects of changes on a geohydrologic system is difficult. This concept suggests that prediction
is more reliable when it is based on a field test that is a prototype
of the real activity, rather than when it depends on testing that
bears little similarity in scale or approach to the real activity.

In geohydrology there are essentially two ways to predict the
behavior of a system when a change is made to it. The first is to
conceptually take the system apart, test the components for their
geohydrologic parameters, re-assemble the components (usually in a
numerical model), and analyze the system using that model. This
approach involves testing small representative portions of each of the
materials in the general mine area, with greater attention to the
materials close to the mine elements. With respect to permeability,
for example, this approach would generally be carried out by measuring
the hydraulic conductivity of small quantities of all the materials of
interest, either in the laboratory on drill samples, or by packer
testing relatively small sections of drill holes in rock materials.
Experience suggests that there are serious drawbacks to this approach
to prediction. The principal difficulty is that this approach does
not provide any direct understanding of the way that the flow system
works. The analyst is left to develop a conceptual model of the flow
system from the geology and the material parameters. While this can
be done with experience, failure to identify and take account of
critical features (for example a small, vertical, clay-filled fault
zone, which might segment the flow system) may invalidate the predictions made.

The second approach to prediction is to perform a large scale test, located as near as possible to the area of interest and reproducing as many of the features of the full scale activity as possible. For example, if an underground mine is proposed and the impact on groundwater pressure is at issue, measuring the impact of the installation of a pilot mine for bulk sampling by monitoring heads in exploration holes and the flow from the mine, is likely to be easy to accurately extrapolate to the full scale mine, even if it is fairly difficult to obtain "exact" parameters from such a test. Experience suggests that the latter approach to prediction in geohydrology is generally feasible, cost effective, convincing, and accurate.

An example of this approach is the investigation that was done by the author for a dewatering system of a very rich gold prospect in Nevada. An early informal test had indicated that there might be unexpected dewatering problems, involving large quantities of water to be removed from thick alluvial overburden and the underlying rock. Preliminary evaluation indicated that well dewatering would be the only viable approach to groundwater control. In order to develop a prediction of how long dewatering would take, and what kind of a system would be required, a prototype dewatering well was proposed and installed in a location that was considered to be appropriate for the actual dewatering system. This well was tested, and was left running to become the first dewatering well in the system. The locations and designs of subsequent wells in the system were based on analysis of the results of the dewatering of earlier wells, in a leapfrog fashion. In addition to finding the correct answer, the incremental cost of investigation was in this way limited to the cost of a few observation wells, and the dewatering program was able to stay ahead of the overburden stripping needs.

Concept #9: Mining Geohydrology is a Project-Long Activity

Groundwater hydrology evaluation should begin with exploration and extend through operation and reclamation. There are many opportunities to learn about the groundwater system in a mining project either directly or while doing other things. An optimal program will accumulate understanding and information on groundwater matters from inception to completion.

First, most areas of the planet have been explored for water resources. Thus even prior to mineral exploration, there is likely some information on the surface and groundwater system in the orebody area. This information should be collected at the start of the project. Second, early exploration, particularly involving drilling, provides an excellent opportunity for definition of the groundwater system, if the geological team is sensitive to hydrology issues.

Some suggestions:

o   Sample and investigate everything that is drilled through,
     whether it is ore bearing or not. The cost of this addition-
     al sampling is trivial, but the cost of re-drilling the holes
     later to get the data is not. Remember that the barren
     materials above and adjacent to orebodies are generally more
     important to hydrological evaluation than the ore, because
     the ore will probably be removed.

o   If possible drill using a method that extracts water, and
     record the flow rate, quality, and temperature of the water
     as a function of depth. Many recent exploratory programs use
     the reverse air rotary/hammer drilling technique. This is a
     boon to hydrologists, as it effectively "dries" the hole
     during drilling. The flow from the hole can, with care, be
     interpreted to indicate important geohydrological informa-
     tion, particularly as each test can be considered to be a
     mini-pump test. In any event, record all water related
     matters (lost circulation, artesian flow, gas, etc.) whenever
     drilling is done.

o   Complete as many of the holes that are drilled as possible.
     This may require considerable effort, as the stability of
     some holes in gold orebodies is poor. However the author
     considers every drill hole to be a resource, in that it prov-
     ides the opportunity for obtaining data from the subsurface,
     which is not generally accessible by other means. Completion
     of the holes should be as good as the conditions allow. If
     possible, insertion of 50 mm diameter PVC slotted pipe is
     optimal, as it allows a water sample to be withdrawn from the
     hole if desired. Failing this, insertion of at least 20 mm
     PVC pipe, slotted occasionally, is adequate, as it allows
     water levels to be measured. Note that some collar protec-
     tion of these completions is essential in a mining area;
     leaving the steel collar casing in the hole is generally well
     worth the cost.

o   Measure the water levels in the completions reasonably
     frequently, for example weekly. It is astonishing how good a
     picture of the behavior of the groundwater system can be
     obtained in this manner, for remarkably little cost. The
     changes in water levels record the effects of nearby drill-
     ing, barometric effects, rainfall infiltration, mining,
     blasting, and many other factors.

   Once the exploration phase is finished, there will generally need
to be a groundwater program of some kind, for permitting, dewatering,
water supply, tailings pond design, inflow evaluation, acid mine
drainage evaluation, or some other purpose. This program should be
additive to the information already obtained, and should be considerably less expensive than it will be in the absence of a good

exploratory hydrology program. As the mine production progresses, the operation of those facilities that impact or are impacted by the geohydrology system provides further insight into the system, so that less and less effort is needed to design and plan future activities. By the time the project is finished, the geohydrology aspects of reclamation activities should be able to be designed and implemented with confidence, as a result of accumulated knowledge.

Few mine managers would disagree with the desirability of the above approach. However the author has almost never seen a mine implement such a program. Some of the difficulties relate to the changing technical personnel that are involved in a project: first exploration geologists; then mine planners; then operations persons, mine geologists, and environmental specialists; and finally reclamation specialists. It is relatively unusual for a mine to have a professionally trained geohydrologist on staff, and it is much more unusual for that role to be filled from an early period with any continuity. Correspondingly, it is very unusual for a mine to retain a mine geohydrology consultant early in the exploration part of a project and retain that person for the life of the mine. Occasionally a geologist is assigned the geohydrology role at a mine, but it is often regarded as a low echelon assignment, with little career prospect, so there is generally high turnover and little continuity. It is suggested that for many gold mines, groundwater issues are important enough for there to be a separate, continuous activity in this area, supported by management, and integrated with the geological and engineering activities of the mine.

Concept #10: Most Mining Geohydrology Investigation is Unnecessary

The author has been involved in over 100 mine geohydrology studies, as a reviewer or technical consultant. It is the author's observation that a considerable proportion of the work that has been done on these projects has turned out to be unnecessary, in the sense that the results turned out not to significantly assist managerial decisions or significantly improve performance of the project. Correspondingly, in many projects the key pieces of information were not collected in the investigation.

There appear to be three reasons why unnecessary work is performed, and necessary work is not performed. First, the information that is needed to make the required decision is not properly identified. Second, some investigations are performed even though enough information exists to allow the required decision to be made without additional data. Third, some investigations are performed even though the precision that is needed from the testing is not available.

To avoid unnecessary investigation, it is suggested that a pre-analysis be made of the question at issue, using the available information. To be able to complete the analysis, it is likely that some of the information that is needed will have to be estimated.

When this happens, it flags the possibility that there is a need for more information to be obtained. To test this possibility, it is normal to perform a sensitivity analysis to see how sensitive the decision is to the missing information. If the missing information could change the decision, then and only then should further investigation of this matter be considered. It should be performed only if the information can be obtained at the required precision.

To assist managers to avoid authorizing unnecessary work, it is suggested that three questions be asked of the geohydrologist prior to initiating work:

1. Do we need the information to make the decision?

2. Is the information presently unavailable at the precision required to make the decision?

3. Can the missing information be obtained at the precision required to make the decision?

If and only if the answer to all three of these questions is "yes" should an investigation be authorized. Note that a geohydrologist will need to perform a data needs assessment of the type described above to be able to answer these questions properly.

CONCLUSION

This paper has attempted to provide guidance to gold mine managers and hydrology professionals in the difficult area of decision-making in mine geohydrology. This decision-making is generally based on predictions about the future behavior of groundwater systems, either in respect of flow, head, or water quality issues. Such predictions are uncertain, difficult, expensive, and time consuming, and have in the past frequently proven to be incorrect. Ten concepts are advanced and discussed with the objective of assisting managers and geohydrologists to improve decision-making power in this technical area:

1. Groundwater hydrology is part of the hydrologic cycle.

2. The foundation of mine geohydrology is geology.

3. Flow of groundwater is subject to the laws of hydraulics.

4. Mining groundwater flows and heads are mainly steady state.

5. Mining solute transport conditions are usually transient.

6. Evaluations in mining geohydrology are uncertain.

7. Simple evaluations are more convincing than complex ones.

8. Prototypes are the best tests of geohydrology systems.

9. Mining geohydrology is a project-long activity.

10. Most mining geohydrology investigation is unnecessary.

In summary, it is considered that mining geohydrology can in general provide predictions of future groundwater conditions that are sufficiently accurate for the purposes of managerial decision-making.

REFERENCES

Freeze, R.A., and Cherry, J.A., 1979, <u>Groundwater</u>, Prentice Hall, Englewood Cliffs, New Jersey.

Chapter 17

RECENT ADVANCES IN EXPLOSIVES ENGINEERING

Lewis L. Oriard and Andrew F. McKown

President, Lewis L. Oriard, Inc.
Huntington Beach, California

Senior Specialist, Haley & Aldrich, Inc.
Cambridge, Massachusetts

ABSTRACT

Advances in explosives engineering are moving forward in several different aspects. These include improvements and new development in explosives products, the application of these products to blasting technologies, and the study of the peripheral effects of the use of explosives, such as ground vibrations and airblast. Topics of special interest in the last several years include new explosives and detonators, cap timing, explosive casting, high-speed photography, perimeter and slope control, air decking and mathematical modeling of blasting processes.

EXPLOSIVES PRODUCTS.

The greatest change in explosives products came about many years ago with the discovery that a mixture of low-grade ammonium nitrate and diesel oil produced an inexpensive, powerful blasting agent (ANFO). Over the years, formulations of ANFO were varied in order to develop some degree of water resistance. Thus evolved the first slurries and water gels. These products are still undergoing further refinements. They consist of various formulations of the oxidizer salts, such as ammonium nitrate, sodium nitrate or calcium nitrate, to which fuels and sensitizers are added in a liquid. Aluminum and oil are common fuels. The liquid is made more viscous and stabilized by the use of gellants and cross-linking agents. Chemical sensitizers are often added, and additional sensitizing has been accomplished by the entrapment of air bubbles. One of the most important chemical sensitizers is MMAN, monomethylamine nitrate. With such sensitizers, it is now possible to make water gels in small diameters.

Among the most recent stages in the evolution of these products has been the development of various water-in-oil emulsions. In emulsions, the oxidizer salts are suspended as droplets of microscopic size, surrounded by a continuous fuel phase, and stabilized against separation by emulsifying agents. As with the water gels, additional sensitizing can be provided by the entrapment of air bubbles, or the addition of glass spheres, a more recent development. Having the better distribution and contact between oxidizer particles and fuel provides more uniform and efficient detonation. In comparison to water gels, the emulsions appear to have better water resistance, greater stability in storage and less sensitivity to temperature changes. They have a higher velocity of detonation and higher detonation pressure than the water gels.

Continuing improvements in blasting products has led to more replacements for conventional dynamites, some using new formulations of conventional ingredients, some using new ingredients. Certain nitroglycerin-based semigelatins are being replaced with non-nitroglycerin high explosives in small diameters (1 to 3 inches). Water resistance is sometimes improved in packaging, for example with paraffin-coated rigid cartridges. Some of the products in this class are good for underground use, designated Fume Class I, generate minimal smoke and fumes, and do not cause headaches (severe headaches may be caused by nitroglycerin absorbed through the skin). Further information is available in blasting handbooks and technical bulletins from explosives manufacturers (1), (2), (3).

Just as there has been an on-going development and improvement of blasting agents and dynamites, the same is true of detonators and initiating devices. For many years now, users have had available to them a wide range of both electric and non-electric detonators (blasting caps). With the newest range of these products now available, it is becoming increasingly easier to develop more sophisticated blast designs to improve fragmentation and perimeter control, and to reduce vibration.

For users of electric detonators, there are continuing developments in sequential timers, electronically controlled blasting machines (4). The latest of these have programmable timing for each circuit, as well as the ability to use slave units with the primary timer. One can now design an electric blast with literally hundreds of delays. These developments have also been accompanied by improvements in the precision of electric delay cap timing. For the first time, U. S. cap manufacturers now have electric delay caps with timing precision comparable to those which have been available in Europe for many years. More comment will follow later about delay intervals and cap timing.

Down-hole non-electric delays have been available now for many years, but developments continue in a manner similar to those for electric blasting. The first non-electric caps did not contain delays in the cap, and some of the earlier delays were separate surface delays tied between instantaneous caps. Later, delays were built into the caps themselves, as was done with electric caps. Detonation was initiated by solid-core downlines of detonating cord of various sizes

(gradually becoming smaller in size). A significant change came with
the development of plastic tubing containing a thin coating of
sensitive explosive (PETN derivative) on the inner surface.   Upon
initiation, a low-order detonation at about 6000 ft./sec. propagates
through the tube to the cap (5).   By repeating a delay sequence of
such caps, it is possible to design certain types of blasting patterns
with virtually no limit to the number of holes which can be detonated
separately.   To carry this a step further, where even the charge in a
single hole may be too large for strict vibration limits or rock
excavation control, explosives manufacturers have recently developed
more convenient product lines of in-hole delays, such as the down-hole
"sliders" made by several manufacturers, for example (1), (2), (6).
With these, a single line of detonating cord can be used to initiate a
series of delays within a single hole between separated ("deck")
charges.

Another type of non-electric initiation system uses empty plastic
tubing connected to down-hole caps. When a blast has been drilled and
loaded, the plastic tubing is charged with a flammable gas from a
portable stored source immediately before firing the shot.   As with
the system described above, a deflagration propagates (at about 8,000
ft/sec) through the tube to detonate the caps (7). The system provides
the same advantages concerning delays as other non-electric systems,
and is basically an inert system at ground surface until charged just
before the blast.   There is a need for greater care in keeping all
lines and connections very clean to prevent any blockage of the
system, but it can be tested.

Research and development efforts also continue on more exotic
products.   Some of these can be seen to be technically feasible, but
still too expensive for everyday use on construction or mining
projects.   Among these are electric caps with exploding bridge wires,
- which explode with a surge of very-high-voltage current into the cap
(a few thousand volts), and caps which are detonated through radio
control.   Research also continues in the hope of developing low-cost
electronic caps for improved accuracy in detonation timing.

For more information on products, the reader is encouraged to refer
to the handbooks and technical bulletins of explosives manufacturers.
A list of names and addresses of major explosives manufacturers in the
U. S. is included following the list of references.

CAP TIMING.

During the mid-to-late 1940's, and in following years, the use of
millisecond delay timing developed rapidly as an accepted method for
reducing vibrations and improving rock fragmentation.   From that time
to the present, there have been on-going discussions of how and why
delays have been so effective in reducing vibration.   Some authors
have argued that there is just as much possibility of constructive
wave reinforcement as there is of the destruction of seismic energy
through wave interference.   If rock masses were homogeneous and

isotropic, and if a sequence of detonations generated pure sine waves and only one type of seismic wave rather than a family of waves, those arguments would carry more validity, and we would find millisecond delays less attractive. Fortunately, however, there are important theoretical and practical reasons why data continues to demonstrate the beneficial character of short-period delays as well as long-period delays, and that it is reasonable to predict ground vibrations based on the charge weight that detonates at any given instant of time. Although we should not completely ignore the subject, constructive wave reinforcement is not an over-riding consideration in everyday practice. Very rarely do vibrations actually approach harmful levels. An important corollary, however, is that it is often valuable to know how much explosive actually (not merely theoretically) detonates at any one instant of time.

Controlled tests by the Bureau of Mines in the early 1960's established that short-period time delays of 34, 17, and 9 milliseconds were effective in reducing vibrations (8), but 5 millisecond delays were not tested. It has become a fairly standard procedure to consider the number of pounds per delay to be that which (nominally) detonates within any given time interval of 8 ms, rather than considering any shorter intervals to act independently. In certain instances, there is a theoretical validity to this view, in that 8 ms might be less than the rise-time of predominant low-frequency seismic waves, that is less than 1/4 cycle, generated by large-scale blasting operations in a soft geological setting. Thus, incoming pure sine waves of the same period but merely out of phase would, theoretically, be additive (9). Fortunately, however, this normally turns out to be an item of academic interest, not usually an item which would determine the presence or absence of damage. Although some constructive wave reinforcement can occasionally be identified on seismograms, the maximum values are usually obscured in the normal variation that accompanies such data, and are comfortably compensated for within the large factors of safety typically used. Experience continues to reinforce long acceptance of the use of the charge weight that detonates at any given time. (Another popular misconception is that caps in a single delay interval detonate simultaneously). Although seismic waves are more likely to converge to a minor extent at more distant locations, the greater distances provide additional attenuation and correspondingly less concern.

If, for whatever reason, the reader is dealing with vibration intensities that are on the borderline of his limiting criteria, then the above items would achieve some significance and he might want to deal with them in a more precise fashion.

From the very earliest monitoring of delay blasting, in the 1940's, it became very clear that the actual delay times deviated from nominal delay times by approximately 10%, plus or minus. Explosives manufacturers often published nominal firing times as "approximate firing times" so as not to misrepresent this known scatter in the firing times of caps. For example, the 1953 issue of Explosives and

Blasting Supplies, Hercules Powder Co. (10) used this terminology.
Also, reproductions of seismograms clearly illustrated this phenomenon
(11).   In 1968, Langefors (12) published a comparative listing of
blasting caps from the U.S., Canada and Europe, showing typical
deviations from nominal firing times. A similar comparative listing
has been repeated recently by Winzer (13). Before the development of
sequential timers and various non-electric timing options, these
variations in cap timing were used intentionally to advantage for the
further reduction of vibrations, especially for long-period caps (14).
On some projects, it was found that up to 8 holes per delay could be
used in later delay numbers without increasing the vibration over that
from one hole per delay. In such instances, the vibration was related
more closely to charge per hole than to charge per delay. Of course,
the principal can still be put to use, although the present-day
explosives user has more options.

Although some European manufacturers have produced more accurate
caps for many years, the situation has remained fairly static in the
United States until very recently. Now, it seems that there is
sufficient demand from explosives users to justify the additional
manufacturing costs and controls to produce more accurate caps.   In
part, this has been prompted by the continuing growth of larger-scale
operations (particularly surface mines and quarries) in closer
proximity to populated areas, and the ever-increasing social pressures
to reduce the perception of blasting operations. In addition, more
accurate timing offers improved control of the fracturing and
displacement process.

As a consequence, most U.S. explosives manufacturers now produce a
new series of more accurately timed delay blasting caps, usually in
addition to the conventional, less accurate series. Previous series
were designed to lengthen the delay interval as one progressed through
the series, in order to prevent overlap of delays in the detonation
process. For the higher delay numbers for short-period delays, the
interval might reach 200 to 500 ms or more, depending on the
manufacturer. In the newer series, the interval remains constant
throughout the series, usually at about 25 milliseconds for the short-
period caps. The accuracy claimed by most manufacturers for the new
caps is in the range of about 1 to 2%, although no independent testing
has been reported to date of actual firing times.

Rather than rely on timing scatter to provide the equivalent of
additional delay periods, explosives users today have available many
more nominal intervals of choice, at greater precision than before.

MATHEMATICAL MODELING.

From time to time there have been efforts to develop mathematical
models of explosives physical effects. There was increased emphasis
on these efforts during World War I and World War II, for obvious
military reasons. During the last thirty years, these efforts have
been oriented more towards civil interests, such as casting and

improved fragmentation in mining/quarrying, and for the prediction and control of secondary effects (vibration and airblast overpressures). With recent advances in computing capabilities, there is a trend towards increasingly sophisticated models, although there are significant differences between the needs of most explosives users and those in certain associated industries, such as geophysical exploration, earthquake engineering, aerospace and military research industries.

Within the last several hundred years of the use of explosives, it has only been in the last 50-75 years that the more primitive drilling technologies have not been a major handicap in the development of blasting procedures. With slow, expensive drilling, it was natural that blasting was essentially looked on as a simple cratering process. Hence, the earliest theories about blasting were based on cratering. It became clear through trial and error that the volume of a crater produced by blasting could be related to the cube root of the explosive charge weight. Cratering relationships, and other scaling relationships associated with blasting phenomena, work best when the explosives charges can be considered approximately as point sources, and are less satisfactory as geometric relationships (and timing relationships) become more complex.

Once an investigator determines through field trials the maximum crater volume and optimum depth of burial for a given single charge at a given site, scaling theories permit an easy extrapolation to lines of charges and arrays of charges for blasts of any size. Proponents argue that the single-charge field trials automatically take geology (field conditions) into account and are therefore accurate representations. Those who do not favor this approach point out that the average production blast often deviates too strongly from the geometry modeled by the tests, and that the approach does not adequately allow for the frequently changing field conditions.

To the present time, it appears that the majority of blast designers continue to use some form of the "traditional" approach of custom-designing blasts through trial and error, while taking into account the site-specific rock properties that appear to influence the results. These often start with typical rules of thumb regarding spacing/burden relationships, hole diameter, stemming, subgrade drilling, and other geometrical factors. A well-known set of such rules of thumb were documented by Ash, after surveying a number of midwestern quarrying operations, and determining a consensus of practice. These have been summarized by Pugliese (15), and again by Dick (16). Such rules are then modified by site-specific features, such as jointing patterns, rock strength and density, variations in layering (or anisotropy in non-layered formations), and the like. Simply put, these designers consider the problem to be one in which engineering geology is at least as important as the physics involved. The explosive charges are often placed in the location and quantity determined by an examination of the varying geological conditions, then modified by trial and error, and adjusted for geology as the work progresses.

A variation of the above rules of thumb follows procedures recommended by authors such as Langefors and Kihlstrom (12), more popular in Europe and South America, whereby a separate judgment or calculation is made for the bottom charge and the column charge. There are many additional variations combining more than one of these concepts, for example considering the charge to be a combination of a bottom crater charge and a top crater charge.

The more recent efforts to model blasting phenomena make use of the capabilities of modern computers, although there are some serious theoretical and practical difficulties yet to overcome. With simple (point-source), large energy sources in a simple, homogeneous, massive material, one can expect computer codes to be relatively easier to develop and correspondingly more representative of a site, but the modeling becomes more speculative as the spatial and time distribution of the charges become more complex, and/or the rock mass becomes more anisotropic or complex.

It is a formidable problem to develop a detailed true mathematical model of even a fairly simple rock site, as opposed to the simplified or substitute models generally in use. This should come as no surprise to those who have attempted to model rock masses for static stresses, for example modeling slopes for static stability. The physics of the dynamic stresses in a blast are far more complex. Similarly, those who have observed a large number of large blasts such as coyote blasts will appreciate the difficulties of finding, identifying and correctly portraying the influence of either major or minor discontinuities, anisotropies or facies changes. Experienced observers still find it easier to predict the outcome based on experience than to determine the physical properties necessary to put into a true mathematical model. Similar to practice in slope stability, we see many slope failures that were not predicted even though the slope was carefully studied in advance, and we try to improve our knowledge of the subject by back-calculating to determine the most reasonable estimate of physical properties and geometry that would account for the result, even though they were not known in advance. The same philosophy might apply to blasting.

Many blasting operations are oriented towards the production of well fragmented material, for example, aggregate quarries. Of these, the one which probably relies more on uniform fragmentation than any other is the oil shale industry. Uniformity of particle size is crucial to the success of an underground retort. At some sites, it has been possible to core the material, determine various physical properties, then develop reasonably representative models based on observations of test blasts. That is, predictive models can be developed, given enough time to study a given material and observe its behavior under blast loading. This also permits gradual progress to be made towards generalized predictive models, but there does not yet seem to be a practical way to simplify (accurately and quickly) the process of developing a realistic model of each new site or site condition. It

still requires a time-consuming and expensive testing procedure to
determine the in-situ moduli and geometric relationships needed for an
accurate model. Further, it becomes increasingly more difficult to
develop a model to predict what happens after the initial fracturing
and displacement of the in-situ rock mass.

Despite these technical difficulties, there is an increasing
interest in mathematical modeling of the blasting process.
Fortunately for development in this area, a number of universities and
government agencies have been able to fund continuing studies, so that
progress continues. Some efforts are designed to model the total
fragmentation process with a single code, others have separated the
initial fractures generated by the seismic shock from the extending of
these fractures and rock movement generated by the slower-acting
expanding gases. Still others are trying to model the physical
processes that take place only after the initial fracturing and
separation of particles from the in-situ rock mass.

Because of the difficulties inherent in developing true mathematical
models of complex families of detonation and stress waves in hetero-
geneous materials, many efforts are under way to develop simplified,
substitute, empirical (pseudo) models which will provide realistic
guidance to explosives users. This compromise with technology is
comparable to that used in predicting ground vibrations. With a good
empirical data base, one can predict the average and upper bound of
typical blasting data, even though the attenuation formulae are not
actually seismic wave formulae. They are data processing formulae
(23).

Nearly every blast designer uses some type of blasting model.
However, the term "model" or "mathematical model" is more ambiguously
used in the blasting industry than in seismology. As used in relation
to blasting, the model may be anything from the application of the
designer's chosen rules of thumb to a 3-dimensional finite element or
finite difference code (17), (18). That is, on the one hand, it may
be some form of an "expert system" involving selected parameters,
selected judgments about these parameters, and simple calculations.
On the other hand, it may be a code which is intended to model the
entire physical process of blasting, and eliminate judgmental input.
The most complete and sophisticated models of the latter type are
somewhat beyond the capabilities of most personal computers and are
not being used at the present time in the blasting industry, but
research continues. At the same time, there is an on-going increase
in the memories and speed of personal computers. We can foresee the
time when personal computers can handle true models of the physical
process as well as simplified models and expert systems.

With the use of a well-defined set of analytical rules, whether in
the form of an expert system or in the form of a mathematical model of
the detonation and fracturing process, the trial-and-error process can
be shortened. In fact, the field-oriented blast designer who uses
neither calculator nor computer can shorten his period of trial and

error if he keeps detailed and accurate records of each of his trials. Similarly, many blasting models have been successful in improving results and/or in reducing the time involved in the trial-and-error process, whether or not they were purported to be true mathematical models of the blasting process.

As is true of any software for sale on any topic, the buyer or new user is at a disadvantage in evaluating the software until he has tried it himself, or at least seen it used in applications comparable to his own. It is wise to demand such a review, and to demand a description of the parameters used in the software and how they are applied to the blast design. For example, a model that works very well for a quarry in uniform, jointed limestone may not work well for an open pit mine in layered, highly variable sedimentary rock types. The model may have no way to deal with such variations, or to deal with them if, for example, the layers are tilted. Even if it does deal with them, does it do so realistically? Did the author of the software know how to handle that question? Can you make the decision more readily with the help of the model or without it? Does the software allow the user to input his own design judgments? In other words, the potential user would want to make the same type of critical judgments that he would for any other type of software or model.

To the present time, research efforts have not produced a generalized, true and complete mathematical model of the complete blasting process. However, there are now many simplified models which have been developed to assist in blast design, and these can be helpful in reducing the time spent in fine-tuning a blasting operation.   Most of these models are some form of modified expert system.

The authors do not wish to offer a detailed critique of any particular model, since modeling efforts are in a state of flux, and any model might be completely changed by the time these words are seen by the reader. The reader who is interested in this topic is encouraged to review, as a minimum, the references in the last 5 or 6 years of proceedings of the Society of Explosives Engineers, and to make further note of the bibliographies included in those references. Space limitations prevent us from providing such a bibliography here.

HIGH-SPEED PHOTOGRAPHY.

Many persons who have not been satisfied with mathematical modeling or standard rules of thumb, or as a supplement to other tools, have turned to high-speed photography as a useful tool for the study of blast performance. It is used extensively by those involved in large-scale blasting operations such as blast casting, for example casting the overburden in surface coal mines, and by those attempting to refine the fragmentation process or reduce costs on many mining projects.

Because of the extremely high velocities involved in transient blasting processes, the human observer has difficulty following the physical action. High-speed photography is a powerful tool, capable of recording the processes that cannot be followed by the human eye. The films can be played back in slow motion, at which time the investigator can study each step of the process, within the limits of slow-motion observation of the face and top surfaces of the blast area.

CASTING.

Casting overburden with explosives in surface coal mining operations has become a standard procedure. It is also finding application in certain metal mining operations, although it is typical in metal mining that there is more often a desire to fracture the material in place with little displacement, to prevent ore mixing.

Also, casting is finding applications in mine reclamation as a means of cutting down abandoned highwalls and reducing the cost of moving material.

There has been some interest in casting as part of the overall subject of explosive excavation for civil construction, such as rapid dam construction (19), but these techniques have not been utilized to any great extent outside of Russia. The technology was explored in the U. S. for the Plowshare Program (peaceful uses of nuclear energy), then extended to chemical explosives (17), but there was resistance to these concepts based partly on environmental concerns and partly on hesitancies to depart so strongly from established and proved procedures. They have not been used to any appreciable extent in civil construction in the U. S. and Canada.

PERIMETER AND SLOPE CONTROL.

Controlled perimeter blasting is not a new concept, even for open-pit mining (20). However, its use for increased slope control is expanding continually, due to such factors as the need for improved slope stability, decreased maintenance and the ever-increasing need to reduce overburden/waste costs, especially for marginal ore bodies.

Often one hears that "pre-splitting" is being done at a certain mining property. Sometimes there is an inadvertent use of the term "pre-splitting" as a synonym for any and all controlled perimeter blasting methods employed. This tendency is encouraged by common reference to small-diameter cartridged explosives as "pre-split" explosives, regardless of their actual application at the moment. Unfortunately, this ambiguous use of the term leads many persons to believe that true pre-splitting is being done and may encourage them to use pre-splitting techniques when it is not appropriate nor intended by the speaker. Actually, the term should only be used when it is meant that the perimeter holes are detonated in advance of the production holes. When that is done, there is no production achieved with the perimeter blasting. Its only function is to generate a fracture plane along the perimeter. Further, there must be very accurate control over the breakage from the last row of production holes. They must break as far as the perimeter to permit excavation, but if they break beyond the perimeter, they destroy the results achieved with the pre-splitting. True pre-splitting is often appropriate for structural excavations, but is not often the best approach to controlled excavation of large slopes such as those found in surface mining, especially when nearby production charges are very large.

In contrast, there are several advantages to using a form of cushion blasting, where the perimeter charges are detonated last in the sequence. Under these circumstances, the last row of production holes can be kept farther from the perimeter, with less risk of overbreaking beyond the perimeter, because the perimeter holes will produce muck, not merely a fracture. Often, the charges can be lighter than pre-split charges because of the nearby free face. Thus, the results of well-designed cushion blasting for an open-pit mine may be less expensive, more effective and less likely to leave a damaged final wall (20).

Various efforts have taken place to explore effective but inexpensive ways of charging perimeter holes. These efforts have included various approaches to decoupling charges. One of these approaches has been that of using a small-diameter charge in a larger diameter hole, such as might be accomplished by using cartridged explosives, or placing bulk explosives in a smaller diameter container in the field (for example, light-weight plastic pipe). Another approach has been that of using very-low-density, sensitized explosives to reduce the total quantity of explosives in holes that are sufficiently close together to permit neat fracturing along the line of perimeter holes. Loading densities have been lowered in various ways, including the addition of plastic or glass spheres to the explosives mixture.

One important factor that has not gained full recognition is that a significant portion of perimeter blasting has been done with more explosives than really needed. This is more easily demonstrated in small-scale situations, such as structural excavations. In more and more instances, the conventional factory-packaged perimeter blasting products (typically weighing from 0.25 to 0.34 lbs/ft) are being replaced with detonating cord in the range of 150 to 400 grains per foot. Detonating cord is relatively expensive, but there are no comparable small-diameter explosives products currently on the market in the United States. When we scale up to large-scale surface mining, there has been a comparable use of explosives in amounts that have been more than the minimum needed for perimeter control.

Recent developments suggest an approach called air decking that has many of the more desirable features of several previous practices combined. Experience to date is promising for large surface mining operations. More field experience in a wider range of rock types is needed before good "rules of thumb" can be provided reliably. This method provides a technique for reducing the quantity of explosives in perimeter holes, while being fast to use and relatively less costly than the traditional methods described above. More comment follows.

For certain special applications, a method called fracture control blasting is gaining attention. The technique was extensively studied and advanced at the University of Maryland (21). It involves the scribing or grooving of drill holes to provide an easy start and directional control of fractures. Field tests have shown promising results for certain applications (22).

AIR DECKING.

Several independent but synergistic concepts have combined to lead to the use of air decking, a design concept where only the lower portion of a blast hole is loaded, and only the collar zone contains a plug of conventional stemming. An open, air-filled zone remains between the explosives charge and the stemming. Hence the term "air stemming" or "air decking".

It has long been recognized that one can decouple explosives charges from the surrounding rock to reduce the shattering effect of the initial shock wave, and still permit the longer-lasting effects of the expanding gases and remaining pressure pulse to continue doing work on the rock mass. That is, it is possible to do work on a portion of a rock mass which is not in direct contact with the explosives.

Independently, one might observe what appears to be contradictory conclusions about the benefits of stemming. On the one hand, conventional practice does not usually call for the use of stemming in the horizontal holes used for tunneling (exceptions include those cases where it is necessary to reduce airblast overpressures). Extrapolating this tunneling experience, some have eliminated stemming from large-hole blasting operations in surface mines (in areas where the overpressures can be tolerated). In a manner similar to that of tunneling, the work can be accomplished despite the loss of energy through the top of the open hole. Under these circumstances, there is an increase in the useful work when there is good coupling between the explosive and borehole, as occurs with the use of bulk blasting agents. However, it is a simple matter to illustrate through controlled experiments that the presence of stemming increases the useful work, and the use of good-quality stemming increases it significantly. Without stemming, energy is being wasted, and loss of energy must be weighed against the increased cost of stemming.

In surface mines where an upper cap rock overlies weak overburden materials, it has been common practice to load a bottom charge, place stemming through the central part of the hole, and load a top "deck" or "decked" charge in the cap rock. Further, it has been observed that such cap rock can often be broken effectively even when the intermediate stemming and deck charge are left out, but a good stemming plug is placed immediately above the location where the deck charge ordinarily would be placed. Thus has evolved the use of air stemming or air decking.

The physical principle which permits this design to be effective is the propagation of a tube wave upward from the explosive charge, which is then reflected at the bottom of the stemming plug. This newly reflected energy is comparable to that developed by a modest deck charge.

This practice has been extended to perimeter control where it has been shown to be similarly useful. The main difference is simply the quantity of explosive used in a given hole. It has an advantage of being very fast to implement, and provides an unlimited choice of explosives quantities for the perimeter holes.

GROUND VIBRATION AND AIRBLAST.

It could reasonably be expected that a non-specialist reading today's literature could become quite confused regarding peripheral effects of blasting, such as ground vibration and airblast overpressures. On the one hand, there is a strong trend towards ever more restrictive limitations on blast effects. To many readers, this would imply that it is necessary to keep below highly restrictive limits in order to avoid physical damage of various kinds, and/or that both the intensities and the physical effects are unpredictable. Such conclusions are erroneous and tend to foster needless fear, or result in needless excess costs or prohibitions of work which is quite safe and simple. Of course, it is well recognized by everyone that explosives must be handled and used with proper knowledge and care. Carelessness and inadequate training can be causes of accidents, just as is true with vehicles or mechanical equipment.

On the other hand, there has been an ever-growing base of data and experience from the earliest use of explosives to demonstrate that vibration intensities and effects are easily predictable to the degree of accuracy needed for daily use, and that ground vibrations and airblast overpressures very rarely cause real damage to nearby structures, even when conventional vibration limits are exceeded.

The main problem has always been, and will continue to be, that people can easily feel very small vibrations and easily hear small secondary sounds associated with blasting, are very inaccurate in their observations, and that most are poorly informed about the silent environmental forces that are the causes of so much damage to structures and other facilities. For example, a change in outside temperature of 27 degrees Fahrenheit can generate strains in a residence which are comparable to those generated by ground vibrations with particle velocities in the general range of about 8 ips (23).

For most mining operations, the primary concerns about vibration (or airblast) relate to the structures and facilities on the mining properties themselves, but more and more mines find themselves within perceptible range of the public. Even for those far removed from the public, it is common to fail to recognize the effects of environmental forces and equally common to attribute many items of environmental damage to vibrations.

VIBRATION MONITORING.

There has been a significant change in vibration monitoring equipment in the last few years. The majority of portable seismographs used for the monitoring of blasting operations are now microprocessor controlled, digital seismographs. A typical instrument of this type has 4 channels (3 for ground motion, one for airblast overpressures), is seismically triggered at a selectable threshold, and has a digital memory capable of storing about 7 seconds of data which is sampled at 1024 bps. Unfortunately, the 7-second memories are not capable of storing all the data from some blasts using long-period delays. Under those circumstances, it is more common to use oscillographic analog recorders or digital tape recorders. However, it is expected that with the rapid advances being made in digital memory devices, the memories of the digital seismographs will be expanded in the near future.

As is true with other types of electronic monitoring devices, it is also becoming less expensive and easier to combine remotely located sensors with central recording and processing facilities (telemetering). Several instrument manufacturers now offer such capabilities in their product lines.

REFERENCES.

1. Atlas Powder Company, "Explosives and Rock Blasting", subsidiary of Tyler Corporation, Dallas, Texas, 1987.

2. E. I. duPont de Nemours & Co. (now ETI), "Blasters Handbook", 16th edition, Wilmington, Delaware, 1978.

3. Ireco, Incorporated, Technical Information Data Sheet D-02A-01-88-2K, "Iredyne 365 Non-Nitroglycerin High Explosive (formerly Hercodyne 365), Salt Lake City, Utah.

4. Research Energy of Ohio, "REO Technical Data Bulletins", regarding various models of sequential blasting machines, Cadiz, Ohio 43907.

5. Ensign-Bickford Company, "NONEL Primadet Technical Bulletins", Blasting Products division, Simsbury, CT.

6. E. I. duPont de Nemours & Co.(now ETI), Technical Information Data Sheet E-78079,12/85, "Detaslide", Wilmington, Delaware.

7. Ireco Incorporated, "Hercudet Nonelectric Delay Detonator System", Technical Information Data Sheet A-10-05-87-2K, Salt Lake City, Utah.

8.   Duvall, W.I., Johnson, C.F., Meyer, V.C. and Devine, J.F., "Vibrations From Instantaneous and Millisecond-Delayed Quarry Blasts", U.S. BuMin RI 6151, 1963.

9.   Oriard, L. L. and Emmert, M., "Short-Delay Blasting at Anaconda's Berkeley Open-Pit Mine, Montana", AIME Preprint No. 80-60, February, 1980.

10.   Hercules Powder Co.,"Explosive and Blasting Supplies", Wilmington, Delaware, 1953, p. 33 (see, also, Leet, p. 15).

11.   Leet, L. Don, "Vibrations From Blasting Rock", Harvard Univ. Press, Cambridge, Massachusetts, 1960, p. 120.

12.   Langefors, U. and Kihlstrom, B., "The Modern Technique of Rock Blasting", John Wiley & Sons, 1968, p. 143.

13. Winzer, S. R., "The Firing Times of MS Delay Blasting Caps and Their Effect on Blasting performance", NSF, APR 77-05171, Martin Marietta Labs, Baltimore, MD, 1978.

14. Oriard, L. L., Ewoldsen, H. M. and Perez, J. Y., "Blasting Techniques and Safeguards Used in Enlarging the Underground Powerhouse at Salto de Villarino, Spain", Rapid Exc. and Tunneling Conf., ASCE/AIME, 1974.

15. Pugliese, J. M., "Designing Blast Patterns Using Empirical Formulas", BuMines IC 8550, 1972.

16.   Dick, R.A., Fletcher, L.R. and D'Andrea, D.V., "Explosives and Blasting Procedures Manual", BuMin IC 8925.

17.   U. S. Army Engineer Waterways Experiment Station, Explosive Excavation Research Laboratory, "Annotated Bibliography of Explosive Excavation Related Research", Technical Report E-72-32, Livermore, California.

18.   Heusinkveld, M., Bryan, J., Burton, D., and Snell, C., "Controlled Blasting Calculations With the TENSOR74 Code", Lawrence Livermore Laboratory, University of California, Livermore, California, 1975.

19. Oriard, L. L., Ewoldsen, H. M. and Mahmood, A., "Rapid Dam Construction Using the Directed Blasting Method", 2nd Iranian Congress of Civ. Engrg., Pahlavi University, 1976.

20.   Oriard, L. L., "Blasting Effects and Their Control in Open-Pit Mining", Proc. 2nd Intl. Conf. on Stability in Open Pit Mining, Vancouver, 1971.

21. Fourney, W.L., and Dally, J. W., "Grooved Boreholes for Fracture Plane Control in Blasting", Mechanical Engineering Dept., Univ. of

Maryland, 1977.

22. Oriard, L. L., "Field Tests With Fracture Control Blasting Techniques", SME of AIME, Lib. Congr. Catalog Card No. 81-65517, 1981.

23. Oriard, L. L., "Close-In Blasting Effects on Structures and Materials", ASCE National Meeting, Boston, October, 1986.

LISTING OF MAJOR U.S. EXPLOSIVES MANUFACTURERS

Atlas Powder Company, Subsidiary of the Tyler Corporation, 15301 Dallas Parkway, Suite 1200, Dallas, Texas 75248.

Austin Powder Company, 3690 Orange Place, Cleveland, Ohio 44122.

Ensign-Bickford Co., Blasting Products Division, Simsbury, Connecticut 06070

Explosives Technology International (ETI), formerly duPont USA and duPont Canada, Wilmington, Delaware 19898.

Hercules Powder Co. (now Ireco).

Ireco, Inc., Crossroads Tower, Salt Lake City, Utah 84144.

(Lack of space and time prevents us from listing major manufacturers in other countries besides the U.S. Please accept our apologies.)

Chapter 18

VENTILATION CONSIDERATIONS IN UNDERGROUND GOLD MINES

A. E. HALL P.ENG

DEPARTMENT OF MINING & MINERAL PROCESS ENGINEERING
UNIVERSITY OF BRITISH COLUMBIA
VANCOUVER, B.C., CANADA

INTRODUCTION

Underground gold mining operations are found in South Africa, U.S.S.R., Canada, the United States, Zimbabwe, Ghana , Brazil, Australia, China and many other countries. The methods used for extraction differ from highly mechanized trackless mining in the developed countries to the labour intensive systems of the third world nations. The climatic conditions in which mines are situated vary from equatorial to arctic. The rock temperatures and the depth of workings also vary significantly from shallow to extreme depth. The Western Deep Levels mine in South Africa is the deepest and exceeds 4,000 metres in depth with strata temperatures above 55°C.

These major differences between mining fields result in significantly different ventilation problems in underground gold mines. The ore types also vary and some deposits have high silica contents while other mines extract sulphide ores. Methane may also be encountered in underground gold mines and can have a large influence on the ventilation practice.

Several topics are currently of concern to operating mines and are reviewed here:-
      a) Air circulation and fans
      b) Mine cooling
      c) Permafrost operations
      d) Mine heating
      e) Control of diesel exhaust
      f) Gases, fires and explosions

AIR CIRCULATION AND FANS

Small mines use simple ventilation systems, usually circulating 20 to 50 cubic metres per second (cu m/s). Such mines use either a single forcing or exhaust fan to move the air through the mine. Forcing fans are common in shallow mines with intake air heaters, especially where a ramp or decline haulage system to surface is used. The ramp or decline can then be ventilated by exhaust air from the workings which removes the diesel exhaust emissions directly and leaves the ramp free from ventilation doors or fans. This layout is good for transportation as vehicles are not delayed by doors or airlocks. Damage to doors by vehicles is prevented with the consequent ventilation leakage and short circuits. The disadvantages of the system are that the air can cool significantly in the ramp and produce fog if the mine has high humidity (Golder Assoc., 1983), and if the forcing fan fails the air pressure in the mine decreases which can cause more methane or radon to be emitted into a gassy mine.

Large mines use several separate circuits with multiple main fans on surface at the different shafts and ventilation raises, plus underground booster fans in the individual circuits. The circulated air quantities can be large:- Western Deep Levels circulates 1,000 cu m/s and the Homestake Mine in South Dakota circulates 450 cu m/s. Most of the South African gold mines use systems of this type with the main fans sited on the exhaust shafts, which leaves the intake shafts clear of obstructions and airlocks. Hoisting of men and materials is then done through the intake shafts which can create localized dust problems, but these are easier to deal with than the excessive corrosion which would be experienced in the exhaust shafts.

Many of these mines use refrigeration machines and this results in the exhaust air being saturated with moisture and its temperature often exceeding 30 °C. The reduction in atmospheric pressure in the exhaust shafts causes water condensation as the air becomes supersaturated and droplet clouds form in the shaft. Unless the air velocity is kept below 7 metres per second (m/s) or above 12 m/s, these clouds remain in suspension in the shaft and significantly increase the resistance to flow of the circuit (Lambrechts, 1956). The No. 3 shaft at Anglo American's Vaal Reefs Mine was extremely wet and the water collected in the fan drift required two pumps, each connected to a 50 mm diameter pipe running full, to remove it. When a pump failed, the water would spill back into the shaft and cause the whole ventilation system to stop. The only way to restore the ventilation was to stop one of the twin surface fans, which caused the water cloud to drop down the shaft and enabled air to flow through the system. the second fan could only be restarted when both pumps were running again. This shaft has subsequently been sealed to reduce its resistance (van Wyngard, 1987).

The pressure loss through a mine circuit can be as high as 15 kiloPascals (kPa) and this requires high pressure main fans plus the use of boosters in the circuit. Homestake is a North American mine

which is similar in layout to South African practice. This mine uses
air cooling and has been described in detail (Struble, Marks and
Brown, 1988). The mine produces 1.7 Mt/a of ore and uses three
independent ventilation circuits to draw 450 cu m/s through the mine
at pressures up to 4 kPa. The mine has three surface centrifugal fans
of 225, 520 and 560 kW power respectively, plus a two stage 932 kW
axial fan and a 560 kW back-up fan. There are three large underground
booster fans of 300, 300 and 220 kW power and secondary ventilation
is handled by seven 30 kW booster fans. Auxiliary ventilation
requires 307 smaller fans. Total refrigeration output is 3.7
MegaWatts of cooling.

Homestake reviewed their ventilation system and decided to install
a new centrifugal surface fan to ventilate the mine below the 1800 m
level. The fan is designed to handle 226 cu m/s at a pressure of 6.5
kPa. The cost was just over $1 million, with the following
breakdown:-

Fan, motor and associated controls	$450,000
Fan installation, motor, electrics and ducts	$461,000
Underground excavations	$149,000
Total cost	$1,060,000

This represents a capital cost of approximately $4,690 per cubic
metre per second circulated.

AIR COOLING

The deep South African mines have major problems in keeping their
ambient underground air temperatures below 30 °C to prevent heat
stroke and heat exhaustion in their labour intensive workings. The
Vaal Reefs operation of Anglo American Corporation employs
approximately 36,000 underground workers and they must be provided
with a suitable environment to work in. The increase in air
temperature at depth because of autocompression requires the majority
of these mines to refrigerate their ventilation air. The amount of
air which can be circulated through the mine is restricted and the
use of underground cooling plants or refrigeration machines resulted
in many problems and restrictions. The air temperature leaving
underground heat rejection cooling towers can exceed 55 °C and this
restricted access to return airways for inspection and repairs. In
several mines, underground booster fans were installed to overcome
the resistance caused by rock falls which partially blocked and
restricted airways, in preference to shutting down refrigeration
machines to access the airways with consequent stope production
losses.

During the 1970's the mines examined the cooling problem and a
major change in refrigeration technique was introduced. Large surface
refrigeration machines were installed to cool water on surface and
circulate it through the mine workings. Figure 1 shows a typical
installation of this type. The circulation of water in deep mines in

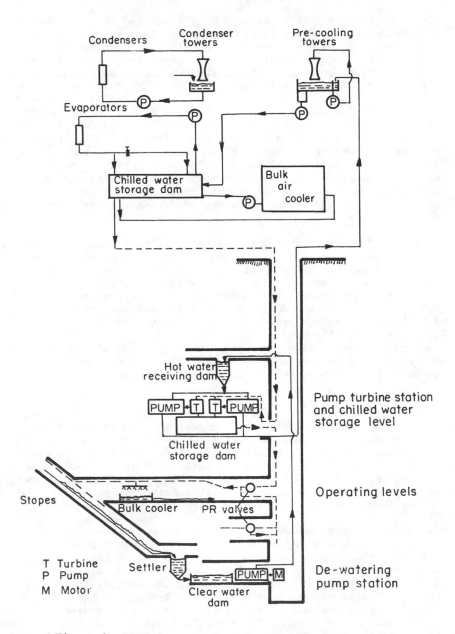

Figure 1. Surface Refrigeration Reticulation System

open circuits resulted in high pumping costs and energy recovery
turbines were introduced to recover energy from the incoming water to
power pumps and reduce costs. One installation of this type at
Buffelsfontein mine circulates 300 l/s of water to produce 3,830
kiloWatts of power from a 1,500 metre head at an efficiency of 86%.
A system of this type is in use at the Lucky Friday mine in Idaho.

The installed cooling capacity in South Africa exceeds 1,000 MW of
refrigeration and has grown from 250 MW in 1975 and 600 MW in 1980.
The continued circulation of water for refrigeration is therefore
extremely expensive and alternative methods of using the water are
being investigated. One alternative is to use the hydropower to
operate rockdrills and other machinery. Several other experiments
have been conducted with ice circulation. This is attractive because
only about 25% as much ice as water requires to be circulated for the
same cooling rate.

Figure 2 shows a pilot ice plant used at East Rand Proprietary
Mines Ltd to study aspects of conveying ice underground. Ice is
injected into the conveying pipeline by three specially designed
pneumatic power vessels which are arranged in parallel and located
adjacent to the ice storage bunker. After injection into the pipe,
the ice travels at speeds of 25 m/s as a succession of plugs
separated by the conveying air. Ice has been conveyed underground at
rates of up to 15 tonnes per hour using this system. A large deep
mine would require 800 tonnes per hour to provide 100 MW of cooling.
The problem is to develop machines capable of producing the large
quantities of ice required by these systems (Sheer, Burton and Bluhm,
1984)

A recent development is the long-cycle hydro-lift system which
uses the chilled water being sent underground to force hot water back
to surface, thus greatly reducing the power requirements for pumping.
All the chilled water requirements for an entire shift are sent
underground to a sealed excavation before the shift begins. The
chilled water forces hot water out of the excavation and returns it
to surface over a number of hours. This method is now being studied
by the Chamber of Mines of South Africa (Chamber of Mines,1987)

The only North American gold mine using refrigeration is
Homestake, with a capacity of 3.7 MegaWatts, which is planned to be
extended to 11.8 MW by 1993. Other mines using refrigeration are the
Lucky Friday mine of Hecla Mining in the Couer d' Alene district of
Idaho and Teck-Cominco's Polaris mine in the Arctic. The Polaris
installation is unusual because the mine is in permafrost and the
system is provided to cool intake air in summer to prevent melting of
the permafrost. Melting would result in ground instability and this
justifies the cost and power consumption of the refrigeration units.
This is despite the limited power generation capacity at the mine and
the high cost of diesel fuel for power generation because of
transportation and storage costs.

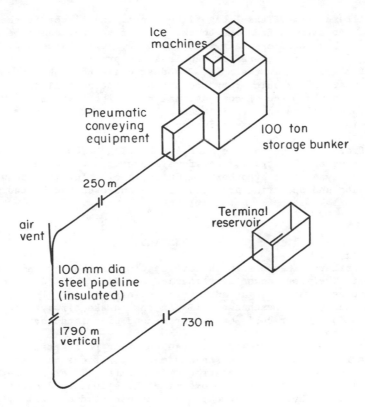

Figure 2. Schematic of Ice Reticulation System

PERMAFROST OPERATIONS

Echo Bay's Lupin Mine also operates in permafrost which extends to a depth of 540 m below surface. All mining is carried out by trackless equipment and it is impossible to use normal water flushed in-the-hole drills. All production operations are carried out using hydraulic drills which are flushed with brine to prevent freezing and the drills are fitted with external flushing to prevent corrosion.

Dust control and allaying is extremely difficult in these conditions and if operations such as raise boring have to be performed without the use of water, then personnel may have to be removed from the downstream airways. This is necessary to prevent excessive dust exposure. Stoppage of other operations can be very expensive and may not be acceptable to management. One alternative is to limit the dust creating operation to non-production shifts but this reduces the utilization of expensive machinery. A possible solution is to reduce the air passing the borer to an amount which can be filtered by a flannel bag filter unit. The filter is then positioned downstream from the borer. Size of the units may be a problem as they require an air velocity of 0.1 m/s at the bag. The filtering of 5 cu m/s therefore requires the provision of 50 square metres of filter bag in the unit. The dust collected can be removed from a hopper in the unit or by the use of disposable filter cartridges in the place of conventional bags. The flannel bag filters usually work well in cold environments because there is minimal moisture in the air to cause blinding of the bags.

Permafrost mines do not use mine air heating because this would cause melting of the strata, therefore measures have to be taken to protect employees from hypothermia. Suitable clothing must be worn and vehicle cabs and operating areas should be enclosed and heated where possible.

MINE HEATING

Most northern mines use intake air heating if they are not in permafrost. The most common systems used are direct fired burners using propane or natural gas and steam boilers heating glycol, which is circulated to heat exchange coils in the intake airways. The costs can exceed $3,000 per cubic metre per second circulated over a 6 month heating season.

Increases in propane costs have caused mines to re-examine their heating strategy to prevent waste and many operations cut back their heating at weekends and on non-production shifts. In areas where cheap electrical power is available, some mines have installed electric heaters. Ontario Hydro have encouraged the use of dual fuel systems where the electric heaters are used at times of low power demand. This technique is known as valley-filling and the mine receives cheap power while the utility company stabilize their 24 hour load. A computer program analysis is available from Ontario Hydro for potential customers who wish to investigate the potential for a dual fuel system. The key to success is the valley-filling as this keeps the electric cost down. The drawback of the method is that if the heaters increase the mine maximum demand then the operator will pay in excess of 7 cents per kiloWatt-hour.

A lot of attention has been paid to potential heat recovery systems to decrease heating costs. Unfortunately most mine waste heat is at relatively low temperature and is therefore classed as low grade. The installation of heat recovery coils in the return airways has been tried by several mines and investigated by many more. The problem is that the return air is usually at a temperature below 10 °C and is very humid. Cooling the air below 0 °C causes ice to form on the heat exchanger surface and this prevents further effective heat exchange. In mines where the outside air is at -30 °C and the intake air is heated to about 0 °C, this means that only the heat picked up in the mine itself is recovered. This low recovery is a major disadvantage of the systems and they are not likely to be used unless the air is dried before it reaches the exchanger.

The ice-stope method of mine heating has received attention from Inco and has possible application for gold mines with large excavations available for ice storage. In this system water is sprayed into cold intake air and changes into ice. The formation of ice causes the latent heat of fusion to be transferred to the air and heats it. The ice is then stored in the mine until the spring/summer period when it melts due to the warmer air and rock temperatures. The resulting water is pumped out of the mine and the excavations empty ready for the next winter heating season.

A possible method for recovering waste heat is to monitor the quality of the exhaust air from the mine. Most mines use a large amount of diesel equipment and the ventilation quantity is set by the installed underground diesel power. The operation of this diesel equipment is not continuous and it has been found that the exhaust air quality varies significantly over the working day. During the periods when the return air quality is good, a significant percentage can be recirculated to the intake. This reduces the intake air quantity and therefore the heating costs without reducing the airflow rates in the mine workings. Trials of this method are being conducted by U.B.C. and several papers are available (Hall, 1985 and Hall, Mchaina and Hardcastle, 1988). The method is not suitable for mines subject to the presence of radon gas because it increases the air residence time in the mine. Problems with the method have been the lack of a reliable on-line dust detection instrument to provide effective monitoring of dust levels and the problem of increasing fine dust load in the recirculated air. Experiments in South Africa have shown that the coarse dust rapidly settles out of the air leaving the minus 5 micron fraction. Recent experiments have proved the effectiveness of water spray chambers in removing the fine dust and this may assist in developing a working system in Canada. There is potential for applying the system in the North West Territories under the Mineral Development Agreement (MDA), if a mining company is prepared to take a lead role. The Manitoba and Saskatchewan mining legislations have also been very supportive of the tests conducted to date at Ruttan and in the potash industry.

A novel system of heat recovery is being used by the Macassa mine in Ontario. The system is shown in Figure 3. The warm water pumped from underground is used to cool the mine air compressors which increases the water temperature to 34 °C. This hot water is then used to heat the intake air to the mine. The system is suited to mines with a significant amount of water being pumped from underground, which is of adequate quality to be used for compressor cooling.

If the use of water spray chambers is effective in cleaning exhaust air so that it can be recirculated then the use of water pumped from the mine is feasible. This would warm the air further and add to the amount of heat recovered by a recirculation system.

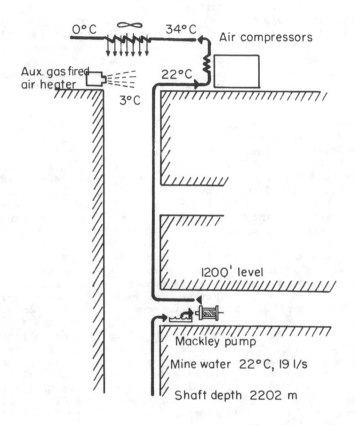

Figure 3. Macassa Mine Air Heating System

DIESEL EXHAUST

Historically the mining legislations concentrated on the carbon monoxide and dioxide levels in the air. The recent French report (French & Associates, 1984) has changed the emphasis and it is possible that the air quality index (AQI) may be introduced as a legislative standard. The index is most sensitive to the levels of NO_2 and SO_2 in the air and the presence of respirable combustible dust (RCD). The introduction of legislation to control RCD levels would force many mines to install exhaust filters to remove solid emissions from the air.

Small mines with restricted ventilation may find it beneficial to fit diesel equipment with fume diluters to prevent smoke clouds from being generated in the vicinity of large machines. The speed of diesel trucks and scooptrams should always be controlled relative to the air speed. It is bad practice to operate at the same speed because the diesel emission cloud will travel with the machine and generate excessive pollutant concentrations. The use of augmented ventilation when diesels operate in dead end drifts can be useful. An auxiliary fan is started by the driver when he enters the drift and stopped when the vehicle exits. This ensures adequate ventilation without wasting air when operation ceases. Spreading out the diesels around the mine helps control the pollutant levels and diesels should not work in the same ventilation split unless this is unavoidable. In this case particular attention should be paid to the ventilation quantity to ensure that is adequate for the two machines. Electric traction can be considered as an alternative to diesels where air quantity is limited, particularly in small mines using rail transport.

GASES, FIRES AND EXPLOSIONS

Most gold mines do not have gas problems with two notable exceptions. Some Ontario mines are subject to the presence of radon gas and several South African mines experience large quantities of methane.

Radon creates problems if it decays to produce the radon daughter products. The risk is minimized by reducing the time that the gas is resident in the mine. Ventilation circuits are arranged to reduce the air residence time as much as possible and to prevent uncontrolled leakage and recirculation from the return to the intake.

Several South African mines experience methane, especially in the Evander and Orange Free State goldfields. There have been several explosions caused through this, the most recent being at St. Helena gold mine in Welkom in late 1987. In order to prevent such occurrences the Mine Managers' Association produced a guide for dealing with methane(Chamber of Mines of S. Africa, 1973). This publication is currently being revised by the Association and should be available in the near future. Precautions that can be adopted range from bans on naked lights and the implementation of regular methane testing to full fiery mine regulations requiring explosion proof electrical equipment and the use of flameproof diesel engines.

Sulphide ores are also subject to ignition depending on their mineralogy and sulphur content. There have been several ignitions in Canada but none so far in gold mines. It is advisable to test sulphide ores in advance of mining to ensure that they are not liable to ignition. Low sulphur can prevent ignition but Falconbridge experienced an ignition with 16% sulphur ore at Sudbury. Sulphide ignitions usually occur after blasting. The first hole raises a cloud of sulphide dust which is ignited by a following hole. The explosion is usually of low power but can stall fans and blast out ventilation seals or destroy brattice and tubing. The SO2 liberated from the ignition can then be drawn into intake airways because of the damage to the ventilation system. There have been fatal exposures of workers including one at Geco in 1985. Most of the mines experiencing such ignitions withdraw all labour from underground during blasting, but this is expensive and delays production. There is a current research effort to develop methods of preventing these ignitions and the current state of the art was presented in a special session of eight papers at the CIM annual general meeting in Edmonton in May, 1988. Normal precautions include the use of water or limestone stemming and washing down the face area thoroughly before blasting.

Fires in any mine can be extremely serious and gold mine fires are no exception. Serious loss of life occurred at both Bracken and Vaal Reefs in South Africa from burning explosive accidents. The Kinross accident caused 177 deaths and resulted from a cutting accident with oxygen and acetylene bottles igniting an insulation foam applied to the walls and roof of the airway. The use of flammable materials in mines should be controlled to prevent the formation of toxic gasses. Explosive transport should take place between shifts or when there are fewest people in the mine. Cutting and welding operations should be tightly controlled and should stop at least one hour before the end of the shift with a subsequent inspection of the cutting area at the end of the shift to check for smoldering waste.

Diesel fuel is flammable and the storage areas and refuelling points should be classified as high risk areas subject to regular inspection. Fire fighting equipment should be provided and inspected to ensure that it works and regular clean-up of spilt fuel should be done. Electric equipment is also a fire risk and all cables should be carefully protected when there is blasting in their vicinity. Electrical power supplies into blasting areas should be cut off if they are not required. Most South African mines now isolate their stope power supplies for lighting and slusher winches during blasting and have found that this reduces stope fires.

All mines should draw up an action plan for fires. Planning should account for all possible eventualities and the procedure to be adopted in each case should be determined. These plans should be updated regularly particularly when changes are made to the ventilation layout. In a fire situation the initial response is critical and all employees should be trained in the actions required in the event of a fire. Refuge bays and stench alarms are vital parts of fire protection systems. The survival of the coal miners in the recent fire at Borken, West Germany stresses the importance of adequate refuge facilities.

Other measures which assist in fire prevention are good housekeeping, the removal of flammable waste material from the mine and control of smoking materials. The mine water reticulation system should be designed to supply large quantities of water to all parts of the mine in emergency conditions. The provision of fire doors at strategic points in the mine can reduce the ventilation flow to the area of the mine experiencing a fire, thus giving more time to bring it under control before it gets out of hand. A system of on-line gas or combustion product monitors can provide early warning of a fire in the mine and should be considered for installation.

CONCLUSIONS

Gold mines are too varied in size and layout to generalize their ventilation requirements. A brief overview of the important considerations in determining the requirements for individual mines has been given. The actual system selected will depend on the geographic location of the mine and the type of mining method used.

Gold mines are subject to problems of heating and cooling and to dilution of diesel exhaust products. Some mines are subject to gas emissions, particularly radon and methane. The risk of fire and explosions in gold mines can be high and mines should establish good fire prevention procedures and provide adequate facilities to deal with such accidents. Planning for emergency conditions and training of employees in fire fighting can assist in reducing the severity of these occurrences.

REFERENCES

Chamber of Mines of S. Africa, 1973, "A Guide to Managers to Assist in Combating Methane in Metal Mines", Johannesburg

Chamber of Mines of S. Africa, December 1987, "Hydro-lift System-Saving Pumping Costs", R&D News.

French & Associates, April 1984, "Health Implications of Exposure Of Underground Mine Workers to Diesel Exhaust Emissions", Ministry of Supply & Services, Canada Report No: M 38-17/1-1985E.

Golder & Associates, June 1983, "Study to Investigate the Problem of Atmospheric Fogging in Underground Mines", Energy, Mines and Resources Canada Contract Report: OSQ82-00091.

Hall, A.E., 1985, "The Use of Controlled Recirculation Ventilation to Conserve Energy", 2nd. U.S. Mine Ventilation Symposium, pp.207-215, Balkema, Boston (Ed., P. Mousset-Jones)

Hall, A.E., Mchaina, D.M. & Hardcastle, S.G., July ,1988, " The Use of Controlled Recirculation to Reduce Heating Costs in Canada', 4th International Mine Ventilation Congress, Brisbane.

Lambrechts, J. deV., 1956, "The Value of Water Drainage In Upcast Mine Shafts and Fan Drifts', J. Chem. Metall. Min. Soc. S.Afr.,Vol 56.

Sheer, T.J., Burton, R.C. & Bluhm, S.J., 1984, "Recent Developments in the Cooling of Deep Mines", S. Afr. Mech. Eng., Vol. 34.

Struble, G.R., Marks, J.R. & Brown, A.B., 1988, "Installation of a New Surface Fan at the Homestake Gold Mine", Mining Engineering, p.111-15.

van Wyngard, A.L., 1987, "Sealing Fissure Water in the Vaal Reefs No.3 Main Vertical Upcast Shaft", Assoc Mine Managrs. S. Afr. 2/87, p.92-105.

5

Processing

Chapter 19

DEVELOPMENTS IN GOLD PROCESSING TECHNOLOGY:
A COMPARISON OF METHODS

T. V. Subrahmanyam and K. S. Eric Forssberg[2]

Laboratory of Mineral Processing, Departamento de Geologia,
CCE/UFRN, Campus Universitario, 59.000 Natal-RN (BRAZIL)

[2] Division of Mineral Processing, Luleå University of
Technology, S-951 87 Luleå (Sweden)

ABSTRACT

The present paper deals with the important technological developments
in gold processing and where possible , the data is presented in the
form of tables. Gold processing methods have evolved to such an extent that low grade and complex ores once considered uneconomic   are
now becoming viable. Constant efforts are underway to innovate technologies in order to improve the efficiencies and cut down the capital and operating costs. The methods are discussed.

INTRODUCTION

Economic, technological and environmental considerations
have always been associated with innovative   technologies
and the metallurgy of gold is not an exception to   these.
During the last few years gold processing methods   have
evolved to such as extent that to-day low grade ores   are
viable, thanks to the application of heap leaching.   With
the depletion of free milling ores necessity arises to develop methods for processing the refractory ores, for example, bio-heap leaching to highly refractory gold   bearing
arsenopyrite ores. While the economic factor is a major incentive, the environmental considerations of the   modern
world impose limitations and consequently the need for cost
effective alternative technologies, for example, efforts to
substitute sodium cyanide by other lixiviants like   thiourea, organic nitriles, chlorine, etc.; bio-oxidation, pressure oxidation in preference to traditional roasting   for

265

refractory gold ores.

Gold is an exception to the economic crisis that is being faced, in general, by the minerals and   metallurgical
industry all over the world. Probably no other metal   in
the crude form enjoys such a high value to-day as gold does!
The reasons are evident-its bullion value and its   utility
in jewellery. Both in war and in love, at some stage   or
other, gold has been instrumental to the conqueror.

From a peak value of US$ 860/oz at one time in   1980,
gold's price to-day stabilized at US$ 450/oz. In view   of
the prevailing gold prices (compared to low capital   and
operating costs) and the developments in gold   processing
technology, a continued growth of gold search activity and
metallurgy can be expected. The trend is also   encouraging
to small miners because of the portable gold   processing
plants.

The objective of the present paper is to briefly   review
and compare the important methods of gold processing   and
therefore, no attempt is made to describe the   procedures
and process details since they are found in several   papers
(Henley, 1975; Lloyd, 1978; Fleming, 1983;   Wall, et   al.,
1987; Haque, 1987; Dayton, 1987; Jha, 1987).

TYPES OF ORES AND PROCESSES:
AN OVER VIEW OF ASSOCIATED PROBLEMS

Gold occurs under widely varying conditions and the   mineralogy of gold ore is complex. The methods utilized   to
recover gold from its ores vary from primitive to the   most
sophisticated, depending on the nature   of gold's   association. The properties that are exploited for   concentrating
gold are: high specific gravity, wettability by   mercury,
response to flotation collectors and solubility in   dilute
cyanide solutions.

The gold ores may be broadly classified into: ores   containing free gold, non-refractory or readily leachable ores
and refractory ores. Each type of ore presents   problems
specific to cyanidation and the efficiency of   cyanidation
depends on the effective contact between the cyanide   solution and the gold particle. While the grain size of   gold
and the concentration of free cyanide ion influence the kinetics of gold dissolution in free milling ores, the   mineralogy and the nature of gold association are important for

gold recovery from refractory ores. The recoveries   generally low from refractory ores. The low recoveries   are
attributed to physical, chemical and mineralogical   factors
(Subrahmanyam and Forssberg, 1988). Table I shows   the
effect of mineralogical impurities on gold recovery in cyanidation. In summary, the free cyanide ion concentration is
reduced as a result of the reaction of several other   minerals with cyanide thus inhibiting or retarding the dissolution of gold.

Ores containing free gold

   Gravity concentration methods are still a small   miner's
magic formula to recover gold from ores containing   coarse,
free gold. The recoveries are generally low (around   60%).
Fig. 1 shows the effect of particle size on gold   recovery
by some gravity concentration equipment. In order to minimise gold losses in leach residues it becomes essential   to
adopt a preconcentration step ahead of cyanidation or amalgamation. For example, a native gold particle 90% Au   and
10% Ag of 150 µm size will need more than 44 hours for   dissolution in cyanide solution (Graham and Watenweiler,1924;
Barsky, et al., 1934), and with lesser leach times   coarse
particles may be lost to leach residues.

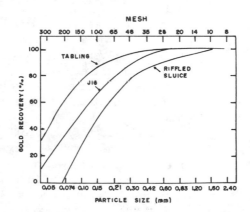

Fig. 1. The effect of particle size
        on gold recovery by gravity
        concentration (taken from
        Wang and Poling, 1983).

TABLE I. The effect of impurities on gold recovery
(reference: Subrahmanyam and Forssberg, 1988)

impurities/species present	effect
antimony (stibnite)	compounds such as thioantimonites are formed and react with oxygen of the cyanide solution to form antimonites.
arsenic (realgar, orpiment)	forms compounds such as thioarsenites which react with oxygen of the cyanide solution to form arsenites.
arsenopyrite	consumes large amounts of cyanide and forms ferrocyanide complexes.
copper	dissolves in cyanide causing excessive cyanide consumption. copper cyanogen complexes formed indirectly affect dissolution of gold. Influences the precipitation of gold by zinc.
covellite, sphalerite, pyrite, etc.	alkaline sulphides or thiosulphates or thiocyanates are products that are formed as a result of decomposition or dissolution in cyanide solutions. These products adversely affect the dissolution and precipitation of gold.
manganese	weathering of manganese results in higher oxidation products which form refractory minerals of silver and manganese, if manganese occurs with silver ores.
nickel	does not affect the dissolution of gold but has detrimental effect on precipitation of gold.
pyrrhotite	a common refractory mineral and its decomposition forms ferrocyanide which removes free cyanide from cyanide solution.
zinc	cyanogen complexes are formed but their effect on the dissolution of gold is less pronounced than those of copper. Causes problems in the cyanidation of silver.
carbonaceous	precipitant of gold in cyanide solutions. Ores containing organic carbon adsorb the dissolved gold before the leach solutions are recovered("preg robbing").
clays	causes solution percolation problems.

quartz, chert, jasperoid	if gold is encased by impervious matrix like quartz the leach solution cannot complex the metal.
water	impurities or dissolved salts in water like bi-carbonates, magnesium sulphate, organic matter, heavy metal salts, chlorides etc. affect the cyanide solution.
flotation products	reagents used for the flotation of gold ores such as xanthates, dithiophosphates, frothers, etc. retard the dissolution of gold in cyanide solution.

Non-refractory or readily leachable ores

For non-refractory ores the methods of leaching are heap, dump, in-situ and vat. Several papers (Potter, 1981; Clem, 1982; Chamberlain and Pojar, 1981; Kappes, 1979) describe the basic operational differences among these methods and heap leaching is most commonly applied to low grade and high tonnage ores; the ore may be run-of-mine or crushed to a size as fine as 10 mm. In view of the varying ore characteristics a detailed metallurgical testing for scale up and an assessment of the amenability of the rock to heap leaching process are essential and are described elsewhere (Potter, 1981). Heaps of 5000- to 2 million tons grading as low as 0.03 oz/st have been successfully leached for gold (Lewis, 1983). In heap leaching, the cyanide solutions (0.5-1.0 g/l NaCN; pH 10-11) applied on top of the heap (at a rate of 6 L/hr/m^2 of heap surface) dissolve the gold during percolation through the stacked ore. The pregnant solutions reach the impervious pad at the bottom of the heap and flow to a collection pond from where the solution is processed through carbon adsorption columns, desorption circuits and finally electrowinning (fig. 2). The heap leaching method is simple and flexible but a major drawback is a low recovery 60-75%, compared to high recoveries like 90-99% frequently reported for ores treated by carbon-in-pulp or counter current decantation methods. Among the problems that may lead to low recoveries in heap leaching operation are: oxygen content, solution percolation and temperature for which the suggested solutions are air injection (Worstell, 1987), agglomeration (Lastra and Chase, 1984), and solution heating (where the temperatures fall below 50º F), respectively.

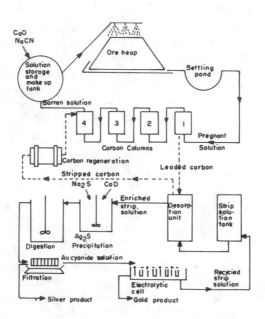

Fig. 2. Heap leaching, carbon adsor-
         ption-desorption system (dia-
         gram of Challenge Mining Co.)
         (taken from Kappes, 1979,).

Refractory ores

    The refractoriness of ores may be due to associated sul-
phide minerals or carbonaceous material or gold   inclusion
in silicious gangue. Gold associated with sulphides   like
pyrite, arsenopyrite, etc., does not yield to direct cyanidation unless roasted. But roasting of large quantities of
ores involves huge costs and also poses environmental problems . Therefore, many plants incorporate a flotation step
in the circuit to produce gold rich sulphide concentrate to
be subsequently treated by amalgamation (if the bulk of the
gold is free), cyanidation or roasting and cyanidation. As
alternative to traditional roasting are bacterial leaching,
pressure oxidation and high pressure/low alkalinity cyanidation (Table II). The bacterial leaching is claimed to be
advantageous from the view point of flexibility,   presents
no pollution problems and cost effective compared to   pressure leaching and roasting. A flow sheet for   bacterial
leaching of a gold bearing sulphide ore   concentrate   is
shown in fig. 3. Several organizations offer the   process

TABLE II. Methods alternative to roasting for refractory minerals

	bio-oxidation	pressure oxidation	high pressure/low alkalinity cyanidation
oxidation reaction effected by	bio-oxidation bacteria like thiobacillus ferrooxidans or thiobacillus thiooxidans	oxygen	oxygen
temperature	$35-40^{\circ}C$	$180^{\circ}C$	–
pH	1.8 (no pH modification is necessary since H_2SO_4 is produced in the reaction by bacterial activity, for example, $$4\ FeS_2 + 15\ O_2 + 2\ H_2O \longrightarrow$$ $$2\ FeS_2(SO_4)_3 + 2\ H_2SO_4$$	1.8-1.9	7.0
pressure		2200 KPa	pressure of oxygen into reactor : 12 MPa
time	from 24 hrs. to 5 or 6 days	1-3 hours	treatment time per batch 135 min.
solids concentration	optimum 15-20% and 30% if the sulphur content is low	–	–
recovery	90-95%	92-93% compared to 5-80% using direct cyanidation.	82-87% compared to very low i.e. 1% extraction by cyanidation.
equipment	open tanks; ambiental condns.	autoclaves	reactor

as an alternative to other treatments of refractory   ores
and pilot bacterial leach plants have been successfully demonstrated. High pressure/low alkalinity cyanidation   has
particular commercial application in the treatment of highly refractory antimonial products, for example: stibnite
concentrates.

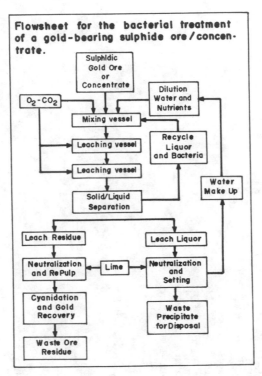

Fig. 3. Bacterial leaching of
        refractory gold ore
        (taken from Pooley, 1987).

BACTERIAL LEACHING

An important development as as alternative to expensive
roasting or pressure oxidation methods, is the application
of bacterial leaching to refractory sulphide gold ores. The
oxidation reactions are effected by bacteria like thiobaci-
llus ferrooxidans, which oxidizes all metal sulphides   to
sulphate and elemental sulphur compounds to sulphuric acid
(see Table II). A recent review on the application of biotechnology in metal extractions deals with the   mechanisms

of bacterial oxidation of refreactory ores (Torma, 1987).

From the photomicrographs the bacterial oxidation was found to occur mainly along the grain boundaries and around inclusions with a faster leaching to the centre of the particle. It was observed that even a 10% oxidation of the sulphide was enough to achieve a recovery of 85-90% in subsequent oxidation; with gold inclusions smaller than 0.5 μm 100% bacterial oxidation was found to be necessary (Pooley, 1987). The extraction of gold associated with pyrite was found to be directly proportional to the bacterial oxidation of pyrite; for example, a pyrite sample without pre-oxidation resulted in a 25% gold extraction and with a pre-oxidized sample the extraction was more than 90%. Similarly pilot scale tests on an arsenopyrite concentrate (10% As) subjected to double stage bacterial leaching showed 90% extraction of arsenic in 100 hours,which followed by solid-liquid separation and cyanidation yielded 92.5% gold extraction (Polkin, et al., 1975). When the gold-arsenopyrite concentrate was directly leached by KCN the final extraction was about 7-10%, and calcination of the concentrate ahead of KCN leaching resulted in a gold extraction of 77%.

It is estimated that the operating costs are equal to pressure leaching and less than roasting and capital costs half those of pressure leaching. Further, it is claimed that the cost of bio-leaching decreases with increasing plant size (Bruynesteyn and Hakle, 1986).

REAGENTS FOR GOLD LEACHING

Table III gives the leaching systems for gold. Cyanidation continues to be economical and a standard practice for gold recovery. However, other alternatives investigated or under investigation have both advantages and disadvantages (see Table IV). Among the alternative lixiviants to cyanide, thiourea is gaining ground since the kinetics of leaching is much faster, less toxic and less affected by interfering ions. Thiourea also reacts readily with copper minerals and hence the derivatives of thiourea for gold bearing copper ores are being investigated. Fig. 4 shows comparative dissolution rates of gold, silver and copper ores in thiourea (0.5% $(NH_2)_2CS$, 0.5% H_2SO_4, 0.1% Fe^3) and cyanide (0.5% NaCN, 0.5% CaO) solutions.On cost consideration, thiourea is about 25% more expensive than cyanide and its rapid degradation leads to a higher reagent consumption.

TABLE III. Leaching Systems for Gold
(reference J. T. Woodcock, 1988)

leaching reagent	oxidant used	gold complex in solution	general conditions
Alkaline Systems			
Cyanide	O_2	$Au(CN)_2^-$	pH $>$ 10
Ammonia–Cyanide	O_2	$Au(CN)_2^-$	pH $>$ 10
Organic Nitriles	O_2	$Au(CH(CN)_2)_2^-$	pH $>$ 7
Alpha-hydroxynitriles	O_2	$Au(CN)_2^-$	pH $>$ 10
Calcium Cyanamide	O_2	$Au(NCN)_2^{3-}$	pH $>$ 10
Alkali Cyanoform	O_2	$Au(C(CN)_3)_2^-$	pH $>$ 10
Neutral Systems			
Thiosulphate	O_2	$Au(S_2O_3)_2^{3-}$	pH $>$ 7, 60 C
Bromocyanide	Br CN	$Au(CN)_2^-$	pH 7
Bromine	Br_2	$AuBr_4^-$	pH 7
Acid Systems			
Chlorine (aqueous)	Cl_2	$AuCl_4^-$	pH $<$ 2, Cl^-
Ferric Chloride	Fe^{3+}	$AuCl_4^-$	pH $<$ 2, Cl^-
Aqua regia	HNO_3	$AuCl_4^-$	pH $<$ 0, Cl^-
Thiocyanate	Fe^{3+}, H_2O_2	$Au(SCN)_4^-$	pH $<$ 3
Thiourea	Fe^{3+}, H_2O_2	$Au(NH_2(SNH_2)_2^+$	pH 1-2

Further, more work is necessary on the recovery of precious metals by activated carbon from the leach solutions of thiourea. Since ion exchange has been shown to work well   for gold recovery from acidic thiourea solutions and pulps, some developments may be expected in this direction in future. The application of thiourea in heap leaching for gold and silver values may bring down the period of leaching due to   faster dissolution rates.

TABLE IV. Properties and applications of leaching reagents for gold

leaching reagent	properties and applications
Alkaline Systems	
Cyanide	universally used.
Ammonia cyanide	used for oxidized gold-copper ores but not persisted with.
Organic nitriles	malanonitrile and alphhydroxynitrile are less toxic than cyanide; expensive and slower leachants; function in alkaline or near neutral solutions; malanonitrile is claimed to be more effective than cyanide at pH 7-8 and may be effective on some base metal ores.
Alpha-hydroxynitriles	function by slow hydrolysis in alkaline solution releasing cyanide ions and the parent ketone. The slow release of cyanide may be effective in dissolving gold while minimising side reactions with other minerals.
Calcium cyanamide and alkali cyanoform	hydrolyse in alkaline solutions to form stable low toxicity carbanions capable of complexing gold;
Neutral Systems	
Thiosulphate	is non-toxic and inexpensive; no commercial scale ore operations are reported. Neutral thiosulphate solutions dissove gold slowly from various products; a mild oxidant and elevated temperatures are needed; copper additions accelerate the dissolution rate.
Bromocyanide	expensive; provides both oxidant and complexing reagent; has been used on sulpho telluride ores to dissolve gold.

Bromine
leaches gold as well as platinum at acid, neutral or alkaline pH; faster leaching kinetics when a protonic cation (NH_4^+) and an oxidant are added; among the disadvantages are the high reagent costs and need for reagent recovery and regeneration.

Acid Systems

Chlorine
gold and silver are leached rapidly at acid pH. Acidified chlorine was found to be worth testing for ores containing less than 0.5% sulphur, beyond which level chlorine consumption is high. High reaction rates achieved when gold is oxidized to Au (III) state and complexed with chloride, bromide or iodide; these reagent combinations are aggressive to other minerals and materials of construction.

Thiocyanate
can be used in conjunction with Fe (III) at pH 2 to dissolve gold and uranium Gold and unused thiocyanate can be recovered by ion exchange .

Thiourea
low toxicity and dissolves gold readily (leach rates of 256 mg $Au/cm^2/hr$, i.e more than 100 times that of gold in an 0.1% NaCN solution); suffers from high reagent consumption which is caused by oxidative degradation; addition of SO_2 is claimed to minimise oxidative degradation. However, it has also been reported that minimal amounts of thiourea were saved at the expense of substantial quantities of sulphite, oxidant and acid. Thiourea also reacts readily with copper.

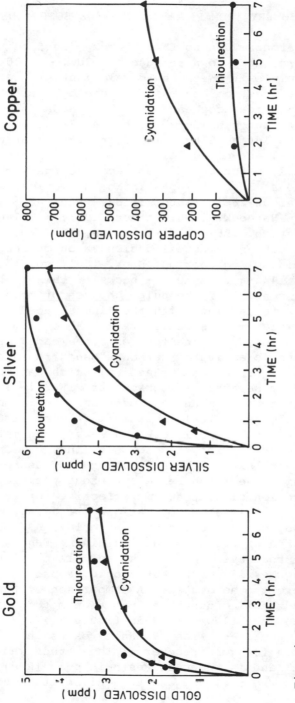

Fig. 4 Comparative dissolution rates of thioureation and cyanidation of gold, silver and copper ores (taken from Chen, et. al., 1980).

GCLD RECOVERY FROM PREGNANT CYANIDE SOLUTIONS

Activated carbons, due to their large   surface   areas
(about 1000 m²/g) are known to adsorb gold (up to 30.000 ppm)
in a cyanide complex . With the development of   efficient
stripping procedures, carbon based technology for   recovering gold gained wide acceptance. The methods for gold recovery from pregnant cyanide solutions are: carbon-in- pulp
(CIP), carbon-in-leach (CIL), zinc dust precipitation,   resin-in-pulp (RIP) and direct electrowinning. Among   these,
the CIP is widely practiced; it is reported that   presently
there are at least 30 CIP plants in North America,   10 in
Australia and about 28 in South Africa. Several   papers
(Mc Dougall and Hancock, 1980; Potter, 1981;   Clem, 1982;
Lewis, 1983; Dayton, 1987; and Michaelis, 1987)   describe
the details of CIP and a detailed discussion on the technical aspects (Mining Engineering, Sept. 1981, pp. 1331-5 and
Oct. 1981, pp. 1441-4). Some drawbacks of this method are:
the need to screen all of the pulp, the use of more cyanide
and gold losses to tailings with the fine abraded   carbon.
An innovation which merits consideration to overcome   the
problem of gold losses in the form of fine loaded carbon is
the use of magnetic activated carbon- "Magchar" (a combination of activated carbon and magnetite) which can be   recovered by magnetic separators (Herkenhoff, 1982). This process dates back to 1940's and on cost basis the   magnetic
separators need more capital investment in comparison   to
the screens presently used in CIP. However, the method may
find potential application for slow settling,   slimy ores.
In carbon-in-leach set up, both leaching and adsorption are
carried out using the leach vessels without the use   of a
separate adsorption circuit and is effective for carbonaceous ores. Many heap leaching operations use either   carbon
or zinc dust method for recovering the gold. As   pointed
out by one of the experts (Potter, in Mining Eng.   Sept.,
1981, p. 1334) that every gold plant and silver plant,   no
matter if using Merrill-Crowe but should be using a   carbon
column somewhere in the system. A comparison of the carbon
in pulp versus the zinc dust precipitation methods is given
in Table. V and the flow sheets in figure 5. Yet   another
development is the use of ion exchange resins in place   of
activated carbon for gold recovery. The method was   developed by Mintek and is known as resin-in-pulp (RIP); it is
claimed to be economical and an attractive alternative   to

TABLE V. Carbon-in-Pulp versus Zinc Precipitation

carbon-in-pulp	zinc precipitation
Gold recoveries are higher; lower capital and operating costs. Large volumes of low grade solutions mainly containing gold may be economically treated in carbon columns.	Higher gold losses. Preferred for rich silver bearing solutions of small volume with a large daily silver production.
Activated carbon added directly to cyanided pulp which obviates expensive filtration, clarification.	Gold and silver are precipitated on powdered zinc. Clarification of the cyanide solution is necessary to eliminate suspended constituents that can coat zinc particles and retard precipitation of precious metals. Elimination of dissolved oxygen from pregnant solution is essential to prevent redissolution of gold.
Activated carbon retains excellent ability to scavenge gold and silver even in the presence of Cu, Ni and Fe.	Sensitive to alkalinity and free and other common constituents of cyanidation liquors i.e. sulphide salts, cyanide complexes of Cu,Ni, As, Sb, all of which decrease gold recoveries.

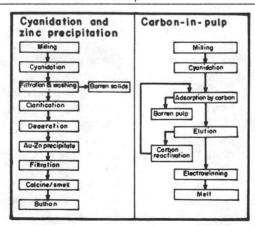

Fig. 5 Gold recovery by   zinc
precipitation and by CIP.

the carbon-in-pulp process (Ryan, 1987). The gold loading
rates were found to be higher when resin fibres were used
instead of resin beads. However, the loading capacities
with resin fibres were low. The resin fibres were developed based on a polyacrylonitrile (PAN) derivative (PAN-
imidazoline) and a polybenzimiadazole (PBI) derivative
(PBI-alkylamine) (Wall, et al., 1987). The technical feasibility of the use of ion exchange resins in a continuous
plant is yet to be confirmed. Table VI gives a comparison
of the CIP and the RIP methods.

When the ore contains high silver to gold ratios, the
leach effluents are treated with sodium sulphide to precipitate the silver sulphide for subsequent removal by filtration. The solutions are processed further through activated carbon columns for gold recovery. Small volumes of
rich gold and silver bearing solutions are processed by
direct electrolysis.

GOLD STRIPPING

The standard method (Zadra Method) to recover gold from
activated carbon is with a solution containing 1% NaOH,0.1
-0.2% NaCN at a temperature of 90°C. The gold is recovered from the effluent solutions by a series of electro-
lytic cells. The resulting solutions after electrodeposi-
tion are returned to the stripping cycle. Usually the time
required for a stripping cycle is 48-72 hours. This stripping time can be reduced to 6-8 hours by adding 10% etha-
nol or propanol to 1% NaOH, 0.1-0.2% NaCN at 60°C.   But
sufficient care is necessary since the system is subject
to fire hazards due to the use of alcohol. To overcome
the problem of fires the use of nonflammable ethylene or
propylene glycol (as a substitute to alcohol) to an atmospheric zadra solution was suggested. The stripping time
was reduced by about 50% by the addition of glycol.   But
ethylene glycol is harmful to activated carbon due to the
presence of corrosion inhibitors and the cost of glycol is
twice that of alcohol. It is claimed that this option may
be viable for smaller and low capital budget operations
(Fast, 1987). With a 1% NaOH, 0.1-0.2% NaCN solution at
120-130°C under a pressure of 482 KPa, the stripping can
be achieved in 2-6 hours. In the AARL (Anglo American Research Laboratory) process or modified Zadra method (Davidson and Duncanson, 1977), the loaded carbon is conditio-

TABLE VI. A comparison of carbon-in-pulp versus resin-in-pulp technologies

	CARBON-IN-PULP	RESIN-IN-PULP
Adsorbant	activated carbon (590 μm - 1680 μm); coconut carbon shell preferred.	ion exchange resins
Technology/adaptability to alternative reagents.	widely accepted; more work is necessary on the adsorption of gold from thiourea solutions.	a suitable resin for gold recovery is yet to achieve widespread acceptance; ion exchange has been shown to work well for gold recovery from acidic thiourea solutions and pulps.
Regeneration	requires expensive and troublesome thermal regeneration of carbon which is essential for good gold adsorption, either in CIP or CIL systems. This requires removal of scale deposits from the carbon usually by acid washings as well as thermal reactivation to remove adsorbed organic species.	less prone to poisoring by organic compounds and could prove useful if the preg robbing characteristics of ores can be overcome by adding diesel ahead of cyanidation.
Interferences	adsorption of gold not inhibited in presence of soluble sulphides copper iron.	the recovery of gold difficult when more than or very small amounts of cyanide soluble copper exists.
Losses	more generation of fines and therby gold losses.	less attrition losses of resin beads.
Solution flow	typical upflow carbon columns operate 40-60 $m^3/m^2/hr$.	at the resins are less dense and smaller than activated carbon and so upflow fluid bed adsorption columns would have to be operated at less solution flow velocity.

ned with a caustic cyanide solution (1.5-5.0% cyanide, 0.5-
2.0% Caustic soda) at 90° C and gold is eluted with water at
150° C. The kinetics of elution was found to increase   with
water purity. The disadvantages of this method are the high
quality water requirement and a high chemical consumption.

CONCLUSIONS

From the foregoing it is evident that two important developments that have taken place in gold processing are: the
application of heap leaching to low grade gold ores and the
development of carbon based technology for gold   recovery
from cyanide leach solutions. With the depletion of   free
milling ores   necessity arises to treat the refractory (or
difficult to treat) ores. Thanks to innovative technologies
there are several options but the adaptability or choice of
a method is based on various factors like environmental and
cost considerations--capital and operating,   metallurgical
efficiencies, flexibility of the process, etc. Undoubtedly
the traditional gold processing methods are still   widely
accepted. However, major developments can be foreseen   in
the following areas:

  * bio-hydrometallurgy- bacterial leaching of refractory
    gold ores, bio-heap leaching,

  * substitution of activated carbon by ion exchange resins

  * thiourea as an alternative to cyanide.

REFERENCES

Barsky, G., Swainson, S. J., and Hedley, N., 1934, "Dissolution of Gold and Silver in Cyanide Solutions", Trans. Am.
Inst. Min. Metall. Engrs., 112, pp 660-667.

Bruynesteyn, A., and Hakle, R. P., 1986,   "Microbiological
Effects on Metallurgical Process", Clum, J.A.,and Haas, L.A:
eds., TMS AIME, Warrendale, pp 121-127.

Chamberlain, P. G., and Pojar, M. G., 1981, "The status   of
Gold and Silver Leaching Operations in the United States",in
Gold and Silver-Leaching, Recovery and Economics, Chap. 1,
Proceedings of SME AIME, Schlitt, W. J., Larson, W.C.,   and
Hiskey, J. B., eds., pp 1-16.

Chen, C. K., Lung and Wang, 1980, "A Study of the leaching of gold and silver by acid thioureation", Hydrometallurgy, 5, pp 207-212.

Clem, B. M., 1982, "Heap Leaching Gold and Silver Ores", E& MJ, Vol. 183, Nº 4, Apr., pp 68-76.

Davidson, R. J., and Veronese, V., 1979, "Studies on the Elution of gold from activated carbon using water as the eluant", Journal of the South African Inst. of Mining & Metallurgy, Oct.

Dayton, S. H., 1987, "Gold Processing update", E& MJ, Vol. 188, No. 6, Jun., pp 25-29.

Fast, J. L., 1987, "Glycol Stripping: A viable option for recovering Gold from Carbon", E& MJ, Vol. 188, No. 6, Jun., pp 48-49.

Fleming, C. A., 1983, "Recent Developments in Carbon- in-Pulp Technology in South Africa", Hydrometallurgy-Research, Development and Plant Practice, Osseo-Assare, K., and Miller, J. D., eds., Proceedings of TMS AIME, pp 839-857.

Graham, K. L., and Watenweiler, F., 1924, "A Search to determine the size of gold particles in the Witwatersrand banket ore and their rate of dissolution in cyanide, Journal of Chem. Metall. Min. Soc. S. Afr., 24, pp 285-292.

Haque, K. E., 1987, "Gold Leaching from Refractory Ores- literature survey", Mineral Processing and Extractive Metallurgy Review, Vol. 2, pp 235-253.

Henley, K. J., 1975, "Gold-ore Mineralogy and its relation to Metallurgical treatment", Mineral Science and Engineering, Vol. 7, No. 4, Oct., pp 289-312.

Herkenhoff, E., 1982, "Magchar: An alternative for gold plants using Carbon-in-Pulp Systems", E& MJ, Aug., pp 84-87.

Jha, M. C., 1987, "Refractoriness of certain gold ores to cyanidation: Probable causes and possible solutions", Mineral Processing and Extractive Metallurgy Review, Vol. 2, No 4, Dec., pp 331-352.

Kappes, D. W., 1979, "Heap leaching is small miner's golden opportunity", Mining Engineering, Vol. 31, No. 2, Feb., pp 136-140.

Lastra, M. R., and Chase, C. K., 1984, "Permeability, Solu-

tion Delivery and Solution Recovery: Critical Factors   in
Dump and Heap Leaching of Gold", Mining Engineering, Vol.36
No. 11, Nov., pp 1537-1539.

Lewis, A., 1983, "Leaching and Precipitation Technolgy for
Gold and Silver Ores", E& MJ, Vol. 184, No. 6, Jun., pp 48-
56.

Lloyd, P. J. D., 1978, "Maximising the recovery of   gold
from Witwatersrand Ores", Minerals Science Engineering, Vol
10, No. 3, Jul., pp 208-221.

McDougall, G. J., and Hancock, R. D., 1980, "Activated Carbons and Gold - A Literature Survery", Minerals Science Engineering, Vol. 12, No. 2, Apr ., pp 85-99.

Michaelis, H. V., 1987, "Recovering Gold and Silver   from
pregnant leach solutions", E& MJ, Vol. 188, No. 6, Jun., pp
50-55.

Polkin, S. I., et al., 1975, Proceedings, XI   Internationl
Mineral Processing Congress, Cagliari, Italy, pp 901-923.

Pooley, F. D., 1987, "Use of Bacteria to enhance   recovery
of gold from refractory ores", E& MJ, Vol. 188, No. 6,   pp
67.

Potter, G. M., 1981,"Design Factors for Heap Leaching operation", Mining Engineering, Vol. 33, No. 3, Mar., pp 277-
281.

Ryan, B., 1987, "Mintek Researching RIP as an   alternative
to CIP", E& MJ, Vol. 188, No. 6, Jun., pp 59.

Subrahmanyam, T. V., and Forssberg Eric, K. S., 1988, "Recovery Problems in Gold Ore Processing with emphasis   on
Heap Leaching", Mineral Processing and Extractive Metallurgy Review,(submitted and accepted).

Torma, A. E., 1987, "Impact of Biotechnology on Metal Extractions", Mineral Processing and Extractive   Metallurgy
Review, Vol. 2, No. 4, Dec., pp 289-330.

Wall, N. C., Hornby, J. C., and Sethi, J. K., 1987,   "Gold
Beneficiation", Mining Magazine, Vol. 156, No. 5, May,   pp
393-401.

Wang, W., and Poling, G. W., 1983, "Methods for recovering
fine placer gold", The Canadian Mining & Metallurgical Bull.

Woodcock, J. T., 1988, "Innovations and Options in gold Metallurgy", Proceedings XVI International Mineral Processing Congress, Forssberg Eric, K. S., ed., Elsevier, Part A, pp 115-131.

Worstell, J. H., 1987, "Enhance Heap Leaching Rates   with Air Injection into the Heap", Mining Magazine, Vol. 156, No. 1, Jan., pp 40-41.

Zadra, J. B., Engle, A. L., and Heinen, H. S., 1952, "Process for recovering gold and silver from activated   carbon by leaching and electrolysis", US Bureau of Mines, RI 4843.

Chapter 20

DESIGN AND OPERATING RESULTS OF THE NITROX PROCESS®

by Gus Van Weert, Ken J. Fair and Vicken H. Aprahamian

Hydrochem Developments Ltd.
16-266 Rutherford Road South
Brampton, Ontario   L6W 3X3

A consortium of Canadian mining companies, engineering firms and the Ontario Government has supported Hydrochem's efforts in the last year to design and cost a 10 tpd NITROX PROCESS® demonstration plant. A 5 kg per hour mini plant was installed at the Hydrochem facilities in Brampton, Ontario.

This paper describes design testwork on Dickenson Mines arseno-pyrite concentrate and application of the findings to the demonstration plant flowsheet. The control of nitrates in discard solids and stack gases is discussed in detail.

Nitric acid oxidation of arsenopyrite results in significant elemental sulphur formation, enough for it to be detrimental to subsequent cyanidation. Flowsheet variations aimed at dealing with this process characteristic are discussed and testwork results on one option (hot liming) are described.

INTRODUCTION

The increase in the gold price, which occurred in the seventies, has resulted in renewed interest in treating refractory gold bearing arsenopyrites by means other than the conventional roasting with air.

As described earlier (1), mining companies and prospectors have generally avoided "refractory" gold and silver ores by walking away from such deposits or filing the details away under "Future Activities (for somebody else)". As long as conventional deposits could easily be found in cooperative jurisdictions, this made sense.

With an increasing world population and increasing volume of trade activities and monetary speculation, total demand for gold has grown, resulting in an all time high price, certainly in inflation adjusted Canadian dollars (2). This has resulted in a search for new deposits, reworking of old deposits, especially tailing dumps, adoption of new techniques (heap leaching) and dusting off these "Future Activities" files.

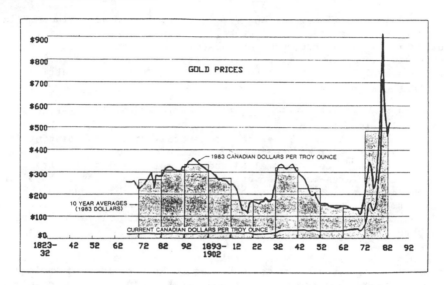

Fig. 1. Historical Gold Prices

Canadian metallurgists have largely spearheaded the treatment of refractory gold and silver ores. The majority of refractory gold and silver ores involve sulphides and arsenides, with pyrites representing the bulk of these (3). The following remarks refer to pyrite and arsenopyrites containing concentrates, and specifically to concentrate produced by the Arthur W. White mine at Balmertown, Ontario. This mine is in the Red Lake area of Ontario and is wholly owned by Dickenson Mines Ltd. of Toronto. Its concentrate is totally refractory; it is produced from the cyanide leach tailings.

As described in detail elsewhere (4), Hydrochem has brought a number of Canadian mining companies, engineering firms and the Ontario Government together for a detailed study of the capital and operating cost of a 10 tpd demonstration plant. The final report was submitted to the subscribers of the study early September for their evaluation. The study included a budget for operating the 10 tpd facility for a year, a site evaluation, an environmental assessment, a limestone availability and cost study, and testwork results based on Dickenson concentrate.

REFRACTORY SULPHIDES

Treatment of refractory sulphides generally involves complete
oxidation of those sulphides and arsenides, which contain the gold
and silver values. Where such oxidation is incomplete, subsequent
gold extraction tends to be incomplete (5, 6). This has also been
found true with the NITROX PROCESS® on samples tested to date.

The logical oxidation agent for arsenopyrite is air, since it is
free (at least at the plant intake). Processes not using air, but
other oxidation agents, such as H_2O_2 or Cl_2 derived principally
through consumption of electrical energy, have not found industrial
acceptance. Oxygen production from air also requires electrical
energy.

Oxidation of refractory sulphides involves their contacting with
oxygen or air. This can be done dry or wet, i.e. pyro- or hydrometallurgically. The former method generally involves roasting in
multiple hearths or more commonly, fluid bed roasters. Some low
arsenic concentrates are smelted and the molten sulphide matte
oxidized in Pierce-Smith converters. This normally involves tolling,
which can be attractive where the producing site is close to a base
metal smelter, the grade of the concentrate is high, no deleterious
elements are present in the concentrate and sulphuric acid can be
marketed.

This paper deals with the situation where one or more producers
wish to process their own concentrates. The roasting option in most
countries now requires sale of sulphuric acid and acceptable collection and disposal of arsenic trioxide. Mineral concentrates often
produce a lower quality acid. Scrubbing of the roaster gases prior
to acid production may yield sludges which are problematic in
disposal. Production of an arsenopyrite concentrate involves a reduction in gold recovery from the ore body. To satisfy the roaster
heat balance, the concentrate should contain at least 16% S, a requirement which may further reduce recovery. The alternative is fuel
injection into the roaster, which may effect acid quality if carbon
is generated and carried in the gas stream.

Arsenic trioxide collection and disposal is even more problematic
and costly. An electrostatic precipitator is required. Quality
specifications for the arsenic trioxide are becoming more stringent
since it is a buyers market, especially with Chile increasing its
production to at least 35,000 tonnes by 1989 (7).

Pyro-oxidation of arsenopyrite (at roasting temperatures of approximately 525°C) proceeds very fast. Achieving similar rates in hydrometallurgical systems is impossible; neither is it necessary.
Roasting requires complete utilization of all oxygen in seconds, to
maintain the calcine at temperature and, especially in a two stage
system, manipulate the chemistry. In the alternative hydrometallurgical systems the air or oxygen can be depleted nearly at leisure,

it's basically a matter of equipment volume. It is important to understand that in hydrometallurgical systems, the oxygen is consumed in solution, but is stored in the gas phase. Transfer of oxygen from the gas phase to the liquid is an important step, if not the limiting step, in sulphide oxidation. In the pressure autoclave system this is dealt with by the use of high pressure oxygen. This increases solubility in, and oxygen transfer to the process slurry. It makes oxidation in the pressure autoclave very energy intensive.

Surprisingly, bacterial oxidation faces an even greater challenge in this regard. The concept is simple: use specially selected or modified bacteria for oxidation of refractory sulphides in agitated slurries. Small amounts of nutrients are also added. Since the oxygen solubility in water at sea level and room temperature is only 9 ppm, the process liquor has to be constantly exposed to air to make the oxidation of arsenopyrite proceed. This is achieved by air sparging and liquor agitation, similar to practices in the pharmaceutical and fermentation industries. Oxygen transfer from air under such conditions measures at best 1 to 2 kg O_2 per HP-hour, under laboratory conditions, but is proportional to $HP^{.50}$ for viscous solutions and $HP^{.43}$ for water in agitated vessels (8). Hence, the larger the vessel, the lower the O_2 transferred per HP-hour. In addition, solids, surface active agents and higher temperatures all tend to reduce the oxygen transfer.

Recently, Demopoulos has reviewed the various aqueous oxidation methods for refractory gold (9). He concluded for the NITROX PROCESS that "Among the advantages of the process are the operation under atmospheric pressure, the use of air instead of O_2, leaching tanks made of stainless steel, fixation of As as basic ferric arsenates (high Fe/As ratio) and high Ag recoveries. Potential drawbacks are the production of $S°$ and its interference during cyanidation on one hand and the effective reduction of nitrate level of the effluent streams on the other. This paper will discuss in particular how the NITROX PROCESS® has evolved to deal with nitrates and elemental sulphur production.

THE NITROX CYCLE

One of the fundamental attractions of the NITROX PROCESS® is the low cost at which oxygen units out of air are made available to the mineral slurry. This is done through an intermediate, i.e. gaseous NO_2, which has a very high solubility in water, even at 85-95°C, the operating temperature of the NITROX PROCESS®.

Not only does NO_2 readily dissolve in the process solutions, it also makes the oxygen readily available for refractory sulphide oxidation. This happens because NO_2 reacts with the water to form nitric acid.

$$3\ NO_2 + H_2O \longrightarrow 2\ HNO_3 + NO \qquad (1)$$

Nitric acid will readily react with sulphides and arsenides. The chemistry has been discussed elsewhere (1), and this chemical reaction can be simplified for sulphur:

$$S + 2HNO_3 \longrightarrow H_2SO_4 + 2NO\uparrow \qquad (2)$$

or: $$S + 3NO_2 + H_2O \longrightarrow H_2SO_4 + 3NO\uparrow \qquad (1+2)$$

Simplistically, the 3 NO_2 enter the liquid phase, oxidize one sulphide sulphur to sulphuric acid and form 3 NO. The gaseous NO has low solubility in water (like oxygen), hence it immediately leaves the process solution and re-enters the gas phase.

NO is not inert however, it readily reacts with the oxygen in air to form NO_2. This fresh NO_2 is promptly absorbed in the process solution, forming nitric acid. It in turn reacts with the refractory sulphides, releasing NO. In this way the NITROX cycle is established resulting in very efficient transfer of oxygen units to the refractory sulphide slurry.

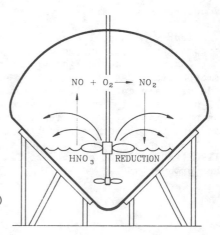

Written as a chemical reaction the oxidation of NO gives

$$3\ NO + 3/2\ O_2 \longrightarrow 3NO_2 \qquad (3)$$

or, for the overall reaction:

$$S + 3/2\ O_2 + H_2O \longrightarrow H_2SO_4 \qquad (1+2+3)$$

Fig. 2.   The NITROX CYCLE

It will be noted that neither nitric acid, NO, nor NO_2 appear in the overall reaction. The quantities of nitrate in the process solution are such, however, that the nitric acid must be seen as a reagent or intermediate in the NITROX PROCESS®, not as a catalyst.

CONTROL OF NITRATES

Having introduced nitric acid, and all its subspecies such as nitrites, NO, NO$_2$, etc., in the process to achieve the low cost oxygen transfer from air via the NITROX cycle, the economic and environmental realities now dictate rigid control of nitrates and all subspecies. Fortunately nitrates are extremely soluble and are not contained in the solids discharged from the NITROX PROCESS®. They are however contained in the free liquid content of these solids.

One of the obvious answers to the problem of nitrates in the spent oxidation liquor is the chemical removal of the nitrates. This can readily be done by adding excess concentrate, as shown in earlier published work by Kennecott personnel (10). This introduces the countercurrent concept, which tends to complicate the process. When treating arsenopyrites, it also may generate arsenites in the barren oxidation liquor. Arsenites cannot be precipitated as environmentally acceptable compounds. The NITROX PROCESS® therefor maintains a high oxidizing potential throughout the reactor, which results in a significant nitrate content in the spent oxidation liquor.

Displacement of nitrate bearing liquids contained in filter cakes requires nitrate free water. This water cannot be brought in from outside the process, otherwise the solutions would dilute and tanks fill up and eventually overflow into the drains. Containment of nitrates requires therefor strict control of the water balance, an unusual discipline for mill operators, but one that base metal refinery operators accept as a given.

Nitrate free water must be generated in the process if it cannot be brought in from an outside source. Earlier versions of the NITROX PROCESS® showed a stream of water vapour exiting the stack. Condensation of this water vapour could provide wash water. More recent testwork has indicated that acceptable wash water can be condensed from the gas stream after the NITROX reactor at a rate of about 2 tons of water per ton of Dickenson concentrate. This change in the flowsheet allows other economies and has therefore been adopted as the preferred alternative. It is shown in the flowsheet, shown as Fig. 3.

The reason that water can be condensed from the gas stream is simply that water evaporation is used to control the NITROX reactor temperature. The air supplying the oxygen for the reaction sweeps away water vapour at atmospheric pressure. There is no heat exchange

° C	kg H$_2$O per kg air at 100% saturation
53	0.1
71	0.3
81	0.6
84	0.8
88	1.0
93	2.0

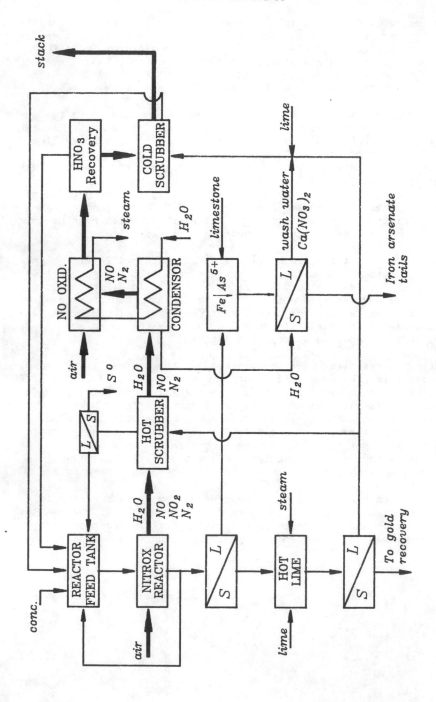

Fig. 3. Schematic flow diagram for Dickenson concentrate

equipment required, in fact the process would not work with such
equipment since the water balance could not be maintained. In a
gypsum saturated circuit, this must be considered as another fundamental attraction of the NITROX PROCESS®.

The 10 tpd demonstration plant will consume 22 tpd air and
evaporate 23 tpd water. If all the air was passed through the NITROX
reactor, the slurry temperature would be 92 °C.

CONTROL OF NOx IN THE STACK GASES

Nitric acid and its subspecies are very much represented in the
gas stream and need to be removed before venting. Removal occurs in
two steps, i.e. through nitric acid production and, subsequently,
through lime (stone) scrubbing. As explained under the earlier discussion of the NITROX Cycle, the bulk of NOx must be converted to
nitric acid to allow the sulphide oxidation to proceed. Scrubbing
with lime produces calcium nitrite from which nitric, or nitrous acid
must be regenerated with the sulphuric acid formed by oxidation reaction (2).

$$NO + NO_2 + \ Ca(OH)_2 \longrightarrow \ Ca(NO_2)_2 + H_2O \qquad\qquad (4)$$

$$Ca(NO_2)_2 + H_2SO_4 + 2H_2O \longrightarrow \ CaSO_4 \cdot 2H_2O + 2\ HNO_2 \quad (5)$$

Since only a limited amount of free sulphuric acid is formed,
lime consumption must be controlled within that limit and lime
scrubbing can only be done on the tail gases. The bulk of NOx must
go to nitric acid. Its regeneration in the NITROX PROCESS® has been
discussed in detail (1).

THE NITROX PROCESS® FLOWSHEET

The NITROX PROCESS® has been under development for over four
years, which is not particularly long for a new hydrometallurgical
process. This particular process has some features, however, that
have allowed fast track development. The most important one is the
separate extraction of the valuables (gold and silver), in a proven
process, i.e. cyanide extraction. Unlike other hydrometallurgical
processes, such as nitrate assisted oxidation of chalcopyrite for
instance, the NITROX PROCESS® makes no attempt to solubilize the gold
or silver, or remove gold or silver from the feed material during the
oxidation reactions.

During the initial process development, Hydrochem became convinced
that a demonstration stage would be very prudent, prior to the construction of a full scale facility. Hence, last year a consortium of
Canadian mining companies, and engineering firms was established

Fig. 4.   The bench scale NITROX Reactor

Fig. 5.   The 5 kg/h miniplant under construction

under the auspices of Hydrochem and the Northern Ontario Development
agency of the Ontario Government to design and cost a 10 tpd NITROX
PROCESS demonstration plant. At the Hydrochem facilities testwork
was stepped up from a bench scale reactor to a 5 kg per hour mini
plant.

Hydrochem also obtained a promise of supply of 10 tpd arsenopyrite
concentrate for the demonstration facility from Dickenson Mines Ltd.
Its Arthur W. White mine in the Red Lake district of northwest
Ontario scavenges a 1 oz Au/ton concentrate from cyanide leach tailings prior to discard. The concentrate is shipped overseas. Both
net back and recoveries have been low (11). In a separate
development, Hydrochem installed novel flotation equipment in the
White Mill, which has increased concentrate output substantially,
making supply of arsenopyrite concentrate to the demonstration plant
feasible without disrupting other commitments by Dickenson (12, 13).

For the flowsheet a Dickenson concentrate of the following
composition has been assumed:

Fe	As	S	Au
27%	10.8%	23.2%	0.93 oz/st

The flowsheet for this particular concentrate at a Northern
Ontario location would approximately follow the block diagram given
as the NITROX Flowsheet (Fig. 3). A description follows.

Arsenopyrite is slurried in calcium nitrate bearing filtrate, then
contacted with part of the reactor exit solution, containing ferric
arsenate and sulphuric acid. Gypsum is precipitated and the slurry
becomes acidic with nitric acid:

$$3Ca(NO_3)_2 + 3H_2SO_4 + 6H_2O \longrightarrow 3CaSO_4.2H_2O + 6HNO_3 \quad (7)$$

The NITROX PROCESS® uses the generated sulphuric acid to recon-
stitute nitric acid from the calcium nitrate in the filtrate.
Essentially, this allows nitric acid to exit the reactor as it must
to prevent formation of arsenites, enter the precipitation, convert
to calcium nitrate and be reconstituted to nitric acid in the reactor
feed tank.

The oxidized concentrate, now greatly reduced in weight since
iron, arsenic and sulphur are all solubilized, exits the NITROX reactor system and is filtered. It has been found that the precipitated gypsum greatly assists in the filtration of what essentially
is a gangue, elemental sulphur and gold residue. The solution goes
to ferric arsenate precipitation, where remaining sulphate and nitrate units are converted to more gypsum and calcium nitrate,
respectively. The calcium nitrate liquor is used to scrub the tail
gases before returning to the reactor feed tank.

The gas stream is also simple. Only part of the air needed for
oxidation enters the reactor system. This allows depletion of the
oxygen and NO_2 in the gas stream. After hot scrubbing of vaporized
nitric acid and trace amounts of NO_2, low nitrate water is condensed
from the gas stream to be used as wash water in the system as discussed earlier under "Control of nitrates". The balance of the air
is then introduced to recover additional HNO_3, which is also returned
to the reactor feed tank. Finally, cold scrubbing of the tail gases
with lime (stone) brings the NOx level to below that allowed for
stack emission.

THE NUISANCE OF ELEMENTAL SULPHUR FORMATION

The one single process characteristic, which necessitated extensive research and in fact, extended process development for the
Dickenson concentrate by a year, is the elemental sulphur formation
in the nitric acid oxidation of arsenopyrite, simplistically represented by the following reaction:

$$16FeAsS + 64HNO_3 \longrightarrow 16FeAsO_4 + 8H_2SO_4 + 24H_2O + S_8 + 64NO \quad (8)$$

Reaction (8) suggests half of the sulphur reports as S_8, a very
stable compound with a ringed structure. The production of elemental
sulphur from arsenopyrite is thought to fluctuate, depending on process conditions. Earlier work has reported as much as 70% of the
contained sulphur reporting as elemental sulphur (14). It probably
does not form as S_8 rings during actual oxidation, since this
requires a high redox of the mineral (15) but forms subsequently at
a lower rate in the oxidation residue.

The formation of elemental sulphur has several connotations for
the NITROX PROCESS®, the most important being a tendency to tie up
gold, or slowing down its extraction from the residue with cyanide
liquors. As important is its effect of increased cyanide consumption
through the formation of thiocyanates.

Rather than searching for process conditions to minimize elemental
sulphur formation, it was decided to deal with the nuisance and turn
it to advantage, if possible. This decision was based on the recognition that the NITROX PROCESS® must work for all feedstocks and this
might not be compatable with a narrowly defined nitric acid oxidation
regime.

It has been shown possible to remove the elemental sulphur from
the oxidation residue by both hydrometallurgical and physical
methods. Some of these will be described in detail. The physical
method of flotation removes both sulphur and gold, and the flotation
concentrate would require separate treatment. This could be hot
liming, or any oxidation method which would turn the sulphur in this
very high grade material into sulphate. Since S_8 is only stable up
to 159°C, high temperature oxidation was considered, but rejected as

inconsistent with the objective of keeping the NITROX PROCESS® as
simple as possible. Thiourea extraction was tried, but appeared
unable to break the gold/sulphur bond. Extractions were no better
than the cyanide ones, although further testwork is still underway.
Melting of the sulphur, followed by hot filtration, or solvent
extraction of the oxidation residue were also considered.

The simplest solution could be burning of the elemental sulphur.
In the NITROX PROCESS® this would take the form of a small, ex-
ternally heated rotary kiln filled with sand or mill tailings,
where the wet S bearing flotation concentrate would flow co-current
with air, be heated to 300-400°C, freed of sulphur and discharge into
the cyanide leach tanks. The air, with a small amount of SO_2, would
enter the NITROX reactor, where the SO_2 would convert to sulphuric
acid (as per the lead chamber process) and the heat would contribute
to water evaporation, i.e. slightly increase the amount of condensed
wash water. The combustion procedure has been tested, but time did
not allow evaluation of the effect of SO_2 in the NITROX reactor. The
effect of this calcining on the gypsum possibly contained in the
flotation concentrate also needs to be assessed. The present flowsheet
therefor incorporates a hot liming step to hydrometallurgically remove the elemental sulphur, since it was most extensively researched.

ELEMENTAL SULPHUR FORMATION WHEN TREATING PYRITE

Earlier references (10) indicated that nitric oxidation of pyrite
produces only a trace of elemental sulphur.This means that the NITROX
PROCESS® for a pyrite concentrate is much simpler.

Our testwork on commercial concentrate samples indicated this to
be correct, but to be certain the following testwork was carried out.

Coarse pyrite specimens were ground to 75% - 75µm and 200 grams of
the ground material were slurried in 2 1 of NITROX PROCESS® liquor
from previous work on Dickenson concentrate:

Fe^{+++}	$SO_4^=$	Redox	pH
39. g/1	27.1 g/1	+750 mV	0

The redox potential was measured against a calomel electrode, and
correlates with approximately 100 g/1 NO_3^-. Concentrated nitric acid
and ground concentrate were added at 85 - 90°C and a constant redox
potential of 750 mV. The last of the 326.1g HNO_3 addition was 45
minutes after the last concentrate addition and 2 hours before the
end of the experiment. In other words, oxidation simulated
Dickenson concentrate process conditions as closely as possible. The
200 g pyrite analyzed at 42.3% S and 46.5% Fe and yielded 15.09
grams of residue at 52.5% S^T, 18.4% S° and 35.3 % Fe, indicating
3.3% elemental sulphur yield and 94.3% iron dissolution.

The 3% elemental sulphur yield is identical to that measured by
the Kennecott team (10), indicating that Dickenson concentrate process liquor does not contain trace elements in sufficient quantity to
effect the pyrite oxidation mechanism. Silver ions have been shown
capable of modifying the nitric acid dissolution of Ni_3S_2 (16).
There is some indication from our work that silver may also effect
the sulphur yield from pyrite under NITROX PROCESS® conditions.

Due to the difficulty of obtaining pure arsenopyrite specimens,
similar work has not been carried out on this material.

ELEMENTAL SULPHUR FLOTATION EXPERIMENTS

Flotation at 80°C under NITROX PROCESS® conditions has been
shown capable of separating elemental sulphur from the oxidation
residue/gypsum slurry. The following describes the testwork.

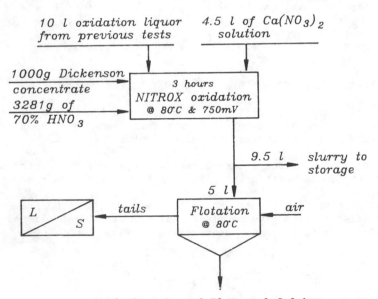

Acid Flotation of Elemental Sulphur

Flotation was carried out in a 150 x 150 x 300 mm stainless steel
flotation cell. Air at a rate of 9.5 1/minute was introduced through
a stainless steel bubble cap type sparger underneath a 50 mm diameter
hydrofoil impeller rotating at 1900 rpm. Froth was scraped every 5
seconds. After 15 minutes 96% of the sulphur and 90% of the gold was
collected in a 49% S° and 10.8 oz Au/ton concentrate. Test results
were as follows:

Wt. g	%S	oz Au/t	Minutes	
9.16	48.5	10.7	1	concentrate
2.76	57.4	11.3	3	concentrate
2.16	49.3	11.8	5	concentrate
0.85	27.0	8.7	15	concentrate
27.63	0.96	0.7	15	tails

From these results it is clear that gold is very extensively concentrated in material floatable under NITROX PROCESS® conditions. The results do not prove the direct link between gold and elemental sulphur, e.g. gold could be cemented on the few remaining sulphide particles or be free floating.

Separate smaller batch experiments were also carried out to investigate the removal of elemental sulphur from NITROX residues which, after acid filtration, were first cyanide leached and then subjected to flotation. This was done to eliminate flotation of free gold as a test interference, since all free gold would likely have been removed during cyanidation.

The cyanided oxidation residue was subjected to two successive flotations, the first after adding frother followed by a second flotation after conditioning with Na_2S, $CuSO_4$ and sodium isobutyl xanthate. The following diagram highlights the results.

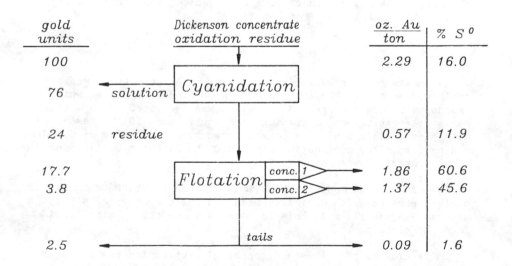

Cyanide Residue Flotation

Even after cyanidation, 90% of the residual gold follows the floatable fraction of the residue, which is largely elemental sulphur. In the next section it will be shown that removal of elemental sulphur allows near complete extraction of gold by cyanidation.

HOT LIME TREATMENT

The dissolution of elemental sulphur in hot lime solution was already known to the ancient Greeks. The resulting liquid was called "thion hudor" i.e. "sulphurous water" by Zosimos (A.D. 250) (17).

To dissolve sulphur away from oxidation residue was found to be fairly easy. To prevent gold from dissolving at the same time without using excessive amounts of lime required a good deal of testwork. The following describes one such experiment.

A three stage NITROX oxidation was conducted with recycle of solutions over a period of six days. Oxidation residues produced during the latter part of the test were estimated to contain approximately 18% elemental sulphur. Hot lime treatment at pH 10.0 -10.5 and 80°C produced residues with less than 0.1% S°. Gold dissolution during hot lime treatment was 6.6% and 4.3% for day 5 and 6 residue respectively. Subsequent 48 hour cyanidation at pH = 10.5 and 1 g/l cyanide extracted 97.5% and 98.0% respectively of the contained gold. Similar cyanidation without the hot lime step extracted only 75% of the gold in 48 hours. Increasing the cyanide level to 5 g/l increased gold extraction from 75% to 85%.

SUMMARY

Extensive testwork was carried out over the past three years on applying the NITROX PROCESS® to Dickenson Mines' Arthur W. White Mill arsenopyrite concentrate. A flowsheet was developed to form the basis of a capital and operating cost study for a 10 tpd demonstration plant. This study was commissioned by a consortium of Canadian mining companies, engineering firms and the Northern Development Fund of the Ontario Government.

One of the fundamental attractions of the NITROX PROCESS® is the minimal cost at which oxygen is made available to the refractory material in the slurry. This is done by introducing nitric acid into the circuit.

In developing the flowsheet, particular attention was therefor paid to control of nitrates and NOx in the solid and gaseous effluents, respectively. For instance, the NITROX PROCESS® was designed to generate wash water within the plant perimeters to control the water balance and prevent dilution of the process liquors.

Elemental sulphur generated by the action of nitric acid on arsenopyrite proved a particular nuisance, since it tied up gold and increased cyanide consumption. A number of process options to deal with this elemental sulphur are discussed. The hot lime dissolution of elemental sulphur prior to cyanidation overcomes these problems and is incorporated in the flowsheet. Low temperature combustion of this sulphur in NITROX PROCESS® air deserves further consideration since it promises to be an even simpler, and hence, lower cost option.

ACKNOWLEDGEMENTS

The authors thank the NOR-DEV organization, the National Research Council, the COSCH Group and the members of the Pyrite Processing Consortium, including Fenco-Lavalin, for their financial support. Special thanks go to Dickenson Mines Ltd. for the supply of concentrates, to Malcolm Downes, Leslie Hendry and Peter G. Turylo for their assistance with the experimental work and John C. Schneider for his counsel and encouragement.

REFERENCES

1. Van Weert, G., Fair, K.J. and Schneider, J.C., 1987, "The NITROX PROCESS® for treating gold bearing arsenopyrites", Presented at the 116th Annual TMS/AIME Meeting, Denver, CO.

2. Cranstone, Donald A., "The history of gold discoveries in Canada - an unfinished story." CIM Bulletin, October 1985, pp 59-65.

3. Springer, Janet, Invisible Gold; 1983, The Geology of Gold in Ontario, A.C. Colvine, ed., Ontario Geological Survey, Miscellaneous Paper 110, pp 240-250.

4. Van Weert, G., 1988, "An update on the NITROX PROCESS®", Presented at the Randol Gold Forum '88, Scottsdale, AZ, January 23-24, 1988, Randol Gold Forum '88, pp 209-210.

5. Robinson, P.C., "Mineralogy and treatment of refractory gold from the Porgera deposit, Papua New Guina", Trans. Inst. Min. Metall., vol. 92, June 1983, pp. C83-C89.

6. Weir, D. Robert and Berezowsky, Roman, M.G.S., "Gold extraction from refractory concentrates", paper presented at the 14th Annual Hydrometallurgy Meeting, CIM. Timmins, Ontario, October, 1984.

7. Crozier, Ronald D., 1988, "Arsenic and the sulphuric acid story", in press, Industrial Minerals (London).

8. Henzler, H.-J. and Kauling, J., "Scale-up of mass transfer in highly viscous liquids", Paper presented at the 5th European Conference on Mixing, Wurzburg, West Germany, June 10-12, 1985.

9. Demopoulos, G.P., 1988, "Liberation of refractory gold by aqueous oxidation methods", Lecture given at the Met Soc. CIM/Laval University, Short course "Possible technologies for gold and precious metals", June 2-4, 1988.

10. Prater, J.D., Queneau, P.B. and Hudson, T.J., "Nitric acid route to processing copper concentrates", AIME Trans., vol. 254, (1973) pp. 117-122.

11. von Michealis, H., 1987, "Shipping concentrate to smelter", Gold and Silver Recovery Innovations - Phase III, vol II, Randol International, Golden, CO, pp. 1105.

12. Connell, L.J, Schneider, J.C. and Sass, G.D., 1987, "Performance of Prochem's new flotation cell at the Arthur W. White Mine in Red Lake, Ontario". Presented at the 89th Annual General Meeting of the Canadian Institute of Mining and Metallurgy, Toronto, Ontario, May 3-7, 1987.

13. Van Weert, G., Contini, N.J., and Knopp, R., 1988, "Startup experiences with the Hydrochem flotation column on auriferous arsenopyrite". Presented at the 90th Annual CIM meeting, Edmonton, Alberta, May 8-11, 1988..

14. Fair, K.J., Schneider, J.C. and Van Weert, G., 1987, "Options in the NITROX PROCESS®," Proceeding of the International Symposium on Gold Metallurgy, Salter, R.D. ed., Pergamon Press, Toronto, pp. 279-291

15. Peters, E., 1988, private communication.

16. Mulak, W., 1987, "Silver ion catalysis in nitric acid dissolution of NI_3S_2", Hydrometallurgy, 18, 1987, pp. 195-205.

17. Partington, J.R., 1954, General and Inorganic Chemistry, MacMillan, London, p. 693.

Chapter 21

CENTRIFUGAL CONCENTRATION AND SEPARATION
OF PRECIOUS METALS

by Byron V. Knelson

Inventor, Manufacturer and Distributor
of the Knelson Concentrator

ABSTRACT

Although the concept of centrifugal recovery has been around in one form or another for many years, it has not been until recent times that it has become clearly understood and properly put into perspective. Simplistic in theory, it is when one begins to examine the variables involved that the true complexities of this matter begin to emerge. In creating an enhanced g force concentrator, there are many aspects both positive and negative that are encountered. Many conditions that are inherent to all conventional 1 g concentrators do not apply to centrifugal concentrators and therefore very few common factors exist between them. In developing this technology, virtually all principles were proven by method of trial and error. Among the topics discussed are the effect of g force enhancement on particulate behavior with respect to size and shape, density, enrichment, fluidization, and the "teeter principle". In addition, we will discuss the physical effects of g force enhancement with respect to manufacturing in the areas of abrasion, bearings, and vibration. In discussing the recovery parameters of centrifugal concentrators we will draw direct comparisons to various other recovery methods, outlining the conditions most favorable to their application.

INTRODUCTION

In order to do a reasonable job discussing particulate behavior in centrifugal recoveries it is necessary to first discuss a few of the best known methods of recovery. I choose the sluice box first because after the gold pan it is the most basic tool and probably that from which all others emanate.

SLUICE BOX

Basically a sluice box in it's simplest form is a man-made creek bed. In the writer's opinion it falls short of a creek bed in the following ways. It does not have the natural swirls, eddies, drops, shallows, turbulent areas, calm areas, change of flow nor length - all of which will drop various types, sizes and shapes of gold during all the millions of combinations possible with the afore-mentioned factors. Any system with specific gravity as its deciding factor tries to create a laminar flow in order to concentrate the denser particles. In a sluice box, as in most gravity systems, removal of oversize particles is essential to recovery of the finer fractions of valuable product for laminar separation is impossible with much oversize in the gangue. So now what happens to particles entering the system?

1) The large and dense particles (nuggets) have the ability to physically displace smaller particles and larger less dense particles so they drop out almost immediately.
2) The smaller but still very dense particles also drop out very quickly - but generally a little further down the box.
3) While this has been happening the total gangue and water are picking up speed. How many times have you heard the old saw "We know we are catching all the gold because after the first five feet there is practically no gold in the sluice box"?
4) As the speed increases, all laminar flow is destroyed by the accompanying turbulence.
5) At some point speed and turbulance combined stops all concentration and everything continues through the box. At this point it is no longer a sluice box, it is a transfer box.
6) Any particles, whether they are gold or just rock, that do lodge in a sluice box usually very quickly become imbedded with black sand and none but the larger particles of gold can penetrate this non-fluidized bed. A simple rule with almost any 1 g device is that if you are losing black sand, you are losing fine gold.
7) The foregoing are the reasons that virtually 97-98% of all gold caught in a sluice is caught in the first 3 - 5 feet.

Critical to this whole separation system is the shape of the gold particles. As a general rule it runs this way.

The larger the particle - The easier to catch
The lesser the aspect ratio - The easier to catch
The smaller the particle - The harder to catch
The greater the aspect ratio - The harder to catch See Figure 1.

If you have large particles with small aspect ratio, you have to work to lose them. If you have small particles with large aspect ratios, you'd better have a good device to catch them and that does not include a sluice box.

It is the writer's opinion that within a very few minutes of operational start up a sluice box is decreasing its efficiency and shortly thereafter starts to select and retain only on the basis of size (nugget). The greater the black sand content the quicker this happens. If the black sand is accompanied by large amounts of clay, impacting and blinding occurs, which accelerates this loss of efficiency.

Figure 1. Weight vs. Area

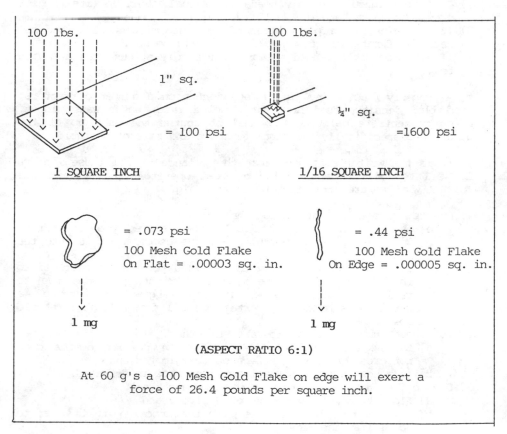

At 60 g's a 100 Mesh Gold Flake on edge will exert a force of 26.4 pounds per square inch.

The start of decrease can be as little as 10 minutes and again it is the writer's opinion that the retention of fine gold is almost non-existent in most sluice boxes after four hours and I am being generous at that.

JIGS

It is often hard for people to understand exactly how jigs work. For what it's worth, I will put forth the writer's understanding of the principles.

Simply stated a jig is a device which has a couple of basic advantages over a sluice box.

1) It is continuous and only needs occasional clean up (its biggest drawing card to hard rock mills).
2) It maintains, **if properly balanced**, an intermittent fluidized bed capable of retaining finer gold than a sluice box.
3) If not properly balanced it can be more disastrous than a sluice box.

A properly functioning jig is dependent upon a number of factors; feed flow density, feed flow rate, pulse rate, hutch water input, hutch concentrate delivery flow and of course the make up of the gangue and specifically the particle shape and size of the gold.

It is the writer's opinion that the balance of all those factors is well beyond the capabilities of most operators. However, lets look at what we are attempting to do in a jig.

1) Create a cross flow on the top that:
    a] is fast enough so that only the desired product and as little as possible worthless gangue accompanies it into the jig hutch and
    b] is slow enough to allow the desired product to settle into the ragging without being rushed right over the top to tails for, once trapped in the ragging, capture of a particle is reasonably certain as all forward motion should now be defeated.
2) Put enough hutch water in to make sure that:
    a] the negative pull of the diaphram or pulsing device does not create a sucking effect on the jig bed and
    b] it's not so much that all settling rate is defeated.
3) Draw off enough water with the hutch product to:
    a] flow it to whatever point of delivery and
    b] at same time balance this with the incoming hutch water to accomplish the aims of 2) above.

If we analyze the foregoing statements, we must realize that if the incoming hutch water rate is too large, then the upward thrust during the positive stroke will be too severe and will push fine gold particles upwards and not allow them to settle out and if the incoming rate of feed and water is too large, then forward motion will be too fast and again not allow fine gold particles to settle out. Most of us know from experience that the settling rate of fine gold is quite slow so we can assume that the settling rate on the negative pulse of the jig must overcome the forward movement turbulance of the gangue; for on the positive pulse, no fine sized gold particles will settle out. We can be quite certain that forward motion takes place on both negative and positive strokes albeit at different speeds depending on the portion of the pulse cycle it is in and depending on where the particle is in relation to the feed entrance and tails exit of the jig cell - for the flow rate speeds up the closer you get to the tails exit and in the same manner as a sluice box at some point the speed of the gange destroys recovery of fine gold. One assumes that forward motion will be much reduced in the very last portion of the negative pulse but that settling rate is ranging from positive to negative in the positive pulse of the jig, and if there is the slightest bit too much hutch make up water the settling rate will go to the negative side sufficiently to stop all fine gold recovery and in the writer's experience, some quite coarse gold recovery (up to 10 mesh in size).

*Interstitial trickling is the redeeming feature of jigs for fine particle recovery.   At the end of each settling cycle, coarse particles will bridge together first, and come to rest.   Finer particles will then trickle through the interstices of the larger one, and come to rest much later.

Flaky particles are generally not well recovered by jigs. Firstly, their terminal velocity is much lower than that of a spherical particle of equal mass.   Secondly, their shape restricts their motion during the trickling phase.

Jigs are often used in gold concentration. Their effectiveness, however, rapidly decreases below 300 um (50 Mesh). This seriously limits their use as sole recovery unit. *LaPlante (1988).

An interesting configuration is the placing of two identical sized jig cells, one behind the other. One of those jig cells absolutely has to be out of sync. If the first one is balanced, it will have sufficient hutch water added that the following combination will be in effect. In jig #2 solids will be correct but liquid content will be far too large, thus the forward speed of the total will be too fast for effective settling. Conversely, if volume and flow are correct for the second jig, then forward motion will be too slow in the first unit and either too much product will go to the hutch or blinding of the ragging and jig bed will take place.   In essence, the same can be said of jigs as of sluice boxes. The large and dense particles will almost immediately drop out of flow and become part of the ragging. Like all 1 g systems the finer the gold and larger the aspect ratio the less likely you are to catch it. Jigs are not a fine gold recovery unit.   Simply stated, unless you can find a way to change the laws of the settling rate in relation to the laws of the flow rate you cannot improve the recovery beyond some fairly definite limits. The laws governing settling and laminar flow are quite absolute.

SPIRALS

The author considers spirals to be one of the better concentrating devices although in truth they are not a concentrator but
rather a rougher. They are quite ingenious in that they have no
moving parts and you can physically see what they are doing.

Spirals are a fine calculation of gravitational pull on the
downward direction and the centrifugal force created by channelling
that downward force in the curve of the spiral. All of this creates
a visible and true laminar flow. No other device will handle such a
large volume of material and retrieve reasonably fine gold while
allowing you to see the actual workings of the unit at the same time
and they are continuous.

Particle behaviour is fairly east to understand. Heavy dense
particles create the most friction; therefore, they move down the
spiral at a slower speed. Consequently, they do not have the energy
or force applied centrifugally to throw them to the outside and will
remain close to the center where the primary cut is taken. Particles
thrown to the outside will be caught in the middling or tails cut.

However, classification is absolutely essential to recovery with
spirals. 10 mesh is pretty well maximum and although I stated they
have no moving parts - that is only if you don't count the miriad of
electric motors, pumps and impellors required to keep the various
flows accurate, as flow rate and density are critical. At any rate
they are a flowing film separator and LaPlante has this to say about
them:

*Flowing film separators can achieve large capacities, especially
when film thickness is much larger than particle thickness and
separator geometry optimized. The best known unit is the Reichert
cone, which has capacities of 60-90 t/h for the 2m unit (surface:
2.58m²) and up to 400 t/h for the 3.5m unit (surface: 8.84m²). The
drawback of such units is their relative inefficiency, both in
rejecting light particles and recovering heavies. Many separation
cycles have to be used. This difficulty is partially alleviated by
grouping individual cones (up to 12) in units which yield relatively
good recoveries, but at a low upgrading ratio. *LaPlant (1988)

As with all 1 g systems if you lose black sand you are losing
fine gold. Spirals are subject to the same laws as other systems
but will generally retrieve finer gold than jigs or sluice boxes
provided proper classification and flow rates are maintained.

Virtually all 1 g systems are subject to high losses if the particles:

Are Fine (below 110 mesh). Because spirals very quickly create a thin veil of water and material - fine gold if allowed to get on the surface will tend to boat its way right out.

Have a high aspect ratio. Very thin flat flakes tend to flip and in so doing will move to the outside and be lost. Also in flipping these flakes can get on the surface and boat again going to tails as once on the surface the centripetal effect of the spiral is lost.

Have a low specific gravity due to inclusion of less dense material in the total particle. With a low specific gravity the particles will move to the outside and again be lost. If there are in fact values other than free gold trapped in the black sand, then spirals are a great way to go as one can retrieve virtually all of the black sand on a continuous basis but if you are after free gold - that is gold that is fine, flaky or contaminated to the point of a low specific gravity - there are better ways to go. You seldom find gold that does not have at least two if not all three of these factors.

TABLES

Suffice it to say that tables are subject to virtually all of the same laws as spirals. Tables also allow you to see and closely guide the delivery of your product and everyone knows they have been used for clean up for many years but in creating this final concentrate they also are incapable of any real volume through-put, and seldom are very effective on gold under 130 mesh.

CENTRIFUGAL UNITS

There are two basic types - single walled rotating cones and centrifuge units with fluidized beds.

Centrifugal units with single walled rotating cones

The first type, which have been around for about sixty years, are single walled rotating cones. There are at least eight different brand names but they all have the following basic specifications:

1) A cone of about 20 inch diameter will spin at approximately 125 R.P.M.
2) Create about 4.5 forces of gravity.
3) Handle about 5 to 6 tons per hour of minus 1/4 inch feed gangue.
4) Should be cleaned up every 1/2 hour maximum. Behaviour of the gangue in these units is fairly predictable. The gangue must be reduced to a maximum size of minus 1/4 inch particles. Reducing to minus 1/8 inch will generally give you a better recovery.

The first material introduced will fill the inner rings and all subsequent material will have to displace light particles with more dense particles in order for concentration to take place. If black sand is present even in reasonably small amounts, they will very quickly concentrate on the innermost 1/8 inch of the material surface. Once this layer is complete these units will increasingly move away from fine gold recovery and select by larger size and density.

It is the writer's experience during a two year period of research with this type of unit that effective recovery seriously down-grades within 10 to 12 minutes of start up and that virtually no fine gold will be recovered below 100 mesh after 2 hours of operation. This two hours can be reduced to as little as 20 minutes in a material with a high content of black sand and fine gold. This type of centrifuge has some ability to recover gold finer than that by jigs, sluice boxes or spirals but their short function re ability both time-wise and volume-wise overshadows any other positive attributes of this type of unit.

Simply stated, compaction in the rings begins almost immediately and continues to impact harder and harder the longer the unit is run. As with all concentrating devices the dense and large particles go to concentration in the first available space - in sluices in the first few riffles, in jigs in the first portion of the jig bed, in centrifuges in the bottom ring.

Centrifugal Units with Fluidized Bed

Now to the second type of centrifugal recovery devise which, just in case you miss it, I will tell you in advance that I'm just a little prejudiced about. The fact that it bears my name of course has no bearing on that prejudice.

*Centrifugal separation of gold experienced a regain in activity with the advent of the Knelson Bowl. This device appears extremely efficient in recovering fine gold, although very little published data are available.

In a mill in the Chibougamau, Quebec area, installing a gravity plant increased overall gold recovery by about 3-4%. Gravity concentration was achieved with pinched sluices and 76cm Knelson bowls, with a significant increase in fine gold recovery. *LaPlante (1988).

Let me briefly cover that point. There is only one true machine of this type - the KNELSON CONCENTRATOR.

On to the forces involved. Let me say that if I had unlimited time I would have explained in detail where the functions in this machine are similar to those in other type of centrifugal units. Since this unit is the one which I am most familiar with I am able to discuss the movement of the gangue in fairly complete detail.

Again, the first material in fills the rings completely. At this point any similarity in the two centrifugal type concentrators begin to separate as the unit greatly enhances and improves the centrifugal factors.

1) A 30 inch Knelson Concentrator spins at approximately 400 rpm's
2) Creates 60 forces of gravity
3) Does not allow compaction
4) Will process up to 40 tph of minus 1/4 inch material
5) In typical alluvial operations will operate up to 24 hours without clean up with no increased loss of gold

In order to understand what happens with a particle we must return to the sluice box for a moment. We realize that as the material heads down a sluice box it picks up speed. In fact, in a rough box, it will get up to nearly 30 mph. The closer it gets to full speed the further it is away from the speed of the container which is zero.

When a particle in a Knelson Concentrator hits the bottom of the cone, it starts to gain momentum and converse to a sluice box, as it gains in speed it gets closer to the rim speed of the cone which happens to be approximately 30 mph. thus if we take the top speed of both as a factor, in the sluice box we have a 30 mph of forward speed with 1 force of gravity. In the Knelson Concentrator we have a forward speed of 30 mph. with 60 forces of gravity.

If both units are reduced to 1 g, then the comparison would look like this:

	Gravity	Forward Speed
Sluice Box	1	30 mph
Knelson Concentrator	1	.5 mph

Conversely, if you raised the sluice box to match the Knelson, the comparison would be as follows:

Knelson Concentrator	60	30 mph
Sluice Box	60	1800 mph

The foregoing shows an immense settling force in relation to the forward speed in the Knelson Concentrator. See Figure 2.

As in all gravity devices, the larger gold particles will go to concentration in the bottom rings or first area available: as the gold particles decrease in size, they will become harder to concentrate and so will be contained further up the rings.

Figure 2.

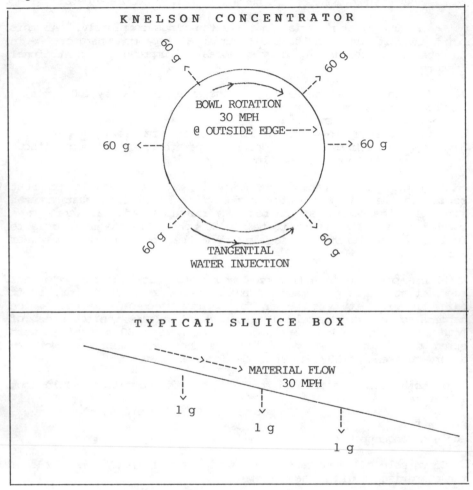

No compaction will take place because of water being injected
from behind the material trapped in the rings. This results in even
the finest particles being able to inject in and displace a less
dense particle of the same general dimensions or volume. Even
extremely fine and flat particles with very large aspect ratios will
inject in and be concentrated. At some point the following situation
will take place. The thin edge of the particle will be presented to
the moving water thus offering little resistance to the water and a
90° thin edge of this same particle will be facing outward toward
the area of concentration. When this happens at 60 forces of gravity,
that particle will go into concentration like a bullet.

Clay balls can be of several types but for our purpose here we can cover it by discussing the most common two.

First: clay particles that have not been reduced to their smallest particle size by whatever wash and scrub system is used. Most one gravity systems will not break this up in the concentrating mechanism and any gold trapped inside will continue on to tails.

Second: (the most annoying) is the plasticene type clay that does, in fact, form into a true ball and in a sluice box actually gather material including gold as it rolls down the box. They are commonly called box robbers. Neither of these types of balls survive in a Knelson Concentrator for the following reasons:

1) Material entering the Knelson Concentrator is travelling vertically but on the horizontal it is dead still. It must both
   change directions and by friction be forced to attempt to attain
   the 30 mph speed of the container. At the edge it changes
   directions again going up the inclined wall of the concentrator
   and as it passes each ring it has a new speed to attain. All of
   this creates the most abrasive material-on-material rub of any
   concentrating device in existence; and, like a centrifugal pump,
   only more so, it will destroy both types of clay balls.

2) The second reason concerns only the plasticene type of clay. At
   60 g forces this type of clay cannot maintain its round shape on
   impact with any solid particles. Result is that these balls
   impact on the wall where they are flattened by the immense g
   force and are then abraded away by subsequent material flow.
   Clay does not affect concentration here as the injected back
   pressure water constantly dilutes the gangue at each ring, so the
   liquids to solids ratio is being increased as the gangue moves
   through the concentrator and no blinding takes place.

An interesting point is that I am frequently asked how a 10 um particle of gold can possibly dislodge a 6 mm particle of rock when the 6 mm particle obviously has much more total weight than the 10 um particle of gold. The answer is: it doesn't. What happens is that any particle of gold only needs to dislodge a particle of like size but lesser density and since gold has a sg of 19.3 and most rock is around 3, it happens fairly easily. Something you should understand is that water is not only injected from behind the material trapped in the rings but it is also injected tangentially in the opposite direction to rotation (see figure 2.). This assists in keeping the concentrate shifting within the container and as long as fluidization and shifting of material takes place, so will concentration of dense particles (gold, platinum etc.).

*Water injected tangentially was a major breakthrough as it helps to keep the material moving within the unit. Most people have done some panning and will agree that if you just gently roll a pan and tip it far enough to wash some of the material over the edge, you

will in effect create the same performance as a stream on alluvial
gravels. You will create a surface enrichment but you will not move
the heavies to the bottom of the pan. If you are to move the heavies
to the bottom of the pan, you are going to have to move that pan
violently enough to make all of the material move within or in
relation to the pan. By doing so, you cause material shifting and
as it shifts, the heavier particles will go to the bottom. This is
the same effect we strive for in the concentrator. We don't want
the material to ever attain the same speed as the cone because if it
does, it will be stagnant, so by injecting the water in the reverse
direction to rotation, we help to keep the material in a constant
fluid motion thereby allowing any particle that is heavier than what
is in the rings to inject in and displace something that is lighter.

In creating tangential injection, we got a fringe benefit we
never even thought about. When you shut the machine down to drop
out the concentrate, you simply turn on the back pressure water and
it immediately blows everything off the rings and you only need to
hose out the bottom ring. This cuts cleanup time from thirty minutes
to five minutes. *Knelson (1985).

With regard to black sand one must understand that this unit does
not recover by seeking a given specific gravity. What it does is
magnify by a figure of 60 the difference in specific gravity between
all particle densities in the gangue. It is this magnification of
the differentiating factors which explain the machine's ability to
effectively recover free gold particles far beyond the parameter of
the aforementioned 1 g devices Consequently, the Knelson Concentrator
can handle virtually any content of black sand or other material
with a high specific gravity and still recover a very high percentage
of the free gold contained in a given product. Because gold and
platinum are two of the materials at the high range of the sg chart,
they will concentrate ahead of mercury, lead, scheelite, black sand
etc. in the Knelson Concentrator.

See Figure 3.

THE TEETER PRINCIPLE AND FLUIDIZATION

Truly, to discuss the teeter principal one must include
fluidization as the two are very closely interacting in the Knelson
Concentrator. If we return for a moment to my earlier remarks about
the use of a single walled rotating ribbed cone and the fact that
concentration only takes place in the first 1/8" on the inner face
of the material in the cone - all material behind this simply packs
because no fluidization takes place. In the Knelson Concentrator
packing is avoided by injecting water through the walls of the cone
and totally fluidizing the bed 100% of the time. This now is where
the teeter principle comes in for if too little water is injected,
then packing occurs and you now have a similar situation to a Knudson
bowl; conversely, if you inject too much water, you will blow out
the material and again concentration is stopped.

Figure 3. Forces of Gravity vs. Weight of Material

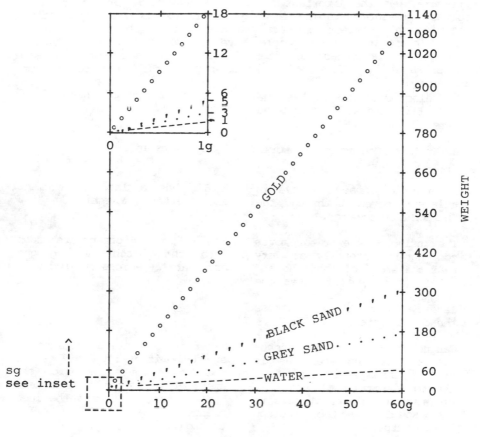

FORCES OF GRAVITY
vs.
WEIGHT OF MATERIAL

This balance was at first thought to be attainable by the use of pressure reducing valves; however, as knowledge of the teeter effect (i.e. correct pressure and the balancing of injected water versus the flotation of the material bed) in the machine was gained it became evident that what was required was a free flowing constant aperture (i.e. ball valve) and a good quality pressure gauge. Once a correct pressure was located for a given material it soon became evident that as heavier material entered the bed a larger pressure registered on the gauge but by using a constant valve opening the water volume entering the machine remained constant. It now became evident that the pressure on the gauge was only an indicatior of the volume of water going by the gauge inlet and that volume, not pressure, was the important factor. We now know that there is a correct pressure (volume) for any material and with very little practice this correct pressure (volume) can be located by even a novice operator. Some people have stated that the Knelson Concentrator is subject to interstitial trickling in much the same manner as a jig. This is not so as several different forces come into play.

A Knelson Concentrator has the bed fluidized 100% of the time.	A Jig is fluidized only 50% of the time.
A Knelson Concentrator works at 60 g forces.	A Jig works at 1 g.
Material is caused to move in a lateral movement through the Knelson Concentrator.	A Jig has a lateral movement and a vertical movement.
The Knelson Concentrator moves concentrate laterally at very near the speed of the cone and at the same time it moves the gangue being treated in a lateral vertical plane across the concentrate face.	A Jig moves concentrate down into the hutch and gangue horizontally across the bed.
Gangue in the Knelson Concentrator picks up speed as it travels through the unit and the more centrifugal speed it picks up the closer it gets to the speed of the Knelson Concentrator.	Material in a jig also picks up speed horizontally but as it does so it moves away from the speed of the jig which is zero and the faster it goes the less chance there is for concentration.

Interstitial trickling is only possible where a bed in effect goes to a passive state and also where the fines can in fact escape through the material.

A summation on the two units would be as follows: a jig allows interstitial trickling to settle out fine particles at some point in the 50% of the time that the bed is passive and before the next positive pulse. A Knelson Concentrator creates material trading (heavy in / light out) 100% of the time as fluidization is constant, balanced and steady and the bed is never allowed to rest so material trading is a constant concentrating function.

ABRASION

Single particles are first subjected to a friction type energy transfer to bring them up to speed and then, after leaving the cone, strike the hull at 30 mph in a slicing tangential motion.   It appeared to be almost impossible to protect the units as we could wear the bottom of the cone out in two weeks and nearly cut the top off the machine in two days. Rubber liners failed to do the job as the silica, water, slicing action combination simply sliced the rubber out. Simply stated, polyurethane of a special design was our savior. It seems to almost be impervious to the foregoing actions and only in the toughest, most abrasive of situations, does it show any signs of wearing at all. We now pre-cast complete abrasion kits and they can be retrofitted to any Knelson Concentrator built since 1983.

BEARINGS

This was a problem because original engineering followed the book and didn't take into account the tremendous fulcrum forces that are applied to a drum suspended at one end from a single shaft. However, the problem was cured when we brought in bearing specialists who utilized their expertise and came up with a double row, spherical pillow block bearing in a one piece housing which has effectively solved our problems.

ROTARY JOINTS

These presented a problem parallel to that of the bearings.   As is often the case, due to the poor quality of the fluidization water and the fact that this water is injected under pressure through a stationary shaft and into a rotating shaft, the seal between these two shafts is highly critical. Our first experience was with ceramic type seals which were quickly destroyed by the abrasives suspended in the water. This problem was solved with the advent of tungsten carbide type seals.

VIBRATION

This was only a problem for a short while. Vibration isolators resembling two donuts resting one on top of the other takes out virtually all vibration that is not removed by balancing.

REFERENCES

Knelson, B.V. "Centrifugal Concentration and Separation of Precious Ores" 17 Annual Meeting of the CMP, Ottawa, 1985, p.p. 346-357

Laplante, A.R. "The Use of Gravity Concentration for Gold Recovery" Part I - Economic Rationale and Basic Principles. Presented at the professional development seminar on small-scale gold projects: Mining, Processing, Economics and Policy Framework. McGill University May 2-6 1988.

Chapter 22

IMPROVED RECOVERIES OF GOLD FROM PYRITE AND CALCINE
BY THE METPROTECH FINE MILLING PROCESS

Keith S. Liddell

Metprotech Pacific
Hong Kong

ABSTRACT

Gold in many ores occurs in sulphides, which, if the gold is
finely disseminated, renders the ore refractory and therefore not
amenable to the simple process of milling and cyanidation.
Traditional methods of treatment of these refractory ores is to
produce a sulphidic flotation concentrate which is roasted to
form a porous calcine from which the gold is recovered by
cyanidation. The roasting process imposes a constraint of a
minimum sulphur grade of the concentrate on the flotation plant,
which may lead to a lower gold recovery being achieved in the
flotation section. Thermal decomposition of the concentrate by
roasting means that the sulphur contained therein has to be
treated and disposed of in an environmentally acceptable manner;
and, if arsenopyrite is present in the concentrate the environmental aspects will be worsened. It has long been recognized
that finer milling of gold-bearing materials improves the
liberation of gold, especially when the gold particles are very
fine.   After extensive laboratory and pilot plant investigations
(Liddell, Patents), a suitable mill has been developed for the
mining industry by Metprotech, which is capable of producing very
fine grinds (down to 90 percent passing one m if necessary). In
the course of the investigations it has been found that a large
number of gold-bearing materials are amenable to fine milling. A
specific advantage that fine milling of sulphidic materials has
over oxidative processes such as roasting and bacterial leaching
is that the sulphides are not decomposed and there is therefore
not a sulphur disposal problem. Furthermore, the milled
sulphides can become a future sulphur feed source, and can be
reclaimed if necessary.   A further feature of the Metprotech
process is that it is possible to add cyanide to the mill feed

318

slurry so that dissolution of the gold occurs within the mill. This results in a large proportion of the gold being dissolved while the slurry is in the mill, which has an advantageous influence on the cost of a subsequent gold recovery plant.

This paper will present the results obtained when a pyrite concentrate and a calcine leach residue were milled in a laboratory mill.

FINE MILLING OF A PYRITE CONCENTRATE IN A LABORATORY MILL

Sample Description

The pyrite concentrate contained 6.89 g/t of gold. The particle size distribution was determined by a Malvern laser diffracto-meter, and the J_{90} (the size at which 90 percent passes), J_{50} and J_{10} were 110, 45.1 and 5.4 m respectively. A cyanidation for 24 hours at a slurry solids content of 50 percent by mass yielded a gold dissolution of 13.5 percent.

Description of the Laboratory Milling Unit

A Metprotech laboratory stirred ball mill was used in a batch manner, with the milling time being varied so as to produce a number of different product size distributions; the mill is shown schematically in Figure 1. The grinding vessel has a height to

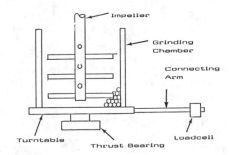

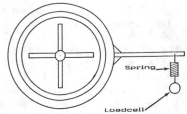

Figure 1. Schematic diagram of the laboratory mill, showing the method of torque measurement

diameter ratio of one, and has a jacket through which cooling water is passed. The vessel is mounted on a turntable which is located on a thrust bearing. The turntable is restrained from rotating by an extension arm which is attached to a load cell; with this arrangement the torque is measured, and hence the energy consumption can be determined. The vessel is filled with grinding medium which is agitated by a vertical shaft that has a number of arms protruding from it. The shaft is driven by a 1.5 Kw motor, through a variable speed gearbox, with the final drive being belt and pulleys; the speed is variable between 70 and 500 rpm.

Experimental Procedure

Four tests were performed, with the milling time being varied, the time periods being 8, 14, 25 and 40 minutes. The grinding medium was ceramic balls, having a top size of 6.4 mm. A slurry of 45 mass percent solids was added to the mill, the volume being sufficient to fill the voids of the media. Sufficient sodium hydroxide was added to the slurry to raise the pH to 11.5.

At the end of the desired milling period the load was emptied out onto a sieve and the medium washed with sufficient water to dilute the solids content to 35 percent. The slurry was then aerated for one hour in a laboratory flotation machine, operated with the air cock open. The slurry was then split into four portions and placed in 5 l winchester bottles to which sodium cyanide was added and leaching initiated. The leach times used were 1, 2, 4 and 24 hours. After leaching the slurries were filtered, washed and dried, and the gold content of the residues determined by fire assay.

A small sample of slurry was removed before aeration and the particle size distribution determined by a Leeds and Northrup small particle analyser.

Discussion of Results

The results are summarized in Table 1, where the product size distributions, gold dissolutions and specific energy consumptions are given. The specific energy consumptions determined from the laboratory mill have been recalculated to correlate with the performance of a 150 l continuous pilot scale mill, using factors determined from tests where other sulphide concentrates were milled in both mills under the same operating conditions. This procedure is necessary to account for differences in the residence time distributions of the batch and continuous mills, and to allow for the fact that the efficiency of the laboratory mill reduces as the milling time increases, due to an increase in viscosity.

TABLE 1. Summary of Fine Milling Test Results for
Pyrite Concentrate

Test No.	Mill Time Min.	Spec Energy KWH/t	Product Size µm			d90 Red. Ratio	After Leaching*	
			d90	d50	d10		Residue g/t	Dissn %
1	8	53	15.8	5.84	1.16	7.0	1.00	85.5
2	14	71	9.75	4.25	0.90	11.3	0.86	87.5
3	25	104	8.01	3.32	0.82	13.7	0.82	88.1
4	40	149	7.13	2.97	0.73	15.4	0.77	88.8

* For 24 hours.

The pyrite concentrate responded well to fine milling, with gold dissolutions in excess of 85 percent being obtained, even with the shortest milling time, and it is apparent that a dissolution of 88 percent can be achieved if the d_{90} is reduced to approximately 10 µm; the specific energy consumption will be approximately 70 Kwh/t. A detailed cost/return analysis will indicate the optimum residence time.

The rates of dissolution are discussed in detail below, where it is shown that fine milling using the Metprotech process results in a very rapid dissolution rate, which has implications for the design of the downstream gold recovery process.

FINE MILLING OF A CALINE RESIDUE IN THE LABORATORY MILL

Sample Description

The calcine residue tested is currently being discarded as tailings to a slimes dam, and contained no dissolvable gold. The gold content of the sample was 28.4 g/t, and the d_{90}, d_{50} and d_{10} were 82, 18.4 and 3.3 µm.

Experimental Procedure

The laboratory mill used to grind the calcine was the same as that used for the pyrite concentrate, except that the measurement of torque was not available at the time.

The solids content of the slurry was 50 percent, and lime and cyanide were added to the slurry before milling so that leaching could take place in the mill. Milling periods of 5, 10, 15, 20, 30 and 40 minutes were used, and after the required time the load was discharged onto a screen and washed with sufficient water to dilute the slurry to a solids content of 35 percent. A portion of the slurry was leached for a further 18 hours, with the remainder being filtered, washed and dried immediately after milling; both residues were than assayed and the gold dissolutions after milling and after milling and leaching determined. A leach time of 18 hours was used for convenience, and the time was not optimized.

The product particle size distribution was determined in the same manner as for the pyrite concentrate.

Discussion of Results

This calcine responds well to fine milling using the Metprotech process, with a gold dissolution in excess of 50 percent being easily obtainable - the yield being greater than 14 g/t, from a material that is being discarded. The results which are given in Table 2 show that the dissolution increases as the d_{90} is reduced

TABLE 2. Summary of Fine Milling Results for a Calcine Residue

Mill Time Min.	Product Size, μm d90	d50	d10	d90 Red. Ratio	After Milling Residue g/t	Dissn. %	After Leaching* Residue g/t	Dissn. %
5	17.0	5.5	1.0	4.8	20.4	28.3	19.7	30.7
10	10.4	4.1	0.8	7.9	17.7	37.7	17.2	39.5
15	8.2	3.3	0.7	10.0	16.0	43.7	15.6	45.1
20	7.1	2.8	0.7	11.6	14.8	47.9	14.8	47.9
30	6.3	2.0	0.6	13.1	13.4	52.9	13.1	53.9
40	5.3	1.6	0.6	15.5	13.1	53.9	13.0	54.3

* For 18 hours.
  Red. - Reduction
  Dissn. - Dissolution

to 6.3 µm, whereafter further size reduction does not improve the dissolution of gold. At this grind, which is achieved after a batch milling time of 30 minutes, the gold dissolution in the mill is 53 percent, and after milling and 18 hours leaching it is 54 percent; this shows that very rapid dissolution rates are obtained when cyanide is added to the mill feed. The rapid rates of dissolution in the mill are attributable to the excellent mass transfer conditions that exist in the mill, the scrubbing action which ensures clean gold surfaces, lack of passivation and the very small particle size.

It has been established that the residence time necessary to achieve an equivalent dissolution in larger continuous mills is approximately two thirds of the batch milling time.

THE EFFECTS OF FINE MILLING ON DOWNSTREAM PROCESSES

The production of very small particles will have both advantageous and detrimental (if not allowed for) effects on subsequent downstream processes and the incorporation of Metprotech milling units into a new or existing flowsheet should not be seen in isolation if maximum benefits are to be obtained. The grinding conditions of the Metprotech mills are designed so that the product has a very narrow particle size distribution, with, for example, the d_{90} and d_{10} of the products having a span of 9 µm. In this way the production of "ultra fine" particles less than 1 µm is minimized, unless liberation conditions dictate otherwise. The effects of fine milling on the rate of gold dissolution and on the activity of activated carbon will be discussed.

Rate of Gold Dissolution

The rate of dissolution of gold can be described by the following second order equation.

$$r_1 = K_r \left((Au)_t - (Au)_\infty \right)^2 \tag{1}$$

where r_1 is the rate of leaching

k_r is the rate constant

$(Au)_t$ is the residue grade at time t

$(Au)_\infty$ is the equilibrium constant (residue grade at infinity)

For quartzitic ores at normal grinds the values of k_r typically vary from 0.7 to 1.4 h^{-1}.

A pyrite concentrate was milled for various times in the
Metprotech laboratory mill to give three products having
different finenesses of grind. Each product was subjected to
cyanidation leaches for one, two, four and 24 hours, and the data
was fitted to equation (1). The results are given in Table 3.

TABLE 3. Leaching Rates for a Pyrite Concentrate Milled to
Three Different Finenesses of Grind

Leach Time, Hrs.	Grind 1		Grind 2		Grind 3	
	Pred. Residue g/t	Pred. Dissn. %	Pred. Residue g/t	Pred. Dissn. %	Pred. Residue g/t	Pred. Dissn. %
0	6.89		6.89		6.89	
1	1.085	84.2	0.860	87.5	0.830	88.0
2	1.053	84.7	0.835	87.9	0.813	88.2
4	1.037	84.9	0.822	88.1	0.904	88.3
24	1.024	85.1	0.812	88.2	0.797	88.4
Grind, d90	15.8		9.8		8.0	
d50	5.8		4.2		3.3	
Rate Constant Kr	107		135		201	
Equilibrium Constant $(Au)_{\infty}$	1.021		0.809		0.796	
R2	0.995		0.993		0.997	

Pred. - Predicted
Dissn. - Dissolution

For all three grinds the rate of leaching is very fast, the
values of k_r being 100 to 200 times greater than those of
quartzitic ores, and the rate increases as the product size
becomes smaller. The equilibrium residue grade also decreases as
the grind becomes finer. The optimum leaching time will be a
function firstly of the economics that determine the most
attractive grind, and secondly of the capital and operating costs
of the leaching section, given the leaching kinetics of the
optimum product fineness; in this case the optimum leaching time
is approximately two hours. If cyanide is added to the mill
feed, then the milled product can pass directly to a carbon
absorption section, further reducing the overall capital cost.

THE ADSORPTION OF GOLD BY ACTIVATED CARBON

To determine whether fine milling of a calcine would result in detrimental effects on the adsorption of gold from solution, a number of tests to determine the adsorption equilibrium and kinetics were performed on a virgin carbon with a slurry made up of finely milled calcine, having a d_{90}, d_{50} and d_{10} of 4.0, 1.7 and 0.5 μm, and water; the pH of slurry was raised to 10.5 with lime and the strength of sodium cyanide in solution was 0.014 percent. The activity of activated carbon which had been conditioned in a slurry made up with the milled calcine was measured by the rate of adsorption of gold and the iodine number, and compared with a carbon that had been conditioned with unmilled calcine.

Equilibrium between Gold in Solution and on the Carbon

The adsorption of gold from solution by activated carbon is a reversible reaction, and an equilibrium isotherm exists between the gold on the carbon and the gold in solution. The operating conditions of a carbon adsorption plant are set so that the carbon loading obtained is lower than the equilibrium value, because the rate of adsorption of the gold decreases as the loading increases towards the equilibrium value. A carbon adsorption plant's operating conditions are designed so that the highest possible carbon loadings are obtained without an unacceptably high gold content in the barren solution; thus, if good equilibrium conditions exist, higher carbon loadings are possible. Four tests were performed for where various masses of fresh powdered carbon were added to the slurry which was then rolled for six days in 100 mL bottles to attain equilibrium. The gold content of the solution was measured and the carbon loading determined. The data was fitted to the Freundlich isotherm, which is,

$$(Au)_c^e = a \ ((Au)_s^e)^b \eqno(2)$$

where

$(Au)_c^e$ is the gold content of the carbon at equilibrium

$(Au)_s^e$ is the gold content of the solution at equilibrium

a, b are Freundlich constants.

The data for the tests is given in Table 4, and, when fitted to the Freundlich equation, the constants a and b are 5242 and 0.325 respectively. The constant "a" is the carbon loading in g/t at that is in equilibrium with a solution concentration of 1 ppm; for quartzitic gold ores the value of "a" typically varies between 7000 and 15000, and for dump reclamation material the

variation is 2000 to 5000 (Johns, 1988). Thus, fine milling of
calcine does not detrimentally affect the equilibrium conditions.

TABLE 4. Data for Carbon Equilibrium Tests

Dry Carbon Mass. g	Solution Au, ppm	Carbon Au. g/t
0.0200	6.5	8904
0.0463	3.8	9033
0.0806	1.8	7397
0.1252	1.3	5116

Rate of Gold Adsorption

The rate of gold adsorption by carbon is realtively slow and is
an important factor for adsorption plants. The rate is given by
an equation which is based on a film diffusion expression, with
the equilibrium described by the Freundlich isotherm (Johns,
1986).

$$r_a - \frac{K_m VA}{\rho} (Au)_s - \frac{b \sqrt{(Au)}c}{a})$$ (3)

where r_a is the rate of gold adsorption

k_m is the mass transfer coefficient

A   is the surface area of the carbon

V   is the solution volume

ρ is the solution density

$(Au)_s$ is the concentration of gold in solution

$(Au)_c$ is the concentration of gold on carbon

a, b are Freundlich isotherm constants.

As the carbon loading increases the rate controlling step changes
from film diffusion to slower intraparticle diffusion, and to
allow for this the mass transfer coefficient, km, is described as

a function of the degree of carbon loading toward the equilibrium value,

$$k_m = k_o \left[1 - \left(\frac{(Au)_c}{(Au)_c^e} \right)^d \right] \tag{4}$$

where k_o is the initial mass transfer coefficient

   d is the fall-away in kinetics parameter.

To perform a batch kinetic test a known mass of fresh carbon was added to the calcine slurry, which was made up to a solids content of 40 percent by mass, contained in a stirred, baffled, 5 litre vessel. Samples of solution were taken at various times and analysed for their gold concentration. The data was fitted to equation (3) by non-linear regression, with an acceptable fit being obtained. The initial mass transfer coefficient k_o was calculated to be 0.29 m/h; under the same conditions typical values for quartzitic ore lie between 0.25 and 0.30 m/h, and for dump retreatment operations the range is 0.04 to 0.19 m/h. The value of d is 1.70, which indicates that the pores of the carbon are not being progressively blocked by the fine calcine. It can therefore be concluded that finely milled calcine does not have a detrimental effect on the rate of gold adsorption by activated carbon.

Carbon Activity

To test whether prolonged exposure to fine calcine would have an effect on the activity of the activated carbon, when compared to calcine before fine milling, samples of virgin carbon were conditioned for eight days in slurries of calcine before and after fine milling, the calcine before fine milling had a d_{90} and d_{50} of 82 and 18 μm. After conditioning the carbon was removed from the slurry by a sieve, washed and subjected to tests to determine the relative degree of fouling.

The rate of adsorption from a solution containing 20 ppm of gold was measured as a function of time, the results being given in Table 5. It can be seen that there is no difference between the carbons conditioned in either calcine slurry, showing that the activity of the carbon is not detrimentally affected by fine milling of the calcine.

TABLE 5. Comparison of the Rate of Gold Adsorption by
          Activated Carbon After Conditioning in Normal
          And Finely Milled Calcine Slurries

Time, min	0	15	30	45	60
Normal Calcine	0	27.5	45.0	56.5	66.0
Finely Milled Calcine	0	27.5	45.0	58.5	66.0

Values are percent of initial gold in solution adsorbed onto
carbon.

The iodine number of a carbon is the quantity of iodine (mg)
adsorbed onto one gram of dry carbon, and is a measure of the
surface area.   The iodine number of the virgin carbon was 1353,
and that of the carbons conditioned in the normal calcine and the
fine-milled calcine slurries were 1122 and 1175 respectively;
again showing that fine milling of calcine does not have a
detrimental effect on carbon activity in comparison with the
calcine that is currently being processed.

The carbons were assayed for iron and calcium, and the values are
given in Table 6.   The assays show that there is a negligible
difference between the iron contents of the carbons conditioned
in the calcine slurries, although iron (and therefore calcine)
has been picked up by both carbons. The calcium contents of the
conditioned carbons are at least twice the iron contents,
indicating calcium fouling. The calcium is likely to reduce the
carbon activity more than the calcine, but is removed by acid
washing during normal operation of a CIP plant.

TABLE 6. Analysis of Iron and Calcium Contents of
          Virgin and Conditioned Carbons

Values in ppm	Fe	Ca
Virgin	116	425
Normal Calcine	2480	6980
Finely Milled Calcine	2520	5550

Simulation of a CIP Plant Treating Finely Milled Calcine

The equilibrium and rate constants that were previously determined for the finely milled calcine were used in a computer program which dynamically simulates a carbon adsorption plant, to demonstrate that acceptable carbon loadings and barren solution grades can be obtained. The gold content of the calcine was 28.4 g/t. and the dissolution after being milled was 60 percent, the relative density of the slurry being 1.42 t/m^3.

The parameters used for the simulation are given below:

Residence time per stage	:	1 hour
Number of stages	:	7
Carbon concentration	:	25 g/L
Slurry bypass	:	5% per stage
Gold content of eluted carbon	:	100 g/t
Activity of eluted carbon	:	100%
Time between carbon transfer	:	24 hours
Fraction of carbon transferred	:	100% per stage

The steady-state carbon and solution profiles resulting from the simulation are given in Table 7; the grades before and after transfer and the average grades are given. The simulated profiles are acceptable, the gold content of the loaded carbon exceeds 11 kg/t, and the gold content of the barren solution has an average value of 0.0015 ppm; the efficiency of the simulated plant is 99.9 percent.

TABLE 7.   Carbon and Solution Profiles Determined
By the CIP Plant Simulation

	Gold on Carbon, g/t			Gold in Solution, ppm*		
Stage	Before Transfer	After Transfer	Average	Before Transfer	After Transfer	Average
1	11 798	9 942	11 090	13.75	10.67	12.49
2	10 560	7,585	9 261	10.84	5.868	8.461
3	7 979	4 525	6 258	5.818	1.954	3.668
4	3 835	1 786	2 685	1.387	0.348	0.754
5	894	433	625	0.1811	0.046	0.012
6	205	143	169	0.022	0.007	0.012
7	112	104	108	0.002	0.001	0.002

* Leaving Specified Stage

The long term effects of fine milling of calcine on a full-scale CIP plant will be known within the next six months. Fine milling of gold-bearing materials provides a number of advantages which can be employed to design cheaper and more compact adsorption plants, details of which will become available in the future.

COMMERCIAL INSTALLATION OF THE METPROTECH PROCESS

The first commercial installation of the Metprotech process was commissioned in August 1988. It is sited at the New Consort Gold Mine in the Barberton district of the Transvaal, and processes calcine produced from two-stage fluidized bed roasting of arsenical and pyritic flotation concentrates. A general view of the installation, showing the Metprotech milling unit is shown in Figure 2. The mill is designed so that tonnage quantities of calcine can be milled to a d_{90} of 8 μm and a d_{50} of 3 μm in one pass before passing to a CIP section; this results in the gold content of the residue being deposited on the slimes dam being reduced by at least 50 percent.

The concept of the process is such that downtime of the milling unit for the removal and replacement of wearing parts is kept to a minimum, and the operation of the process falls under the duties of the personnel of the leaching and CIP sections. This installation is being used to provide scale up information, and energy consumption and operating cost data.

CONCLUSION

The Metprotech process for fine milling of gold bearing sulphidic materials has been shown to be an alternative to oxidative processes, resulting in high recoveries of gold without the environmental problems of having to deal with vapourized or solublized sulphur. Where roasting of concentrates is currently being practised, the Metprotech process can easily be incorporated into the calcine treatment circuit, grinding the calcine before gold recovery, in this way the overall efficiency of the plant can be improved. The rapid dissolution rates obtained in the Metprotech milling unit can be used to the maximum advantage, even eliminating the leaching section of the plant.

The effect of fine milling of calcine was found not to be detrimental for the adsorption of gold onto activated carbon in respect of the equilibrium and kinetic conditions, and the fine calcine had the same effect on the activity of activated carbon as calcine before fine milling.

Figure 2.   General view of a Metprotech milling process
installed in a gold mine.

A full-scale Metprotech process has recently been commissioned to
treat calcine, which will provide data to further optimize the
process conditions and allow operating costs to be accurately
defined.

REFERENCES

Johns, M.W. "Model Application", Chapter 5 of the lecture notes from Carbon School, 8-12 Sept. 1986, The South African Institute of Mining and Metallurgy.

Johns, M.W., Private Communication, 1988.

Liddell, K.S. "Machine for fine milling to improve the recovery of gold from calcines and pyrite". Gold 100. Proceedings of the International Conference on Gold, Volume 2, pp 405-417. SAIMM 1986.

Patent 88/0027   SWA/Namibia

Pat. Appl. 13369/88   Australia
Pat. Appl. 172,650   USA
Pat. Appl. P.I.880/384   Brazil
Pat. Appl. 88/2097   South Africa
Pat. Appl. 88/0023   Venda
Pat. Appl. 88/0036   Bophutatswana

Patent Applications made in other countries.

6

Heap Leaching

Chapter 23

ASPHALTIC CONCRETE LEACH PADS

by G. H. Beckwith[1], G. D. Allen[2]
& A.C. Ruckman[1]

[1]Sergent, Hauskins & Beckwith Geotechnical
Engineers, Inc., Phoenix, Arizona

[2]Sergent, Hauskins & Beckwith Geotechnical
Engineers, Inc., Salt Lake City, Utah

ABSTRACT

This paper describes the design and operational
features of several asphaltic concrete gold heap
leach pads which have been constructed in the United
States over the past 12 years. For these projects,
crushed ore is placed on the pad, leached, removed
and deposited in a waste area. Procedures used in
pavement structure and asphaltic concrete design are
summarized and operational procedures described.
Seepage considerations and the need, if any, for
leakage detection systems and secondary liners are
discussed.

INTRODUCTION

During the past twelve years, several reusable
asphaltic gold heap leach pads have been constructed
in the United States. Several new projects are
currently under design. Details of six leach pads
are shown on Table 1. Structural performance of
these projects has been excellent. Stotts (11)[*]

---

[*]Numbers in parentheses correspond to references at
the end of report.

describes a project in Idaho involving 3 inches of
asphaltic concrete with a secondary PVC liner and
extensive underdrain system.

Unlike "dedicated" pads for which several lifts
are placed and permanently kept on the pad (2),
crushed ore is loaded on asphaltic concrete pads,
leached, washed to remove cyanide solutions,
excavated and hauled to waste disposal areas.

Thickness design methods developed for airport
pavements and asphalt technology widely used for
hydraulic structures have been applied to the design
of asphaltic leach pads. Design and operational
features, design methodology and seepage
considerations are discussed below.

Asphaltic concrete requirements to assure good
performance of leach pads are far more stringent than
for ordinary highway and airport pavements. Thus,
very intense quality control is essential.

DESIGN & OPERATIONAL FEATURES

Pavement structures for gold leach pads designed
by the writer's organization (Table 1) have all
involved two layers of asphaltic concrete with a
asphalt-rubber membrane (4, 5) in between. Where
relatively clean granular soils were present, the
asphaltic concrete has been placed directly on the
compacted subgrade. For weaker subgrades granular
bases have been provided as part of the pavement
structure.

Pads have been laid out in rectangular panels
with the long axis generally parallel to the
topographic contours. Panels have ranged from about
80 to 100 meters in width and to 1000 meters in
length. Pads have been sloped at grades varying from
1.5 to 5.0 percent to V-shaped collection ditches 0.3
to 0.6 meters deep. The ditches are parallel to the
long axis of the pads. For flatter slopes, collection ditches have been provided on each side. For
slopes of 3 to 5 percent, a single collection ditch
has been provided on the lower side. Dams in the
ditches allow collection of solution in a system of
below grade pipes. Separate systems of pipes are
provided for the collection of pregnant solution and
the barren wash solution.

TABLE 1

ASPHALTIC CONCRETE GOLD HEAP LEACH PADS

Project	Location	Year	T_{AC}[f]	T_1	T_2	T_B	Subgrade Soil	Heap Height in Feet
			Component Thickness In Inches					
Smoky Valley Common Operations[g]	Nevada	1976	7.0	5.0	2.0[e]	a	Silty sand, sandy gravel	25 - 30
Smoky Valley Common Operation[g]	Nevada	1985	7.0	5.0	2.0	a	Silty sand, sandy gravel	25 - 30
Round Mountain Gold Corp.[g]	Nevada	1987	7.0	3.5	3.5	12.0[b]	Silty sand (CBR = 55)	40 - 45
Maggie Creek Project	Nevada	1980	7.0	5.0	2.0[e]	7.0[c]	Silty clay, sandy clay	15 - 20
Ortiz Project	New Mexico	1979	7.0	5.0	2.0[e]	a	Silty sand, some gravel	25 - 30
Brewer Gold Project	South Carolina	1988	5.0	2.5	2.5	9.0[d]	Clayey silt (CBR = 3 to 9)	25 - 30

Notes

a. Asphalt concrete placed over compacted subgrade

b. Underlain by 12 inches of compacted subgrade (minimum)

c. Underlay by 38 inches of compacted select granular fill

d. Underlain by geotextile stabilization fabric placed over compacted subgrade.

e. Conventional hydraulic asphaltic concrete

f. Rubber - asphalt layer placed between the two layers of asphaltic concrete.

g. Different pads built at the same project at Round Mountain, Nevada.

h. Year of construction.

Notation & Conversion Factors

T_{AC} = total asphaltic concrete thickness

T_1 = thickness of upper layer of asphaltic concrete

T_2 = thickness of lower layer of asphaltic concrete

T_B = thickness of granular base

1 inch = 2.54 centimeters

1 foot = 0.3 meters

Crushed ores with maximum particle size ranging from about 2 to 8 centimeters have been involved. A 0.3 meter layer of clean crushed ore or crushed rock is placed over the asphaltic concrete as a working surface to protect the pad and enhance drainage. Heap heights have ranged from about 5 to 12 meters.

On several projects, ore has been dumped on the pads with trucks with crawler-type tractors being used to build the heaps. Trucks up to 150,000 kilograms gross weight have been used. For the Ortiz Project in New Mexico, a conveyor system and loading bridge was employed to place ore on the pad. Newer designs have all involved loading of the pads with conveyors and traveling stackers.

Ore is removed with trucks and large front loaders. The cycle length of loading the pad, leaching, washing with barren solution and unloading is typically from 40 to 75 days.

PAVEMENT STRUCTURE

Asphaltic concrete pavements must be structurally designed to resist the wheel loads of front loaders and haul trucks and be essentially impervious to prevent loss of leach solution and release of solutions to the environment.

Special asphalt concrete and various asphaltic membranes have long been used for lining canals, ponds, reservoirs and upstream faces of rock-fill dams (6, 7, 10, 12). Based on this technology, a 5-centimeter thick lower layer of conventional hydraulic asphaltic concrete with an overlying course of asphalt-rubber was selected for the seepage barrier in earlier designs. The upper layer was conventional structural asphaltic concrete. For recent projects, both layers of asphaltic concrete have consisted of special mixes which have achieved both low permeability and high stability.

Thickness Design

The airport pavement design method of the Naval Facilities Engineering Command (9) has been used as the primary method for pavement structure design analysis. Designs are routinely cross-checked with alternative methods such as those of the Asphalt

Institute (1) and Federal Aviation Administration
(3). These methods allow consideration of equivalent
single-wheel loads, tire inflation pressures and
number of load repetitions.

It has been found that the front loader wheel
loads usually control design rather than the haul
truck loads. The equivalent single-wheel loads of
the widely used Caterpillar 966 and 992 loaders are
about 6,800 and 24,000 kilograms, respectively.
Loaders up to 150,000 kilograms gross weight are
being considered for one new design. Up to 9,000
trips per cycle and 300,000 trips for the design life
have been estimated as the pavement design loads.
The subgrades have been characterized with California
Bearing Ratio tests (ASTM D1883).

Asphaltic Concrete

The Marshall method has been used for mix
designs. The strategy used in recent designs is to
employ well-graded aggregates and mineral fillers to
achieve low permeability and good structural properties. Details of three recent designs are given on
Figure 1 and Table 2. Hydraulic conductivities in
the range of 1×10^{-8} to 1×10^{-10} centimeters
per second have been achieved with these mixes.

A matter of great concern is the potential for
"stripping" of asphalt which is subjected to almost
continuous exposure to barren or cyanide solutions.
Stripping involves the loss of bond between the aggregate particles and asphalt cement. If this condition
occurs, then loss of the structural capacity and
increase of hydrualic conductivity may occur.

The predominent test method used to determine the
tendency for stripping is the immersion-compression
test (ASTM D1074 and D1075). An Index of Retained
Strength greater than 75% indicates satisfactory
performance. It should be noted, however, that the
test has a high degree of variability stated in the
precision statement of the ASTM procedure.

Satisfactory performance of leach pads has been
obtained with materials having retained strengths of
less than 50%, although higher rates of maintenance
are involved. Long-term testing suggests cyanide
solutions do not have more severe stripping

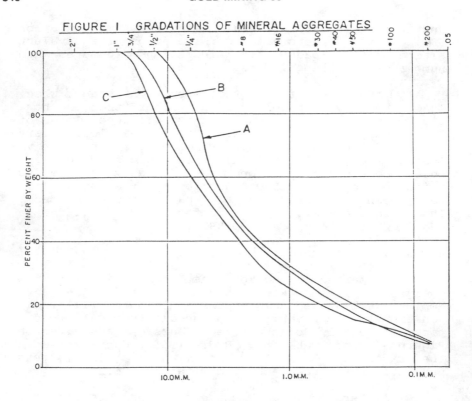

FIGURE I GRADATIONS OF MINERAL AGGREGATES

TABLE 2

IN-PLACE PROPERTIES OF ASPHALTIC CONCRETE MIXES

Mix	Max. Density (lbs/ft)	% Compaction	Marshall Stability (lbs)	V(4)	% Asphalt(5) Cement
A(1)	155.8 (2.49 g/cc)	97 min.	3190 (1450 Kg)	0.2	5.9
B(2)	139.0 (2.23 g/cc)	97 min.	3620 (1640 Kg)	1.3	7.4
C(3)	142.3 (2.28 g/cc)	97 min.	2710 (1230 Kg)	2.2	6.0

NOTES:

(1) Mix included 1% Liquid Anti-Strip Additive

(2) Mix included 1% Hydrated Lime-Mineral Admixture

(3) Mix included 2% Type 2 Portland Cement

(4) Voids in compacted mix (percent)

(5) Percent by weight of total mix

tendencies than water. Hydrated lime, cement, and
CarStab liquid antistrip have been used to mitigate
stripping.

Relatively high mixing temperatures, breakdown
rolling with vibratory rollers, intermediate rolling
with pneumatic rollers and final rolling with static
smooth drum steel rollers is needed to achieve the
high specified densities. As stated before, very
close quality control is critical.

Rubber-Asphalt Membrane & Cover

After careful cleaning of the surface of the
lower layer of asphaltic concrete, the asphalt rubber
membrane, consisting of AC-10 asphalt cement and
granulated reclaimed rubber is applied. A 1.3
centimeter chip seal is placed on top of the asphalt
rubber membrane to protect it during placement of the
overlying asphaltic concrete layer.

SEEPAGE & ENVIRONMENTAL CONSIDERATIONS

Considerabe discussion is presently occurring in
the United States as to whether or not and under what
circumstances secondary liners and leakage detection
systems are needed beneath asphaltic concrete pads.
Extensive systems have been provided in some
designs. The writers do not know of any case where
an asphaltic concrete pad has been ruptured during
operation. Yet, at some sites involving very
permeable granular foundation soils and shallow
ground water, secondary liners and underdrain systems
are clearly required.

For a well designed and constructed pad the asphaltic concrete layers are believed to be as good a
hydraulic barrier as a high quality clay liner of
equivalent thickness. The asphalt rubber membrane is
believed to be equivalent to a geomembrane liner.
Where this type of design is used in the arid and
semi-arid parts of the western United States and a
thick layer of unsaturated soils or rock is present
above the water table, secondary liners and leakage
detection systems are not thought to be necessary.
Partially saturated flow analysis (8) usually demonstrates that minor leakage through the pad would not
reach the ground water.

REFERENCES

1. The Asphalt Institute, 1986, "Thickness Design for Heavy Wheel Loads", Manual Series No. 23, College Park, Maryland.

2. Beckwith, G. H., Hansen, L. A. and Buffington, D. L., 1987, "Considerations in Heap Leach Pad Design", First International Conference on Gold Mining, Vancouver, B.C.

3. Federal Aviation Administration, 1978, "Airport Pavement Design and Evaluation", AC150/5320-6C, Department of Transportation, Washington, D.C.

4. Frobel, R. K., Jemenez, R. A. and Cluff, C. B., 1977, "Laboratory and Field Development of Asphalt-Rubber for Use as a Waterproof Membrane," Report: ADOT-RS-14 (167), Arizona Department of Transportation, Phoenix, Arizona.

5. Frobel, R. K., Jemenez, R. A., Cluff, C. B. and Morris, G. R., 1978, "Asphalt-Crumb Rubber Waterproofing Membrane," Journal of the Irrigation and Drainiage Division, ASCE, Vol. 104, No. IR1, March.

6. International Commission on Irrigation and Drainage, 1985, "Development of Bitumen Concrete and Membranes," ICID Guidelines No. 106, New Delhi, India

7. Internation Commission on Large Dams, 1977, "Bituminous Concrete Facings for Large Dams", Bulletin 32, Paris.

8. McWorter, D. B., and Nelson, J. D., 1979, "Unsaturated Flow Beneath Tailing Impoundments," Journal of the Geotechnical Engineering Division, ASCE, Vol. 105, No. GT11.

9. Naval Facilities Engineering Command, 1973, "Design Manual, Airport Pavements", DM-21, Alexandria, Virginia.

REFERENCES (CONT'D)

10. Sherard, J. L., "The Use of Asphaltic Concrete on the Upstream Face of Earth and Rockfill Dams," Proceedings of the Conference on the Use of Asphalt in Hydraulic Structures, The Asphalt Institute, Pacific Coast Division, Berkeley, California.

11. Stotts, W. G., 1985, "Handling Cyanide at Superior Mining Company,s Stignite Mine", Cyanide and the Environment, Vol. 1, University of Arizona, Tucson.

12. Technical Advisory Committee on Waterdefences, 1985, "The Use of Asphalt in Hydraulic Engineering," Risjks Waterstaat Communications, The Hague, Netherlands.

Chapter 24

DESIGN AND CONSTRUCTION OF CLAY LINERS
FOR GOLD TAILINGS FACILITIES AND HEAP LEACH PADS

Director, Knight and Piesold Ltd.
Vancouver, B.C.

Director, Knight and Piesold Ltd.
Vancouver, B.C.

ABSTRACT

The construction of tailings facilities and heap leach pads for gold mining operations generally requires the installation of one or more impermeable liners beneath the structures to satisfy environmental regulations and prevent the leakage of gold bearing solutions. Liners are usually constructed using clay soils or synthetic membranes, such as high density polyethelene.

Low permeability soil liners can be constructed from natural clays, well graded soils with only a small clay content, or sandy soils blended with a small quantity of imported clay such as bentonite or kaolinite. The permeability of such materials, properly placed with good quality control, generally lies in the range of 10^{-6} to 10^{-8} cm/s, and can be designed to meet minimum prescriptive standards set by regulatory agencies. Important benefits resulting from the use of clay liners include the slow rate of penetration of a saturated wetting front, the decrease in permeability of the liner as it consolidates under the applied load of tailings or ore, and attenuation of potential contaminants in the process solutions. Synthetic liners can be attributed a finite permeability as a result of minute perforations and seaming deficiencies, and possess none of the above properties. Synthetic liners are also, by their nature, relatively thin and even a small head build-up above the liner will result in a large hydraulic gradient across it.

This paper examines the technological and economic benefits of using clay liners. A case history is presented of a drained sub-aerial tailings facility in California, which has been designed to meet current waste discharge requirements, together with the costs of liner construction. Costs of clay liner construction for other tailings facilities and heap leach pad installations are also presented.

INTRODUCTION

The construction of tailings storage facilities can require the construction of some form of liner for environmental protection, should water quality objectives be threatened. For heap leach pads a liner is obviously required for recovery of process solutions and to prevent losses of pregnant solution into the underlying ground. In many instances, designers elect to use synthetic membranes, particularly HDPE, as a quick and simple solution. However, in many locations natural materials exist which can be readily used to construct cheap and efficient liners. The range of materials suitable for soil liners varies widely and a significant clay content is not necessarily required. A deficiency in silt or clay sized particles can often be overcome with the addition of small quantities of bentonite or other imported clay. Current regulations in California prescribe the use of clay liners for certain applications.

CALIFORNIA LEGISLATION AND LINER REQUIREMENTS

Mining activity in California has seen a major resurgance in the last five or six years. The design of any waste management unit in California is regulated by the California Administrative Code, Title 23, Chapter 3, Sub-chapter 15 "Discharge of Waste to Land", which is administered by Regional Water Quality Control Boards. Sub-chapter 15 is an all embracing piece of legislation written primarily for municipal and industrial wastes. It includes some special provisions for the mining industry in Article 7 "Mining Waste Management". Article 7 is currently being updated to, hopefully, simplify the requirements and give special consideration to the large volumes and inert nature of most mining wastes, while maintaining the over-riding objective of non-degradation of groundwater.

Prescriptive standards for liner requirements are outlined in Table 7.3 of Article 7, based on the classification of the waste being stored as either hazardous (Group A), containing non-hazardous pollutants with a low risk, but which exceed water quality objectives (Group B), and inert (Group C). Most tailings storage facilities are either Group B or Group C depending on the presence and concentration of process reagents. Heap leach pads do not fit readily into the regulations since they contain process reagents only during operations and can be readily detoxified after leaching to achieve a Group C classification. For both tailings facilities and heap leach pads, however, the presence of cyanide at concentrations above 5.0 mg/L will require Group A prescriptive standards, and any detectable cyanide below this level will generally require a Group B facility.

Prescriptive standards for a Group B tailings pond or surface impoundment, as laid out in Table 7.3, are for a double liner system under the entire facility, consisting of an outer liner of 2 feet of clay with a permeability of 1×10^{-6} cm/s or less, an inner liner of clay or synthetic membrane, and a blanket drain between the two to monitor leakage through the inner liner and reduce the head acting on the outer liner. This drain is termed the Leachate Collection and Removal System (LCRS). Apart from the obvious cost implications of a double liner system, these prescriptive standards have some in-built deficiencies which can actually result in creating a long-term liability problem for the mine operator. These are discussed in a following section on the design of the Jamestown Mine Tailings Management System, which uses a single clay liner and LCRS as an engineered alternative to the prescriptive standards. Construction of liners is also not the only way of protecting groundwater quality, and numerous tailings storage facilities have been constructed which rely on drainage systems or evaporation for equivalent protection.

CHARACTERISTICS OF SOIL AND SYNTHETIC LINERS

The seepage through any liner system is governed by the engineering properties of the liner, the overall area and the hydraulic gradient across the liner. The quantity of seepage can be calculated using Darcy's Law. The total seepage from a saturated, lined facility may be very similar for a clay liner and a high density polyethylene (HDPE) liner, as a result of the much higher hydraulic gradient acting across the HDPE liner. This illustrates the importance of providing good drainage above the liner to reduce the head build-up on it, which is a basic design requirement for drained tailings facilities and efficient leach pad solution recovery systems. Optimization procedures for leach pad liners and solution recovery systems have been previously published by East et.al. (1987).

Low permeability soil liners can be constructed from natural clays, well graded soils with only a small quantity of clay, or soils blended with additional quantities of clay such as bentonite or kaolinite. The permeability of such materials can vary widely, but typically soil liners can achieve permeabilities in the range of 10^{-6} to 10^{-8} cm/s.

Construction of effective soil liners requires a proper application of moisture conditioning, placing and compaction requirements. Incorrect procedures can render soil liners virtually useless, and a poor understanding of these requirements has contributed to a large extent to previous failures. Due consideration must be given to controlling such factors as dessication, frost-heave, erosion, osmotic consolidation and proper bedding.

In controlling leakage from lined storage facilities, soil liners provide some distinct advantages. Firstly, although initial wetting under unsaturated conditions may be fairly rapid, there is a significant time period for travel of a saturated wetting front of permeant through the liner. This may be tens of years for very low permeant soils. Secondly, the permeability of the soil will generally decrease with increased loading or confining pressure, further reducing the rate of penetration. Thirdly, there may be a significant reduction in the concentrations of waste constituents in the seepage water as a result of dispersion, diffusion and adsorption within the soil matrix. These characteristics of soil liners have been recognized in the drafting of the California Sub-chapter 15 regulations where only clay is permitted as the outer liner.

High density polyethylene liners can also provide a very effective low permeability liner. However, even HDPE has a finite permeability due to pinhole leaks and poor seaming techniques. In most installations, there is of course no way of verifying the actual leakage, which has led to claims of "zero leakage". However, in almost all cases where an effective leak detection system has been installed, some finite leakage is detected. Typical values for leakage from HDPE liners for even the best installations range from 45 to 450 litres per day per hectare (30 to 300 gpd per acre).

One characteristic of leakage through a synthetic liner is that it is likely to be confined to certain discrete areas where defects occur. By installing a synthetic liner in initimate contact with the top of a clay liner, a synergism exists and the characteristic imperfections of each liner are significantly reduced. Seepage through the synthetic liner will occur at the localized imperfections but under a greatly reduced hydraulic gradient due to the underlying clay. Similarly, seepage through the clay will only occur directly beneath the imperfections and the total area of seepage will be greatly reduced. Such a compound liner probably offers the best form of overall protection, but the increased cost of adding the overlying synthetic liner can only be justified for extreme design requirements.

THE JAMESTOWN MINE TAILINGS MANAGEMENT SYSTEM

The Jamestown Mine is an open pit gold mining operation on the original motherlode in California. The tailings management system for the mine was the first to be permitted in California under the then newly adopted Sub-chapter 15 regulations in 1985. The mine came into operation in January 1987.

The tailings from the mill were classified as Group B based on the presence of various flotation reagents and an original proposal to use thiourea as the leaching agent.   During the initial design stage for the tailings management system (TMS) it became apparent that strict adherence to prescriptive standards would result in a large basin of saturated low density tailings that would pose a continuous threat to water quality, would be unstable and difficult to construct a final cover on, and would require extensive long-term monitoring and contingency planning.   An engineered alternative was therefore proposed which consists of a drained tailings storage facility with a single outer clay liner, and a separate double-lined process water pond.   Rotational deposition of tailings is carried out around the perimeter of the storage facility to produce a sub-aerial tailings beach sloping towards the centre of the embankment where surface water is continuously decanted to the process water pond. Seepage from the base of the tailings resulting from consolidation is also continuously removed by the LCRS overlying the outer clay liner.   A schematic section through the facility is shown on Figure 1.

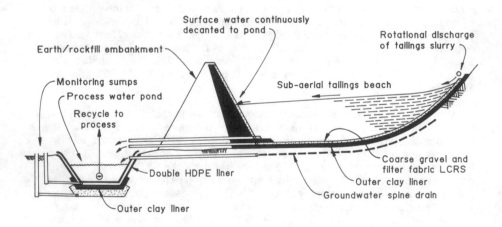

Figure 1.   Schematic Cross-Section.   Jamestown Mine Tailings Management System.

Construction of the first stage of the TMS was carried out in 1986, with subsequent phased expansions in 1987 and 1988. The outer liner over the tailings basin has been constructed using the surface weathered gabbro and serpentinite from within the basin. Both these materials have relatively low fines contents (20 to 40 percent minus #200 sieve) and

Plasticity Indices in the range of 6 to 14. They are well graded and can be readily compacted to achieve a high density, low permeability liner.

Construction of the outer liner required a carefully monitored series of operations to ensure the desired end product was achieved. The material was first excavated to temporary stockpiles and moisture conditioned, then re-excavated to achieve blending and placed in three 200 mm lifts to achieve an overall minimum thickness of 600 mm. Each lift was thoroughly mixed using a disc prior to grading and compaction with a 10 tonne smooth drum vibratory roller to 95 percent of ASTM D1557 maximum density. Detailed operational monitoring at the Jamestown Mine over the first 18 months of operation confirms that the LCRS and outer clay liner are meeting design objectives.

PERMEABILITY TESTING

Sub-chapter 15 requires that the permeability of any liner system be determined primarily by "appropriate field test methods in accordance with the accepted civil engineering practice ...". The principal intent of the statutory requirements for field tests is the measurement of actual construction performance was well as macro-permeability effects which may not be represented in laboratory size samples.

One apparatus for measuring macro-permeability is a sealed double-ring infiltrometer (SDRI). This apparatus measures infiltration from a 5 foot diameter inner ring within a 12 foot diameter outer ring, and allows for calculation of field permeability once steady state seepage through the entire liner has been established. At this stage the hydraulic gradient can be calculated from known boundary conditions on the liner. Tests using this apparatus, while undeniably representative of actual conditions, can take a considerable time for steady state conditions to be established. This could range from several months to more than a year for liner permeabilities in the range of 10^{-6} to 10^{-7} cm/s, and can impose severe limitations on construction scheduling.

An alternative field apparatus called the air-entry permeameter (AEP) has been suggested by Bouwer (1978), and subsequently developed for field quality control testing during construction of impervious liners. This apparatus is identical in general theory to the SDRI, but allows for measurement and estimation of the parameters governing transient flow conditions, so that permeabilities can be calculated for relatively short duration tests (2 to 3 hours). The AEP is essentially an infiltrometer using a 2 foot diameter ring. The size of the ring and the short duration of the test allows for a large number of tests to be carried out, thus giving a large areal coverage of liner for determining macro-permeability effects, as well as providing enough data for a statistically significant mean permeability to be determined. A schematic illustration of the AEP is shown in Figure 2.

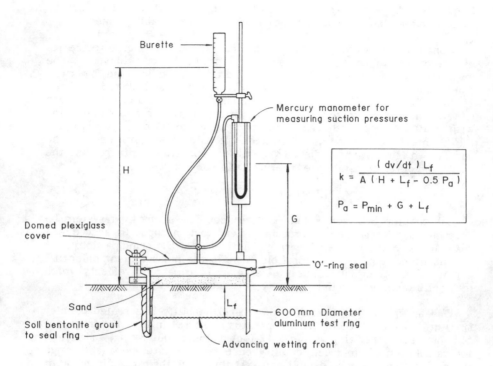

Figure 2.   Air-Entry Permeameter.

The air-entry permeameter was first used for quality control testing on a soil-bentonite liner at the Key Lake uranium mine in northern Saskatchewan, the results of which tests were published by Knight and Haile (1984) and Haile (1985), and at two construction sites in California during 1986.   The first was for the outer liner for the Jamestown Mine tailings management facility described above, and the second was for both inner and outer liners for a heap leach pad for the nearby Carson Hill Mine.   At both locations the Regional Water Quality Control Board required a quality control program involving sealed double-ring infiltro-meter tests, in addition to field and laboratory AEP tests, and laboratory triaxial permeability tests on undisturbed samples. This provided a unique opportunity for comparing the results of four different test methods.

The results of the permeability tests on three test pads and the full scale liner at the Jamestown Mine are given in Table I.

TABLE I. Jamestown Mine Permeability Test Results (Group B Liner)

Test Pads	Coefficient of Permeability (cm/s)		
	#1	#2	#3
In-situ AEP (10 tests per pad)	1.4×10^{-7}	6.2×10^{-7}	1.0×10^{-7}
SDRI	1.2×10^{-7}	3.7×10^{-7}	1.2×10^{-7}
Lab. AEP (10 tests per pad)	2.6×10^{-7}	1.9×10^{-7}	6.4×10^{-8}
Triaxial: (4 tests per pad)			
- Confining pressure 35 kPa	1.6×10^{-7}	4.5×10^{-8}	1.0×10^{-7}
- Confining pressure 840 kPa	1.1×10^{-8}	2.1×10^{-9}	8.1×10^{-9}
Full Scale Liner (13 ha.)			
In-situ AEP (13 tests)		2.6×10^{-7}	
Lab. AEP (15 tests)		2.5×10^{-7}	

At Carson Hill the liners were constructed from a well graded residual soil blended with 30 percent imported kaolinite. The material had approximately 60 percent minus 200 mesh (silt and clay), and an average Plasticity Index of 18. Both inner and outer liners were constructed in three 8 inch lifts each compacted with a sheepsfoot roller to 100 percent of ASTM D698 maximum density. The results of the permeability tests on the initial test pad are given in Table II.

TABLE II. Carson Hill Permeability Test Results (Group A Liner)

Test Pad	Coefficient of Permeability (cm/s)
In-situ AEP (9 tests)	7×10^{-8}
SDRI	9×10^{-8}
Triaxial: (2 tests)	
- Confining pressure 35 kPa	5×10^{-8}
- Confining pressure 840 kPa	1×10^{-8}

On the basis of the test results at Carson Hill, the time for process solution to breakthrough the inner liner was calculated at between 60 and 274 days, depending on the head conditions above the liner. Actual breakthrough occurred after 275 days.

CLAY LINER CONSTRUCTION COSTS

A summary of available construction cost data for various clay liners is presented below in Table III. In all cases the costs include surface preparation, excavation, blending of additives if appropriate, moisture conditioning, placing, spreading and compaction. Costs are in the year of construction.

TABLE III. Clay Liner Construction Costs

Location	Year of Construction	Material	Total Area ha (x 10^6 ft^2)	Unit Cost U.S.$/m^2 (U.S.$/ft^2)
Key Lake, Sask. Outer liner to tailings facility	1982	Sandy till blended with 4% bentonite	45 (4.5)	4.50 (0.45)
Jamestown, Ca. Outer liner to tailings facility	1986	Weathered overburden	13 (1.3)	3.50 (0.35)
Jamestown, Ca. Outer liner to tailings facility	1987	Weathered overburden	10 (1.0)	5.50 (0.55)
Carson Hill, Ca. Outer and inner liners to leach pad	1986	Weathered overburden blended with 30% kaolinite	– –	7.00 (0.70)
Royal Mountain King, Ca. Outer liner to flotation tailings facility	1988	Weathered overburden	24 (2.6)	2.10 (0.21)

SUMMARY

Mining waste management units and heap leach pads can require the construction of some form of liner for environmental protection and recovery of gold bearing solutions. Under current California regulations liner requirements are based on waste classification and favour the use of clay over synthetic membranes for outer liners. This recognizes the inherent advantages of clay liners with respect to rate of solution penetration, long-term durability and the capacity for attenuating potential pollutants. Several facilities have recently been constructed in California using clay liners and the Jamestown Mine tailings management system is of particular significance as an engineered alternative to prescriptive standards. This design incorporates sub-aerial tailings with a complete underdrain (LCRS) overlying an outer clay liner.

In-situ testing of clay liners is of particular significance in determining the actual permeability achieved and the influence of any macro-permeability effects. The air-entry permeability (AEP) has been developed to give a rapid assessment of in-situ permeability and correlates well with other in-situ and laboratory test methods.

The cost of constructing clay liners or modified soil liners is generally in the range of U.S. \$2.00 to 5.00/m^2 (U.S. \$0.20 to 0.50/ft^2). This compares very favourably with equivalent synthetic liners.

REFERENCES

Bouwer, H. (1978), "Groundwater Hydrology", McGraw-Hill, New York, pp. 124-125.

Knight, R.B. and Haile, J.P. (1984), "Construction of the Key Lake Tailings Facility", International Conference on Case Histories in Geotechnical Engineering, St. Louis, Mo.

Haile, J.P. (1985), "Construction of Underseals for the Key Lake Project", 1st Canadian Seminar on the Use of Bentonite for Civil Engineering Applications, Regina, Saskatchewan, March 18, 1985.

East, D.R., Haile, J.P. and Beck, R.V. (1987), "Optimization Technology for Leach Pads Liner Selection". Society of Mining Engineers, Syposium on Geotechnical Aspects of Heap Leach Design, Denver, Colorado, February 24-27, 1987.

Chapter 25

WATER BALANCE EVALUATION FOR HEAP LEACH
OPERATIONS IN HIGH SNOWFALL CLIMATES

Jim M. Johnson, P.E.

Senior Geotechnical Engineer
Steffen Robertson and Kirsten (Colorado) Inc.
3232 South Vance Street, Suite 210
Lakewood, Colorado 80227

INTRODUCTION

The facility water balance and maintenance of sufficient excess storage volume in the solution control system are often amongst the most critical facets of design and operation of heap leach facilities in cold, snowy climates. The purpose of this paper is to summarize and discuss climatic and operational factors which have a significant impact on heap leach water balance evaluations in high snowfall climates. Example water balance simulations are presented for proposed heap leach projects located in areas subject to high snowfall in the northwestern United States. The climatic data input to these simulations represent both net evaporative and net precipitation climates.

FACTORS AFFECTING WATER BALANCE EVALUATIONS

The combined effects of seasonal operations, relatively low evaporative losses, and potentially large inflows to the solution control system from snowpack and spring storms require an enhanced awareness of the factors which impact facility water balance. Unlike operations at mines in more arid climates, seasonal excess water rather than a shortage of make-up water is the rule. A thorough understanding of a number of factors, including regulatory constraints, climatic influences, and impact of operating philosophy and techniques, is critical to successful development of an incident-free heap leach operation under high snowfall climatic conditions. A process water circuit schematic for a typical heap leach operation is shown on Figure 1.

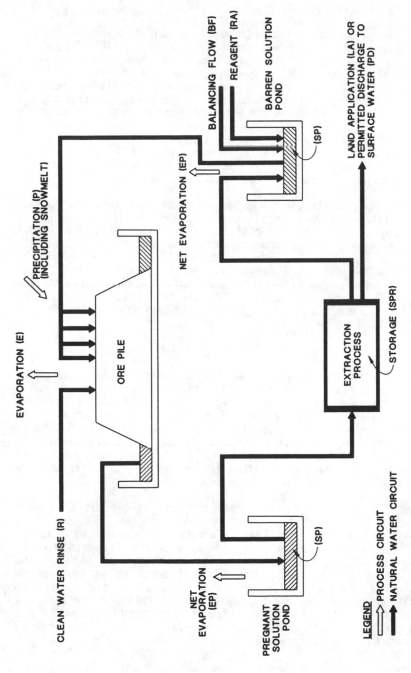

FIGURE 1. Typical Heap Leach Water Circuit

Regulatory Constraints

In areas of the United States where annual evaporation exceeds annual precipitation (net evaporation), the responsible state and federal agencies generally require that heap leach operations be designed and operated as zero-discharge facilities. For these projects, the solution routing and storage facilities must be capable of containing all process solutions circulating through the facility plus all inflows due to makeup requirements and climatic influences. Discharges to either surface water or groundwater are permit violations and are therefore prohibited. The only exception to this requirement, acceptable in many states, is discharge of treated process water through an approved land application system which generates no runoff to nearby surface waters.

For projects located in areas where annual precipitation exceeds annual evaporation (net precipitation), design and operation of zero-discharge heap leach facilities may not be feasible. Under these conditions, excess process solutions may be discharged to nearby surface waters following detoxification. Discharges may be initiated only after application for and approval of a National Pollution Discharge Elimination System (NPDES) permit. The Environmental Protection Agency (EPA) was originally authorized to review and issue NPDES permits. However, in many states, the EPA has delegated this authority to a state agency.

Climate

Climatic influences on heap leach water balance consist principally of temperature, precipitation, and evaporation. As noted in the previous paragraphs on regulatory constraints, the relative imbalance between precipitation and evaporation on an annual basis is typically used to determine the quantity of allowable discharge per year from the processing circuit, if any. When evaluated on a monthly basis, the imbalance between precipitation and evaporation can be used to evaluate anticipated fluctuations in solution storage and/or solution discharge requirements.

Precipitation can enter the solution control circuit in a wide variety of forms, but typically as rain or snow. As rain, the precipitation enters the circuit in a very active and mobile manner and if not evaporated or absorbed by the ore, reports to the solution control ponds relatively quickly. As snow, the precipitation is relatively immobile and inactive as part of an accumulating snowpack. With the exception of moisture lost to sublimation, the accumulated snowpack precipitation reports to the pond during a relatively short time period corresponding to the spring melt. In addition, the snowmelt may be significantly accelerated by spring rainstorms.

Evaporation from the solution control circuit typically consists of a combination of pond surface evaporation and enhanced evaporation in the form of irrigation spray losses. In high snowfall climates, sublimation losses from the snowpack may also be significant.

Operating Philosophy and Techniques

A number of operating techniques are available which allow varying degrees of control over water inflows and outflows and can therefore significantly impact the overall water balance. These techniques include:

- Cellular pad design;
- Spray irrigation (spring, summer, and fall);
- Drip irrigation (winter);
- Snow removal;
- Land application;
- Permitted discharges; and
- Water budgeting.

Depending upon project-specific requirements, all or some of these techniques may be used to maintain an acceptable water balance even under high snowfall climatic conditions.

The cellular design concept is illustrated schematically on Figure 2. The purpose of cellular pad design is to minimize potential inflows to the process water circuit through minimization of the active area of the heap at any time. For projects in areas having very cold, high snowfall climates, the cellular system allows leached and neutralized portions of the heap to be decommissioned and removed from the process water circuit prior to winter shutdown. The cellular system provides the same advantages for facilities which will be operated through the winter using buried drip irrigation. The reduction in potential inflow from storm precipitation, heap draindown, and spring snowmelt must be balanced against additional costs associated with berm construction and marginally larger lined areas. However, the cellular system provides the greatest degree of control over the process water circuit and may therefore provide the most efficient operation and greater potential cost savings over the life of the project.

The use of spray irrigation during the predominantly nonfreezing months provides operators their most valuable tool for controlling process water storage requirements. Spray losses from both leaching and rinsing operations can often be used to offset large storm and snowmelt inflows if the solution control ponds are large enough to store these inflows temporarily. In most instances, spray losses are large enough that fresh make-up water must eventually be input to the system during the peak irrigating period.

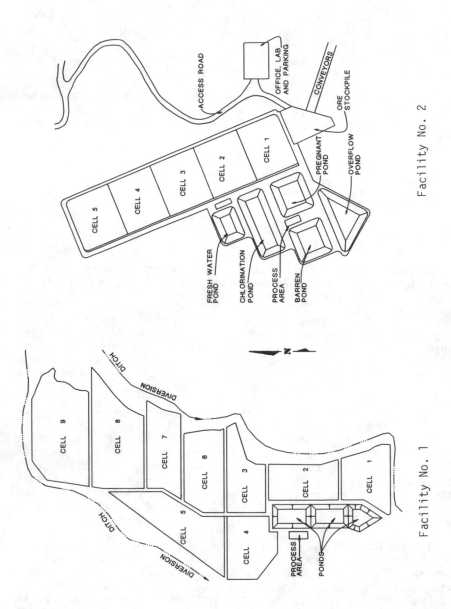

Facility No. 2

Facility No. 1

FIGURE 2. Example Heap Leach Facilities

The use of buried drip irrigation during the predominantly freezing months provides operators a tool for minimizing the volume of process water which will report to the solution control ponds over the winter. Large quantities of solution will be reapplied and temporarily stored in the active heaps rather than draining into the ponds for winter storage.

Under special circumstances, snow removal from the heaps may be an acceptable technique for controlling inflows to the process water circuit. Some states will accept this technique as part of a process water control plan, but only as an emergency action to prevent greater environmental damage. Due to the financial costs and practical difficulties associated with snow removal operations this technique is typically not warranted except in the event of an exceptionally heavy snowpack which will exceed available storage capacity when melted.

Many states will permit land application systems for discharge of treated, excess process water. If properly designed and operated, a land application system will maximize water losses due to evaporation, transpiration, and infiltration, and will generate no surface runoff. However, reliance on such a system requires large solution control ponds to store draindown, storm precipitation, and spring snowmelt until the ground surface is free of snow and dry enough to inhibit runoff during application.

As noted earlier, treated process water may be discharged to nearby surface waters if the project is located in an area with a net precipitation climate and if an NPDES permit is approved. Due to the expense associated with installing and operating the treatment and delivery systems, this technique is typically not warranted unless all other techniques are not sufficient to adequately control the accumulation of process water.

Water budgeting, meaning use and reuse of all available process water rather than relying on liberal infusions of fresh make-up water, is also a valuable technique for controlling accumulation of excess process water. Although this technique seems too obvious to require mentioning, it is often underutilized when freshwater is readily available and the operator's attention is diverted to other pressing concerns.

ENGINEERING EVALUATIONS

Engineering evaluations of facility water balance should consider the following possible inflow and outflow mechanisms:

· Inflow from average monthly rainfall;
· Inflow from design storm rainfall;

- Seasonal inflow from snowpack melt;
- Make-up water inflow;
- Pond evaporation losses;
- Irrigation evaporation losses;
- Permitted losses by discharge to surface water; and
- Permitted losses by discharge through land application.

In addition, inflow to the solution storage ponds in the form of heap draindown in the event of seasonal or unplanned shutdown must be considered.

Inflow Mechanisms

Inflow from average monthly rainfall can most easily be determined using site specific data if available. Otherwise, station-specific and regional data published by the National Oceanic Atmospheric Administration (NOAA) in the United States and by Environment Canada (EC) can be used. Data from nearby stations can be converted using various empirical correlations to estimates of project site conditions. The converted data should be used in conjunction with regional data to determine average annual precipitation and the monthly distribution of precipitation. The precipitation values can then be converted to inflow quantities based on an assumed runoff percentage (typically 100 percent) and a measured catchment area.

The potential surge inflow from a design storm can also be evaluated using regional data published by NOAA and EC. Storm events of 6 hour and 24 hour duration having return periods of 2, 5, 10, 25, 50, and 100 years are available and procedures are provided to determine events of intermediate or very short duration. Most permitting agencies require the use of the 100-yr, 24-hr storm event when determining solution storage pond size requirements.

The seasonal inflow from snowpack melt can be evaluated using station-specific and regional data published by the Soil Conservation Service (SCS) in the United States and by EC. Data from nearby stations can be converted to estimates of project site conditions using empirical correlations. The data can also be analyzed statistically to determine snowpack water equivalents for various return periods. The return period to be used for sizing the solution storage ponds is usually not specified by the permitting agencies and is therefore a function of project-specific conditions. For instance, the ponds can be designed to hold only the one-year anticipated snowmelt if discharge of treated process water to nearby surface waters is anticipated on an as-needed basis. Conversely, the ponds should be designed to contain the ten-year anticipated snowmelt if no discharge of any kind is desired and the project is expected to have a ten year life.

Inflows to the solution storage ponds in the form of heap draindown can be estimated based on the results of simulated leaching and draining tests conducted on representative ore samples in the laboratory. Although heap draindown must be considered in design of the solution control ponds, it represents a transfer of storage from the heap to the ponds rather than a new inflow to the process water circuit.

Anticipated inflows to meet requirements for fresh make-up water can be evaluated for various operating and climatic conditions using computerized water balance simulations.

Outflow Mechanisms

Outflow due to average monthly pond evaporation can most easily be determined using site specific data if available. Otherwise, station-specific and regional data published by NOAA and EC can be used. Data from nearby stations can be converted to estimates of project site conditions using empirical correlations. If the ponds are sufficiently shallow, pan evaporation values or a combination of pond and pan evaporation values may be used. The appropriate evaporation values can then be converted to outflow quantities based on the measured pond areas.

An estimate of outflows due to irrigation evaporation losses can be made based on the anticipated irrigation rate, type and setting of sprinkler head, area to be irrigated, and empirical correlations which incorporate site-specific effects such as average temperature and wind speed. Sprinkler heads which allow the formation of relatively large water drops are generally preferred to minimize the potential for offsite drift due to wind.

Anticipated outflows through the mechanisms of permitted land application or surface water discharges can be evaluated for various operating and climatic conditions using computerized water balance simulations.

EXAMPLE WATER BALANCE SIMULATIONS

Example water balance simulations have been completed for two proposed heap leach operations in the northwestern United States. Facility No. 1, shown on Figure 2, represents an expanding pad system and is based on a proposed project located in a net evaporative area of the Pacific northwest. Facility No. 2, also shown on Figure 2, is based on a reusable pad project proposed for a net precipitation area of the northern Rocky Mountain states. Both facilities are proposed for areas which experience relatively high snowfall. The average annual snowfall for Facility No. 1 is approximately 120 inches and that for Facility No. 2 is approximately 190 inches.

Facility No. 1

As Facility No. 1 is proposed for a net evaporation climate, no discharges to nearby surface waters will be allowed by the regulatory agencies. In addition, land application of treated process water was determined to be inappropriate due to environmental concerns. For these reasons the water balance simulation was completed assuming zero-discharge conditions. The heap leach pads were split into cells to minimize catchment areas and the solution storage ponds were made sufficiently large to store anticipated operating solution, draindown, storm surges, and snowmelt inflows.

The climatic data, facility data, and the results of years 6, 7, and 8 of a 12-year water balance simulation are included on Table 1. The projected pond storage quantities for years 6, 7, and 8 are graphed as a function of time on Figure 3. Years 6, 7, and 8 were chosen as they represent the most critical phase of the water balance for this particular facility. During year 7, projected pond storage as determined from average climatic conditions has accumulated to a peak of 4,629,000 cu ft. After adjusting for design storm storage and a possible ten year snowpack, potential pond storage has increased to approximately 7,429,000 cu ft.

Based on this simulation, total pond storage of approximately 7,524,000 cu ft will be required to contain the anticipated process solutions with a minimum freeboard. However, the simulation also indicates that some of this pond storage will probably not be required until year 7. The operator of this facility may elect to defer some pond construction until year 6 to minimize initial capital expenditures.

Facility No. 2

Unlike Facility No. 1, Facility No. 2 is proposed for an area having a net precipitation climate. Therefore, discharge of treated process water to nearby surface waters will be allowed by the regulatory agencies if required. In addition, land application of treated process water will be environmentally acceptable. For these reasons the water balance simulation was completed based on average rain and snow conditions. It has been assumed that excess process solutions generated during wetter than average years will be treated and discharged as needed, thereby minimizing solution storage requirements.

The climatic data, facility data, and the results of the first two years of a five-year water balance simulation are included on

TABLE 1. Facility No. 1, Water Balance Simulation

FACILITY NO.1 WATER BALANCE

IRRIGATION RATE (CF/SF HR)	0.0401	= .005 GPM/SF
WETTING RATE (CF/TON)	4.0100	= 30 GALLONS/TON
100 YR-24 HR STORM	0.2330	
PRODUCTION RATE (TPD)	4500	
TEN YEAR SWE	250%	
OPERATING DAYS/MONTH	27	
TOTAL PONDS AREA (SF)	542,000	
START-UP VOLUME (CF)	160,400	= 1 MONTHS SOLUTION

PAD #	AREA (SF)	VOLUME MILLION (TONS)
1	580,000	1.89
2	550,000	1.14
3	588,000	1.59
4	632,000	1.77
5	649,000	1.66
6	697,000	1.72
7	501,000	1.08
8	989,000	2.13
9	1,115,000	2.05

POND #	AREA BOTTOM (SF)	AREA TOP (SF)	STORAGE VOLUME (CF)	HEIGHT (FT)
1	108,000	204,400	3,122,000	20
2	101,000	194,250	2,935,000	20
3	49,500	111,300	1,467,000	18
TOTAL	258,500	509,950	7,524,000	

MONTH	HOURS OF OPERATION	LOSS IRRIGAT.	DAYS/MONTH
APRIL	8	0.050	30
MAY	14	0.070	31
JUNE	14	0.090	31
JULY	14	0.090	31
AUGUST	12	0.080	31
SEPT-	10	0.070	30
OCT-	8	0.050	31

Monthly Water Balance Simulation

MONTH (END OF)	PAD(S) IN OPERATION	ORE ON PAD(S) (TONS) X1000	AREA UNDER IRRIG. (SF) X1000	PRECIP. (FT)	EVAP. (FT)	SWE (FT)	PRECIP. SNOWMELT (CF)	SCHED. DRAINDOWN (CF)	IRRIG. EVAP. (CF)	POND EVAP. (CF)
APRIL	6 & 7	2,680	1,102	0.123	0.215	0.520	214,020		530,282	86,835
MAY	6 & 7	2,800	1,147	0.158	0.386	0.120	274,920		998,085	104,345
JUNE	6,7 & 8	2,920	1,246	0.132	0.474		360,228		599,575	129,537
JULY	6,7 & 8	3,040	1,394	0.051	0.598		139,179		2,183,431	158,835
AUGUST	7 & 8	3,040	816	0.075	0.496		152,400		973,795	168,435
SEPT	7 & 8	1,560	944	0.083	0.305		168,556	2,117,280	794,942	93,263
OCT	7 & 8	1,680	1,014	0.163	0.136		331,216		504,201	37,030
DEC	7 & 8	1,800	0	0.205	0.044	0.240	416,560	2,223,144		
JAN	7 & 8	1,920	0	0.254	0.033	0.270				
FEB	7 & 8	2,040	0	0.208	0.036	0.470				
MARCH	7 & 8	2,160	0	0.188	0.061	0.550				
APRIL	8	2,280	1,340	0.123	0.134	0.520	249,936		644,808	85,982
MAY	8	2,400	940	0.158	0.386	0.120	241,898		817,960	120,929
JUNE	8	1,560	989	0.132	0.474		202,092		475,907	136,372
JULY	8	1,800	989	0.051	0.598		78,081	1,117,600	1,549,077	157,566
AUGUST	8	1,920	989	0.075	0.496		119,825		1,180,249	131,323
SEPT	8	2,040	989	0.083	0.305		127,073		832,837	88,883
OCT	8	2,040	989	0.163	0.136		249,553	1,270,368	491,770	36,644
NOV	8	2,130	0	0.205	0.044	0.042	313,855			
DEC	8 & 9	2,250	0	0.254	0.033	0.036				
JAN	8 & 9	2,370	0	0.208	0.036	0.270				
FEB	8 & 9	2,490	0	0.178	0.061	0.470				
MARCH	8 & 9	2,610	0	0.188	0.134	0.550				
APRIL	8 & 9	2,730	1,658	0.123	0.215	0.520	325,458		797,830	91,857
MAY	8 & 9	2,850	1,769	0.158	0.386	0.120	418,068		1,539,331	106,263
JUNE	8 & 9	2,970	1,881	0.132	0.474		349,272		905,137	129,357
JULY	8 & 9	3,090	1,914	0.051	0.598		314,846	1,455,300	2,392,910	158,764
AUGUST	8 & 9	3,210	1,948	0.075	0.496		198,450		2,325,296	132,764
SEPT	8 & 9	3,330	1,992	0.083	0.305		219,618		1,677,463	81,863
OCT	9	1,320	1,037	0.163	0.136		270,091	2,659,833	515,638	49,302
NOV	9	1,440	0	0.205	0.044	0.042	339,685	1,746,756		
DEC	9	1,560	0	0.254	0.033	0.270				
JAN	9	1,680	0	0.208	0.206	0.470				
FEB	9	1,800	0	0.178	0.061	0.550				
MARCH	9	1,920	0	0.188	0.134		911,350		911,350	

MONTH	SNOW EVAP. (CF)	HEAP WETTING (CF)	NET INFLOW (CF)	STORAGE VOLUME (CF)	MAKE-UP (CF)	POND STORAGE (CF)	POSSIBLE INFLOWS 100 HR 24 HR STORM (CF)	100 10 YR SWE (CF)	DRAIN DOWN (CF)	AVAILABLE POND STORAGE (CF)
APRIL		10,746,800	-11,149,897	3,122,000	7,242,529	160,400	405,420		3,546,444	3,411,736
MAY		481,200	-1,308,710	2,935,000	1,308,710	160,400	405,420		3,705,240	3,252,940
JUNE		481,200	-850,085	1,467,000	850,085	160,400	635,857		3,864,036	2,863,707
JULY		481,200	-2,684,287		2,684,287	160,400	635,857		4,022,832	2,704,911
AUGUST			-1,127,450			160,400	473,455		1,905,552	3,857,142
SEPT		481,200	-1,890,750		73,299	1,287,850	473,455		2,064,348	4,825,796
OCT		481,200	2,639,704		691,215	160,400	473,455		2,223,144	4,267,000
DEC						2,800,104	473,455	213,360		4,250,440
JAN						2,800,104	473,455	1,371,600		4,037,080
FEB						2,800,104	473,455	2,387,600		2,878,840
MARCH			1,111,600			3,917,704	473,455	1,676,400		1,862,840
APRIL		9,624,000	-10,104,854		6,347,550	733,777	356,723		3,175,920	1,456,440
MAY		481,200	573,377			160,400	356,723		1,905,552	1,527,948
JUNE		481,200	-891,386		318,009	160,400	356,723		2,064,348	4,942,529
JULY		481,200	-2,109,762		2,109,762	160,400	356,723		2,223,144	4,783,733
AUGUST		481,200	-1,627,947		1,627,947	160,400	356,723		2,381,940	4,624,937
SEPT		481,200	-760,062		760,062	160,400	356,723		2,540,736	4,466,141
OCT		481,200	-760,062			160,400	356,723		2,699,532	4,307,345
NOV			3,013,387			3,173,787	616,513			3,993,490
DEC						3,173,787	616,513	277,830		3,455,865
JAN						3,173,787	616,513	1,786,050		1,947,645
FEB						3,173,787	616,513	3,109,050		924,645
MARCH			1,455,300			629,087	616,513	2,182,050		95,145
APRIL		10,947,300	-11,511,528		7,042,841	160,400	616,513		3,612,609	3,134,473
MAY		481,200	-1,708,726		1,708,726	160,400	616,513		3,771,405	2,975,677
JUNE		481,200	-1,166,422		1,166,422	160,400	616,513		3,930,201	2,816,881
JULY		481,200	-3,502,911		3,502,911	160,400	616,513		4,088,957	2,658,085
AUGUST		481,200	-2,040,908		2,040,908	160,400	386,08		4,247,253	2,499,289
SEPT		481,200	2,364,985			2,525,385	386,08		4,406,549	2,340,493
OCT			2,086,441			4,611,826	386,08	173,985		2,865,778
NOV						4,611,826	386,08	1,118,475		2,526,093
DEC						4,611,826	386,08	1,946,925		2,352,108
JAN						4,611,826	386,08	1,367,025		1,407,618
FEB			911,350			5,523,176	386,08			579,118
MARCH										247,718

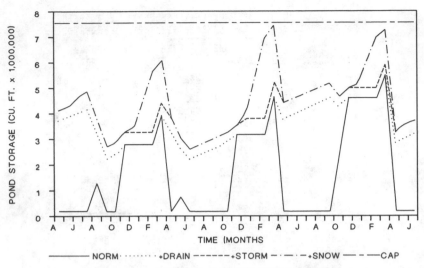

**WATER BALANCE SIMULATION
FACILITY No.1**

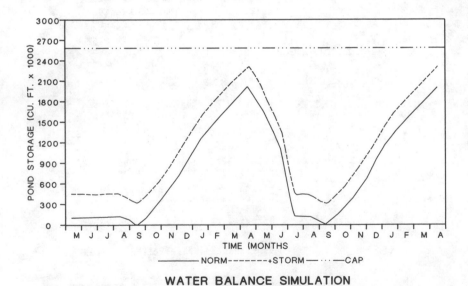

**WATER BALANCE SIMULATION
FACILITY No.2**

FIGURE 3. Example Water Balance Simulations

Table 2. The projected pond storage quantities are graphed as a function of time on Figure 3.

Maximum projected pond storage based on average climatic conditions and no discharges is approximately 2,013,000 cu ft. After adjusting for the design storm, total pond storage has increased to approximately 2,324,000 cu ft.

Based on this simulation, total pond storage of approximately 2,593,000 cu ft will be required to contain the anticipated process solutions with a minimum freeboard. However, total pond storage requirements may be decreased significantly if the treatment and discharge system is designed and operated on a monthly basis in addition to operating during storm and snowmelt surges.

SUMMARY AND CONCLUSIONS

In practice, the design storm and snowpack melt inflow mechanisms in combination with heap draindown, represent the bulk of the solution storage requirements and present the most opportunity for errors in design and facility operation in high snowfall climates. Pond and irrigation evaporation losses and permitted losses through land application or discharge to surface waters in conjunction with careful scheduling of the leaching operations present the best opportunities to control the facility water balance. The benefits to the operator of maintaining tight control over the water balance include reduced capital outlay on solution storage facilities and fewer opportunities for uncontrolled discharge of process solutions to the environment. The latter benefit has the potential for reducing neutralization bond requirements, as well as reducing possible cleanup costs during the life of the operation.

TABLE 2. Facility No. 2, Water Balance Simulation

FACILITY NO. 2 WATER BALANCE — REUSABLE PAD

TOTAL PAD AREA	525,000	SQ FT
TOTAL ORE PRODUCTION	5,000,000	TONS
DAILY PRODUCTION	6,000	TONS/DAY
LOAD CYCLE TIME	10	DAYS
LEACH CYCLE TIME	30	DAYS
RINSE CYCLE TIME	5	DAYS
UNLOAD CYCLE TIME	5	DAYS
TOTAL CYCLE TIME/CELL	50	DAYS
NUMBER OF LEACH CELLS	5	
AVERAGE LIFT HEIGHT	17.5	FEET
MAX TONS/CELL	60,000	TONS
MAX TONS UNDER LEACH/RINSE	240,000	TONS
MAX TONS ON HEAP	270,000	TONS
MAX SPRAY IRRIGATION AREA	497,000	SF BASED ON 3.5/5 (70 %) OF TOTAL PAD UNDER IRRIGATION APPROX. 2500 GPM PUMPING REQUIREMENT
AVG. INITIAL DENSITY OF HEAP	85	PCF
INITIAL ORE MOISTURE	13	(%)
LEACHING MOISTURE	25	(%)
FINAL ORE MOISTURE	17	(%)
ORE WETTING REQUIREMENT	25.13	APPROX. 3.85 CF OF SOLUTION/TON OF ORE
SPRAY APPLICATION RATE	0.005	GPM/SQ FT
SYSTEM CATCHMENT AREA	1,100,000	SQ FT 23.6 ACRES
POND SYSTEM CAPACITY	2,828,000	CF BASED ON 1 CELL (125,000 SF), 0.005 GPM/SF & 24 HOURS OF SOLUTION
100 YR 24 HR STORM MINIMUM POND VOLUME	120,000	CF
START-UP MINIMUM POND VOLUME	3.4	INCHES
MEAN ANNUAL PRECIP	34.75	INCHES
MEAN ANNUAL PAN EVAPORATION	42.40	INCHES APPROX. 31.4 INCHES OF EVAPORATION/YEAR
PAN EVAPORATION COEFFICIENT	74.00	(%)

MONTH	TONS ADDED TO HEAP	NEW TONS WETTED (DURING MONTH)	TONS ON HEAP (AT END OF MONTH)	TONS OF DRAIN DOWN (DURING MONTH)	AREA UNDER IRRIGATION (SF)	PRECIP (INCHES)	EVAP (INCHES)	SPRAY EVAP % OF APPLIC	PRECIP (X1000CF)	DRAIN DOWN (X1000CF)	ORE WETTING (CF)	SPRAY EVAP (X1000CF)	POND EVAP (X1000CF)	MONTHLY BALANCE (X1000CF)	MAKE-UP (X1000CF)	CUMULATIVE STORAGE (X1000CF)	100 YR 24 HR STORM (X1000CF)	CUMULATIVE + STORM (X1000CF)
MAY 89	0	0	0	0	0	3.2	3.8	5.0	0.0	0	0.0	0.0	0.0	0.0		120	312	432
JUNE	180,000	120,000	180,000	0	105,000	3.4	4.3	5.0	312	0	462	152	42	-343	343	120	312	432
JULY	180,000	180,000	240,000	120,000	346,500	1.4	5.9	10.0	128	308	692	1,000	32	-1,289	1,289	120	312	432
AUG	180,000	180,000	240,000	180,000	367,500	1.8	5.3	10.0	165	462	692	1,061	35	-1,162	1,162	120	312	432
SEPT	180,000	180,000	240,000	180,000	367,500	2.0	3.1	5.0	183	462	692	1,061	21	-1,130	1,010	0	312	312
OCT	180,000	60,000	240,000	240,000	262,500	2.5	2.2	5.0	229	615	231	379	17	218	0	218	312	529
NOV	0	0	240,000	0	0	3.2	0.8	2.0	293	0	0	0	0	293	0	511	312	823
DEC	0	0	240,000	0	0	3.8	0.4	2.0	348	0	0	0	0	348	0	859	312	1,171
JAN 90	0	0	240,000	0	0	4.3	0.6	2.0	394	0	0	0	0	394	0	1,254	312	1,565
FEB	0	0	240,000	0	0	2.9	1.5	5.0	266	0	0	0	0	266	0	1,519	312	1,831
MARCH	0	0	240,000	0	0	3.2	3.0	5.0	293	0	0	0	24	269	0	1,789	312	2,100
APRIL	0	0	240,000	0	0	3.0	3.8	5.0	275	0	0	0	51	224	0	2,013	312	2,324
MAY	180,000	120,000	240,000	0	105,000	3.2	3.8	5.0	293	0	462	152	62	-382	0	1,631	312	1,943
JUNE	180,000	180,000	240,000	180,000	367,500	3.4	4.3	5.0	312	462	692	531	40	-490	0	1,141	312	1,453
JULY	180,000	180,000	240,000	180,000	367,500	1.4	5.9	10.0	128	462	692	1,061	32	-1,196	174	120	312	432
AUG	180,000	180,000	240,000	180,000	367,500	1.8	5.3	10.0	165	462	692	1,061	35	-1,162	1,162	120	312	432
SEPT	180,000	180,000	240,000	180,000	367,500	2.0	3.1	5.0	183	462	692	1,061	21	-1,130	1,010	0	312	312
OCT	60,000	60,000	240,000	240,000	262,500	2.5	2.2	5.0	229	615	231	379	17	218	0	218	312	529
NOV	0	0	240,000	0	0	3.2	0.8	2.0	293	0	0	0	0	293	0	511	312	823
DEC	0	0	240,000	0	0	3.8	0.4	2.0	348	0	0	0	0	348	0	859	312	1,171
JAN 91	0	0	240,000	0	0	4.3	0.6	2.0	394	0	0	0	0	394	0	1,254	312	1,565
FEB	0	0	240,000	0	0	2.9	1.5	5.0	266	0	0	0	0	266	0	1,519	312	1,831
MARCH	0	0	240,000	0	0	3.2	3.0	5.0	293	0	0	0	24	269	0	1,789	312	2,100
APRIL	0	0	240,000	0	0	3.0	3.0	5.0	275	0	0	0	51	224	0	2,013	312	2,324

Chapter 26

MANAGING HEAP LEACH SOLUTION STORAGE REQUIREMENTS

by Scott E. Broughton and Robert T. Tape

Mining Engineer, Klohn Leonoff Ltd.
Richmond, British Columbia

Senior Geotechnical Engineer, Klohn Leonoff, Inc.
Kirkland, Washington

ABSTRACT

Heap leaching requires containment and storage of cyanide solutions, the volume of which may increase dramatically following an extreme storm event or extended wet periods. This paper describes a computer operated solution storage model developed to assist in the management of these requirements. It results from the accumulative efforts of various Klohn Leonoff engineers working on a number of heap leach projects throughout the western United States for the past several years.

INTRODUCTION

One aspect of gold/silver heap leaching is containment of the cyanide leach solutions. This includes a substantial amount of effort to prevent leakage from occurring through the containment barriers. Additionally, the leaching facilities need to be designed to satisfy a requirement that the total capacity of all containment areas is adequate to hold the maximum volumes of solution anticipated throughout their operating life. Also, since precipitation is included within this requirement, some form of design upper limit must be established.

Conceptually, the problem is quite simple. Solution volumes, including precipitation intercepted by the containment system, must be stored because discharge from cyanide leaching operations, in the general sense, is not allowed. However, identifying all the elements of stored solutions, establishing design requirements, and organizing their applications is a tedious process; just the type of problem that is well suited to computerized modelling.

This paper elaborates on one such model developed over a period of years by Klohn Leonoff engineers. The authors identify the basic components of the model, including how each of the input parameters are defined, and explain some of the more significant benefits and user-friendly features of the program. Finally, they discuss future improvements that are being contemplated.

HEAP LEACHING

Heap leaching of low-grade gold/silver ores is a relatively new practice. It has been going on for about 20 years and has become widespread throughout the western United States during the past 10 years. Alternately vat leaching; that is, cyanidation of higher grade crushed ores in tanks, has been ongoing for about a century.

The ore, primarily obtained from open-pit mines, is often crushed to some optimum gradation as determined by leach column testing. It may also be agglomerated to maintain a more porous structure, thereby facilitating percolation of solution through the ore. Ore, placed on a leach pad containment system, is irrigated with weak cyanide solutions, adjusted and pH buffered as necessary for the cyanide to dissolve the gold and silver into solution. The pregnant solution is then directed through a process plant where the precious metals are extracted. The barren solution with cyanide added to again facilitate leaching is cycled back over the ore heap.

Various combinations of pad configuration and solution storage facilities occur, with both permanent and load/unload pads being used. As the name suggests, in the former type, the ore is intended to reside permanently on the pad site and thus will be handled only once with continued leaching, year-after-year until spent. The load/unload pads are much smaller in nature since they are designed to hold only a portion of the ore for a limited period of time, after which the spent ore is removed to some disposal area and a fresh charge is placed on the pad. On flat ground terrain, both permanent and load/unload pads tend to be flat and, consequently, have little or no internal solution storage capability, other than solution in circulation during leaching. In these cases, solutions are generally stored in geomembrane-lined ponds constructed near the leach/processing areas.

In hilly or mountainous terrain, heap stability, especially for large permanent pads, generally dictates that a dike be constructed to hold the ore in place. Often this entails placing the dike across the mouth of a local gully or valley, hence the title of "valley-fill heap leaches". Such configurations result in sizeable solution storage volumes upstream of the dikes, within the voids of the ore heaps.

The more common solution storage facilities include:

- in-heap storage within the pore spaces of the ore behind dikes;

- process operation ponds, that is, pregnant and barren solution ponds; and

- emergency-spill or storm-contingency ponds.

A significant portion of the operating solution is normally circulating within the heap, percolating down through the ore during leaching. Following shutdown of leaching, whether planned or not, this solution comes out of circulation and occupies storage space which needs inclusion in design considerations. However, by maintaining circulation both during and after an extreme storm event, the actual total volume of solution that can be contained is substantially increased. Thus, under emergency conditions, the act of circulating solution provides a temporary means of surplus storage.

Typically, leaching is started soon after relatively small amounts of ore have been placed on the pad, a feature of heap leaching that allows fairly rapid return on investment compared to other mining processes. Also, for permanent pads, construction of the dike and pad surface area may be staged over several years. Combined, these features mean that both solution storage requirements and solution storage capacities may increase substantially over time, which can result in a complex balancing process to ensure that the storage capabilities keep up with increasing storage requirements.

WATER BALANCE MODELLING

The concept of managing heap leach solutions in its simplest form may be expressed mathematically by the following expression:

$$S = I - O$$

where S is the total storage requirement and I and O are the sums of all inflows and outflows, respectively. Given that heap leach projects are normally considered as closed systems throughout most of their lives, one might consider outflow as zero. However, there are some losses such as evaporation, and others that will be noted later, which must be included in water balance computations.

Inflows

An inventory of common inflows for storage design is as follows:

.   solution circulating within the heap;

.   precipitation intercepted by the leach pad and
    solution storage ponds;

.   make-up water;

.   transfer of solution from one storage facility or heap
    to another;

.   minimum solution levels and hence base solution
    volumes required for pump operations.

Solution circulating within the ore heap is best accounted for
by dividing it into two components. The first is vertical
percolation flowing down towards the containment surface or the
in-heap static solution level. This often represents a very large
portion of the total storage requirement, especially for fully
developed permanent heaps. Its magnitude depends on ore
characteristics such as porosity, grain-size distribution,
mineralogy, etc; the volume of ore under leaching and rate of leach
solution application.

The amount of solution in vertical circulation, frequently
expressed as a percent of dry ore weight, is determined from leach
column tests which are required for ore leaching evaluations. The
extra data and observations required to determine circulation
quantities are easily obtained. A limited amount of full-scale heap
test data, however, has indicated the leach column tests may
excessively overestimate actual in-heap circulation quantities by as
much as two or three times. This feature should be kept in mind,
with perhaps full-scale field tests being conducted once leach
operations have started, to fine tune the initial design
computations.

The second component of circulating solution is a saturated zone
immediately above the primary liner. It represents flow of solution
parallel to the liner surface which for permanent valley-fill heaps
progressively increases in depth from the outer fringes of the heap
to the inner storage zone. The configuration and, therefore, volume
of saturated-zone circulation varies depending on pad slope and
leaching configurations. In any case, this component may be
reasonably estimated from hydraulic computations given a knowledge of
pad geometry, particularly slope angles, leach rates and areas and
permeability of the porous materials immediately above the liner.

Liner cushioning and solution recovery considerations often result in layered porous media being placed within this zone.

Defining precipitation rates and quantities is done using conventional hydrologic procedures entailing past records and statistical analyses. Since we have no control over precipitation extremes, however, maximum storage requirements need to be based on some form of upper design limits augmented by contingency plans in the event that those limits are exceeded. The more controversial aspects of accounting for precipitation pertain to the severity of the extreme wet events for which storage ought to be provided and, if exceeded, what should be done about it. Regulations and their clarity vary from state to state.

Some authorities consider a 100-year return period (or occasionally even less) is an adequate level of protection with the provision that precipitation in excess of this limit may be treated (de-toxified) and released, generally in the form of land-spray applications. Others reason that should spills occur due to a more extreme event, the large volumes of surface runoff, also attributable to that same event, would readily dilute the solution to such low levels as to eliminate any negative effects of the spill. This particular feature applies to short-term storm events. It does not necessarily apply to 100-year wet years which can often exceed 100-year design storm events. Still other authorities prefer to have storage facilities which are capable of accommodating the probable maximum precipitation (PMP) event. Depending on location and operations plans, this may result in similar or substantially larger facilities than otherwise might be required. Once the level and type of design limits are selected, hydrologists, using generally well established techniques, are able to compute their magnitudes.

Surface runoff from surrounding areas needs to be directed away from the storage facilities as much as possible. Sometimes, when this is not entirely practicable, the runoff from certain adjoining areas has to be included as inflow to the system. At various times of the year and from time-to-time, particularly during periods of heap development when solution losses are greatest, shortages can occur, requiring that make-up water be added into the system. Alternatively, operators of facilities with multiple storage locations, especially with multiple heaps having internal storage, may wish to transfer solution from one site to another. Solution pumps generally require a minimum head for safe operations which translates to a minimum base storage volume. All of these items need to be included in the model.

Outflows

Typical sources of system losses include:

.   evaporation;

.   losses to ore wetting;

.   solution treated and released.

Evaporation from open ponded surfaces is generally determined from the best pan evaporation data available which is adjusted for open water bodies, nominally 0.7 times pan evaporation. Most of the evaporation losses, however, can occur from the spray irrigation works set up for leaching. The actual rate of these losses vary, being dependent upon droplet size, travel distance, wind velocity, cloud cover, vapor pressure, humidity, air temperature, etc. For simplicity, the authors commonly use a factor of 1.0 times pan evaporation for conservative evaluations of storage requirements and 1.5 times pan evaporation for estimates of make-up water requirements. Snow covered heaps, possibly shut down for the winter, will also experience sublimation losses providing warm temperatures do not melt the snow, allowing it to percolate down into the ore. Although many heap water balances benefit from maximized evaporation losses to offset precipitation, solution needs to be preserved at some dry locations, in which case closely spaced trickle pipes are used instead of sprays.

Solution loss due to ore wetting occurs when the ore is leached for the first time. Ore wetted by leaching does not drain down to as dry a state as before leaching. The wetted water content, available from leach column test data, minus the initial dry water content, represents locked-in solution which is lost from circulation. Such losses are limited to those periods when the heap is still under development.

Although not common, heaps may be operated such that solution is treated and released whenever an undesirable surplus occurs. In wet climates this can be on a seasonal basis. In dry climates, such practices will most likely be restricted to the release of rare, extreme inflow events. Seepage losses are not considered since any amount of solution leakage out of heap leach operations that might be considered acceptable practice would be so small as to be totally negligible to water balance computations.

Storage

The in-heap capacity for storing solution within the ore's pore spaces needs to be based on net usable (not gross) porosity; that is, the pore space available after pore capacity used up by ore wetting has been deducted. Leach column tests can be adapted to help in estimating net porosities.

Summary

In summary, heap leach water balance models need to properly account for all of the inflows and losses previously discussed. A reasonable time interval for successive computations appears to be monthly based on data availability and management requirements. Given all the necessary input parameters and the many years of monthly computations required, modelling an ore heap through its operating life requires a large number of repetitive calculations, even before varying ranges of precipitation events and combinations of operating plans have been considered. Heap leach water balance modelling is thus well suited to some form of computerized solution.

COMPUTERIZED MODELLING

While many engineering problems involve repetitive iterations for solution, some solutions are found by 'balancing' or 'accounting' of the data. In mineral processing, metallurgical balances are examples of these types of problems, where a balance of material quantities is required. Similarly, water balances are performed for reservoir design, where monthly hydrologic data is balanced with monthly demands to aid in the evaluation and optimization of reservoir size. Simply stated, a heap leach can be modelled as a reservoir with inflows of barren solution and precipitation, and outflows of pregnant solution and evaporative losses, such that the modelling of heap leach solution is performed by solution balance.

During the development of a computer model a decision is required as to the language or method of programming best suited to the task. If a language is chosen it must be recognized that the 'programming' of the engineering task may represent a small proportion of the program itself, and that a large part of it is code required to make the program usable by others. Any comprehensive program must include modules which make the program user friendly, including editing functions to add or append data, data locating functions, hardcopy printout functions, and graphics.

Financial accountants have long used the common spreadsheet as a computer software tool for balancing their books. This is because the spreadsheet is presented graphically in rows and columns, in a similar manner to a blank ledger. The key to this computerized ledger is that a cell (the intersection of a column and a row) can be programmed to perform a specific calculation based on data in other cells. In this way data can be entered in specific cells and calculations made based on this raw data can be performed and displayed rapidly.

Engineers who are currently programmers of spreadsheets will acknowledge the power of these self-calculating tables, however, the real advantage to engineering balances is realized when changes in conditions are required. Raw data is edited on screen and the table is recalculated under new conditions at high speed.

Some of the benefits found in popular programmable spreadsheets are summarized as:

- ease of programming, including, logical statements and all other common math functions;

- ease of raw data input, ease of data recognition by user, and editing;

- rapid recalculation under partially or entirely new conditions;

- ease of formatted hardcopy output;

- advanced macro programming language for creation of user-friendly and self-running programs; and

- software support of many common hardware configurations.

HEAP LEACH SOLUTION BALANCE MODEL

The computerized model presented herein is a programmed spreadsheet developed to calculate total inflows and outflows of solution associated with heap leach operations. The model uses a Lotus 1-2-3 or Symphony spreadsheet. It has been used as a design and operating tool on several projects to date.

When used for design, a single-heap simulation is performed to calculate the storage requirements and thus ultimate height of impoundment dikes for internal heap storage or the size of ponds required for external storage. The model also allows for staged monthly construction of both ore loading and lined area, such that balanced leaching can be performed during the development of the heap.

For operating purposes, the model can be extended to include multiple-heap configurations. A facility to transfer to another pad or pond permits a "link" for solution management of the entire multiple-heap leaching operation. Summary tables of storage requirements, storage availability, and make-up water requirements are also featured in multiple-heap configurations.

The current model has been successfully extended to balance an operation involving up to six heaps. For this particular model many user-friendly features have been added to the spreadsheet to help, and to expedite the use of the model by the mine operators on site. These features are made possible by the macro language available as part of Lotus 1-2-3 and Symphony, and include:

.   rapid printing of individual heap operations and summary tables with easy to use menus;

.   rapid "go to" with easy to use menus;

.   menu driven editing of operating conditions including adding or deleting a month, and moving data forward a month.

As mentioned in the previous section, Lotus and Symphony support of hardware configurations is a benefit of programming with spreadsheets. Other methods of programming the task, for example using popular micro-based languages, would have put restrictions on user hardware requirements. This model therefore requires no additional graphics cards, monitors, printers, or plotters, other than the configuration on which the user is currently operating Lotus 1-2-3 or Symphony. The ease of expanding the single-heap model into a multiple-heap model is another direct benefit of using a spreadsheet such that customizing the model to simulate a heap leach facility can be easily performed.

The single-heap model consists of basically two tables or pages. The first page is a "model constants page". It was designed to exactly fit the width of the monitor, and is thus easy to edit. Initial conditions are required since the model operates on a monthly basis and start-up conditions are entered into the model here. A typical printout of a single heap is presented on Figures 1 and 2.

Initial conditions are the results of the balance, or, the condition of the heap, at a time of start-up minus one month. They may be considered at any stage of development prior to leaching, or, may indicate tonnages and volume of ore and collection pool storage after a winter shutdown, and prior to resuming leaching.

```
| WATER ACCOUNTING MODEL                   OUR FILE:  Sample         |
|                                          DATE:     24-Jul-88       |
| L'eau D'or Mine                                                    |
| --------------------------------                                   |
|                                                                    |
| HYDROLOGIC CONDITIONS:                                             |
|        - Average annual precipitation =        88.8  inches        |
|        - Storm surcharge calculated for         4.4  inches        |
|                                                                    |
|                                                                    |
| WATER BALANCE PARAMETERS:                                          |
|                                                                    |
|  1. Calculations start in      Apr-90                              |
|  2. Loss to ore wetting =           3% by weight of wetted ore     |
|  3. Porosity (Net Useable)         35% by volume                  |
|  4. Volume of solution in circulation in unsaturated              |
|     percolation zone =          1.3% by weight of ore being sprayed|
|  5. Heap pile density =        15.0 (cu. ft./ton)                 |
|  6. Volume of solution in circulation in saturated                |
|     basal zone =               1.0    ft  x leaching area          |
|  7. Spraying rate for barren solution                             |
|     over heap pile =          0.005  gpm/ sq ft                    |
|  8. Evaporation during spraying                                   |
|     pan evaporation    x       1.0                                |
|  9. Initial conditions                                            |
|     Cumulative ore loaded =                    1000000 tons        |
|     Ore placed but not yet wetted =                  0 tons        |
|     Collection pool reading    30    volume =>    0.70 (Mil gal)|  |
|     Leaching (from barren pond):                                  |
|     - Leaching area =                             0.00 acres       |
|     - volume in unsaturated percolation zone =       0 (Mil gal)|  |
|     - volume in saturated basal zone =               0 (Mil gal)|  |
|     Sprinkling (from recycle or transfer):                        |
|     - Sprinkling area =                           0.00 acres       |
|     - volume in unsaturated percolation zone =       0 (Mil gal)|  |
|     - volume in saturated basal zone =               0 (Mil gal)|  |
| 10. Upslope catchment areas                                       |
|     - contributory area =     0.00  acres                         |
|     - runoff = precipitation x  0.00                              |
| 11. Minimum pumping requirments                                   |
|     - minimum depth =                            10.0 feet         |
|     - minimum storage in collection pool =          0 (Mil gal)|   |
| 12. Collection Pool storage vs. Elevation data                    |
|                                GROSS VOL.  STORAGE                 |
|                   ELEVATION    OF STORAGE  IN VOIDS                |
|                      (ft)      (Mil gal)   (Mil gal)               |
|                   ---------    ----------- --------                |
|                           0       0.00       0.00                 |
|                          20       0.50       0.18                 |
|                          40       3.50       1.23                 |
|                          60      11.00       3.85                 |
|                          70      16.50       5.78                 |
|                                                                    |
| 13. Lined pad area =        13.77 acres                           |
```

Figure 1.   Model Constants Page

WATER ACCOUNTING MODEL : MEAN PRECIPITATION AND EVAPORATION

L'eau D'or Mine   OUR FILE: Sample

MONTH-YEAR	Apr-90	May-90	Jun-90	Jul-90	Aug-90	Sep-90	Oct-90	Nov-90	Dec-90	Jan-91	Feb-91	Mar-91
1 LOADED PAD AREA (acres)	13.77	13.77	13.77	13.77	13.77	13.77	13.77	13.77	13.77	13.77	13.77	13.77
2 LEACHING AREA (acres)	8.26	8.26	8.26	8.26	8.26	8.26	0.00	0.00	0.00	0.00	0.00	0.00
3 SPRINKLING AREA:SOLN FROM RECYCLE/TRANSFER (acres)	0.00	0.00	0.00	0.00	0.00	0.00	0.00	0.00	0.00	0.00	0.00	0.00
4 ORE LOADING (tons/month)	200000	200000	158000	0	0	0	0	0	0	0	0	0
5 ORE PLACED BUT NOT YET WETTED (tons)	200000	200000	158000	1558000	1558000	1558000	1558000	1558000	1558000	1558000	1558000	1558000
6 CUMULATIVE ORE LOADED (tons)	1200000	1400000	1558000	1558000	1558000	1558000	1558000	1558000	1558000	1558000	1558000	1558000
7 PRECIPITATION OVER PAD AREA : MEAN MONTHLY (inches)	4.10	2.40	2.00	1.40	1.80	2.10	3.80	5.70	7.90	8.10	7.20	6.00
8 (Million gallons)	1.53	0.90	0.75	0.52	0.67	0.79	1.42	2.13	2.95	3.03	2.69	2.24
9 SURFACE RUNOFF FROM SURROUNDING AREA (Million gallons)	0.00	0.00	0.00	0.00	0.00	0.00	0.00	0.00	0.00	0.00	0.00	0.00
10 SOLUTION FROM PROCESS PLANT SPRAYED OVER HEAP PILE (Million gallons)	78.82	78.82	78.82	78.82	78.82	78.82	0.00	0.00	0.00	0.00	0.00	0.00
11												
12 SOLUTION/WATER RECIRCULATED WITHIN HEAP PILE (Million gallons)	0.00	0.00	0.00	0.00	0.00	0.00	0.00	0.00	0.00	0.00	0.00	0.00
13 PAN EVAPORATION (inches)	2.80	4.50	4.60	4.80	3.90	2.80	1.40	0.00	0.00	0.00	0.00	1.40
14 HEAP PILE EVAPORATION (Million gallons)	0.00	1.01	1.03	1.08	0.87	0.63	0.00	0.00	0.00	0.00	0.00	0.00
15 SUBLIMATION (inches)	0.00	0.00	0.00	0.00	0.00	0.00	0.00	0.00	0.00	0.00	0.00	0.00
16 (Million gallons)	0.00	0.00	0.00	0.00	0.00	0.00	0.00	0.00	0.00	0.00	0.00	0.00
17 ADDITION TO ORE WETTING (Million gallons)	1.44	1.44	1.14	0.00	0.00	0.00	0.00	0.00	0.00	0.00	0.00	0.00
18 ADDITION TO CIRCULATION IN UNSATURATED PERCOLATION ZONE												
19 - LEACHING (Million gallons)	2.24	0.37	0.30	0.00	0.00	0.00	-2.91	0.00	0.00	0.00	0.00	0.00
20 - SPRINKLING (Million gallons)	0.00	0.00	0.00	0.00	0.00	0.00	0.00	0.00	0.00	0.00	0.00	0.00
21 ADDITION TO CIRCULATION IN SATURATED BASAL ZONE												
22 - LEACHING (Million gallons)	0.94	0.00	0.00	0.00	0.00	0.00	-0.94	0.00	0.00	0.00	0.00	0.00
23 - SPRINKLING (Million gallons)	0.00	0.00	0.00	0.00	0.00	0.00	0.00	0.00	0.00	0.00	0.00	0.00
24 INFLOW OF SOLUTION/WATER TO HEAP PILE COLLECTION POOL (Million gallons)	75.73	76.89	77.10	78.26	78.61	78.97	5.28	2.13	2.95	3.03	2.69	2.24
25 PORTION OF INFLOW REQUIRED TO SATISFY MIN. STORAGE FOR PUMPING	0.00	0.00	0.00	0.00	0.00	0.00	0.00	0.00	0.00	0.00	0.00	0.00
26 PORTION OF INFLOW TRANSFERRED TO ANOTHER PAD OR POND (Million gallons)	0.00	0.00	0.00	0.00	0.00	0.00	0.00	0.00	0.00	0.00	0.00	0.00
27 PORTION OF INFLOW RECIRCULATED WITHIN HEAP PILE (Million gallons)	0.00	0.00	0.00	0.00	0.00	0.00	0.00	0.00	0.00	0.00	0.00	0.00
28 SOLUTION TREATED AND RELEASED (Million gallons)	0.00	0.00	0.00	0.00	0.00	0.00	1.39	2.13	2.96	3.03	2.69	2.24
29 SOLUTION RETURNED TO PROCESSING PLANT (Million gallons)	76.34	76.89	77.10	78.26	78.61	78.82	0.00	0.00	0.00	0.00	0.00	0.00
30 EXCESS PORTION OF INFLOW ADDED TO COLLECTION POOL STORAGE	0.00	0.00	0.00	0.00	0.00	0.16	3.89	0.00	0.00	0.00	0.00	0.00
31 COLLECTION POOL STORAGE (Million gallons)	0.09	0.09	0.09	0.09	0.09	0.24	4.13	4.13	4.13	4.13	4.13	4.13
32 APPROXIMATE COLLECTION POOL ELEVATION (ft)	10.0	10.0	10.0	10.0	10.0	21.3	61.5	61.5	61.4	61.4	61.5	61.5
33 MAKE-UP WATER REQUIRED AT PROCESS PLANT (Million gallons)	2.48	1.92	1.72	0.55	0.20	0.00	0.00	0.00	0.00	0.00	0.00	0.00
34 COLLECTION POOL STORAGE (Million gallons)	0.09	0.09	0.09	0.09	0.09	0.24	4.13	4.13	4.13	4.13	4.13	4.13
35 CIRCULATION IN UNSATURATED PERCOLATION ZONE												
36 - LEACHING (Million gallons)	2.24	2.62	2.91	2.91	2.91	2.91	0.00	0.00	0.00	0.00	0.00	0.00
37 - SPRINKLING (Million gallons)	0.00	0.00	0.00	0.00	0.00	0.00	0.00	0.00	0.00	0.00	0.00	0.00
38 CIRCULATION IN SATURATED BASAL ZONE												
39 - LEACHING (Million gallons)	0.94	0.94	0.94	0.94	0.94	0.94	0.00	0.00	0.00	0.00	0.00	0.00
40 - SPRINKLING (Million gallons)	0.00	0.00	0.00	0.00	0.00	0.00	0.00	0.00	0.00	0.00	0.00	0.00
41 TOTAL OPERATING STORAGE REQUIRED WITHIN HEAP PILE; NOT INCLUDING STORM SURCHARGE (Million gallons)	3.27	3.65	3.94	3.94	3.94	4.10	4.13	4.13	4.13	4.13	4.13	4.13
42 STORM SURCHARGE (Million gallons)	1.65	1.65	1.65	1.65	1.65	1.65	1.65	1.65	1.65	1.65	1.65	1.65
43 OPERATING STORAGE INCLUDING STORM SURCHARGE (Million gallons)	4.92	5.29	5.59	5.59	5.59	5.75	5.78	5.78	5.77	5.77	5.78	5.78
44 TOTAL STORAGE AVAILABLE WITHIN HEAP PILE (Million gallons)	5.78	5.78	5.78	5.78	5.78	5.78	5.78	5.78	5.78	5.78	5.78	5.78
45 STORAGE CAPACITY AVAILABLE IN EXCESS OF OPERATING REQUIREMENTS. NOT INCLUDING STORM SURCHARGE: MINUS SIGN MEANS MORE STORAGE IS REQUIRED THAN IS AVAILABLE (Million gallons)	2.50	2.13	1.83	1.83	1.83	1.67	1.64	1.64	1.65	1.65	1.64	1.64
46 STORAGE CAPACITY AVAILABLE IN EXCESS OF OPERATING AND STORM SURCHARGE REQUIREMENTS (Million gallons)	0.86	0.48	0.19	0.19	0.19	0.03	0.00	0.00	0.00	0.00	0.00	0.00

Figure 2. Model Solution Balance

The critical conditions of the model which require initializing are:

. cumulative ore loaded (tons);

. ore placed but not yet wetted (tons);

. collection pool volume (million gallons);

. leaching area (acres);

. sprinkling area (acres).

As discussed previously, multiple-heap configurations and large single heaps present the opportunity of reducing storage requirements if a transfer or recycling is performed; that is, pregnant solution is pumped and sprayed directly back onto a heap.   In this way solution volume can be reduced by an increase in evaporation losses. Because this operation can be performed on many heaps, a facility to simulate it must exist in the model.   To distinguish between the spraying of barren solution from the process plant and the spraying of pregnant solution from the recycling operation, the terms leaching and sprinkling have been adopted respectively.   The model calculates additions to the percolation and the basal zone from leaching and sprinkling and thus the initial conditions must include leaching and sprinkling areas.

The model constants page also includes heap parameters, previously discussed, used monthly to calculate storage demand and availability.   The constants are displayed on the first page and are easily edited for calibration of the model, they include:

. storm surcharge (inches of storm precipitation);

. loss to ore wetting (measured in percent of wetted ore);

. net usable porosity (measured in percent by volume, or in gallons per ton of ore);

. volume of solution in saturated basal zone (measured in feet above lined leaching area);

. evaporation during spraying (factor of 1.0 typically used, times pan evaporation);

. upslope catchment area (acres);

. minimum pumping requirements (feet);

.   collection pool storage/elevation data (presented in tabular form, this table provides storage/elevation data over the range of the lined area elevation). Storage in voids is gross volume times net usable porosity. The model references this table and interpolates linearly between data point sets to calculate monthly solution levels and available storage.

Page two of the model is the programmed table where monthly balances are calculated. This table has headings divided into seven main categories:

.   basic heap characteristics, such that staged monthly changes may occur;

.   inflows to the heap;

.   outflows from the heap and additions to circulation zones;

.   accounting of collection pool portions;

.   make-up water required;

.   total storage requirement accounting;

.   total storage availability.

The first column of the table has many functions which reference the initial conditions of the model, while all following columns reference previous columns for key data, including storage availability. Monthly precipitation and evaporation data can be entered manually into the table, allowing the user to input actual records or superimpose a wet year at any point in the model. Alternatively, a hydrologic data input table can be created and required information will automatically be stored in the model.

Some of the column cells are highlighted for direct user input or editing while others are "protected" programmed cells that cannot be inadvertently changed. Many programmed cells are in fact complex logical statements performing "if" functions. These cells perform different calculations based on operational outlines input by the user. For example, if the leaching area is given for a particular month, then calculations for inflows to the percolation zone are performed. If not then no additions are entered into the balance.

Column 2 and all subsequent columns to the end of the table are similar in that calculations are always based on the previous column. When the user wishes to add columns the Lotus copy command is used to copy the previous column as a whole to the new extension. A benefit of Lotus spreadsheets is that the copy command will copy equations and change any reference cell addressed relative to the new extension. Absolute cell addresses are assigned to certain equations which require coefficients from the model constants page, which do not change location.

Expansions of the single-heap model to a multiple-heap model are also performed by the copy command, such that two or more single-heap models operate on one spreadsheet. Customizing can then be performed by a competent spreadsheet programmer to produce a user-friendly and efficient, custom solution management model for site use.

Multiple-heap configurations with many heaps and long operating periods occupy a large amount of space on the spreadsheet. These large spreadsheets in turn occupy a large amount of memory and may limit the user to few extensions. A new development of the model will have one single-heap configuration per spreadsheet, such that a multiple-heap link would be achieved by linking separate spread-sheets. This is a task which requires advanced macro programming to be fully automatic and user friendly but will ultimately allow the user to extend and expand to a much larger limit. It was originally thought that no one mine would have more than five to six heaps for balancing, and these configurations have been modelled on a single spreadsheet, however, a desire to model larger multiple-heap configurations for extended periods has initiated these types of new developments of the model.

Figures 1 and 2, as mentioned, present a typical hardcopy printout of the single-heap model. Figure 1 shows the model constants and initial conditions of the example heap. The heap represented is a typical valley-fill heap with large internal collection pool and no external ponds of appreciable size. The heap was operating the previous summer and was shut down for winter such that the collection pool volume is significant. In the spring another lift of new ore is to be placed, staged over a three-month period. Leaching will be ongoing during the placement of the new ore, and will shut down again in the fall.

As with all accounting the most important result of Figure 2 is the "bottom line". Here the available storage is shown with and without storm surcharge added in. For the example case shown, the user found a negative storage capacity when calculations were performed for the design storm event and has decided to perform a transfer to another facility (line number 26) to maintain storage capacity for mean precipitation and the storm event.

DATA RELIABILITY

Precipitation and pan evaporation data are frequently not available for remote mining sites. Therefore, on-site collection of data for these two parameters as early as possible before heap construction and continuing throughout leaching is very desirable. Data for initial studies is generally extrapolated from the nearest available sources.

Most of the ore properties needed for water balance computations may be estimated from leach column and other laboratory tests. However, full-scale field tests are very valuable to reconfirm the laboratory data. This applies especially to the circulation component of solution storage and to estimates of in-heap net available porosity.

It is our view that refinements in the techniques used to estimate the rates of evaporation losses would be useful to the industry. We suspect, however, that very carefully controlled field observations made over several months at numerous sites using a variety of leach irrigation equipment would be needed to produce reliable results. Benefits would result from evaluating equipment which either maximized or minimized evaporation losses.

CONCLUSIONS

Although conceptually simple, heap leach solution/water balance analyses and, hence, solution management is laborious and very sensitive to operational or other changes which frequently occur. Analyses such as these are well suited to computer models, one of which is demonstrated in this paper. Also, as demonstrated, the adaptation of commercially available spreadsheet technology can be very effective. While appreciating that refinements in both the physical and computer models are a natural, ongoing process, Klohn Leonoff's experience with heaps already in service confirms that the model presented is reasonably efficient and reliable. Its usage is equally applicable to the initial design of the containment facilities and to managing solution-balances throughout leaching operations. The benefits of computerized water balance modelling becomes readily apparent when alternative designs are being evaluated and when a heap leach operator, shortly after having experienced a rare storm, is wanting immediate but reliable answers concerning various mitigative action plans.

7

Placer Mining and Dredging

Chapter 27

PRECIOUS METAL RECOVERY USING SLUICE BOXES
IN ALASKA AND CANADA

by James A. Madonna

Associate Professor, Mining Extension
School of Mineral Engineering
University of Alaska Fairbanks

ABSTRACT

Information gathered while studying geological parameters that govern fine gold deposition in the alluvial environment can be applied to the technological development and selection of fine gold recovery systems.

It is recognized that substantial gold losses occur in the minus 1.5 mm (minus 12 mesh) fraction using conventional sluicing systems.

Within the past two decades much attention has been given the design and efficiency of these thick bed concentrators. Studies and comparisons drew information from several foreign and domestic placer operations. These studies concentrated on:

1) the physical and chemical properties of alluvial gold

2) the parameters that control fine gold deposition

3) the geometrical form, components and general working structure of conventional sluicing systems

4) the physical properties which contribute to the overall efficiency of a sluice system

5) the comparison of three major styles of conventional sluicing systems.

A dramatic improvement regarding fine gold recovery and water quality was observed when using modern multiple-channel sluicing systems in place of conventional single-channel sluicing systems.

BACKGROUND

The careful consideration of parameters governing the hydraulic separation of ore minerals from a gravel aggregate has led to the development of more effective concentrating equipment. The modern equipment is capable of processing increased volumes of gravel while simultaneously more efficiently recovering the valuable minerals from placer deposits.

The more effective processing of increased volumes of gravel has compensated for a decreased tenor of deposits and, therefore, a profitable mining of ground which was previously uneconomical.

The three major parameters that control the separation of gold and other minerals from a gravel aggregate in a hydraulic environment include size, shape and specific gravity. The subject here is the profitable extraction of gold but the model developed can be applied to any ore mineral amenable to hydraulic separation.

SIZE:

In regard to separating gold disseminated in alluvial deposits from its associated gravel aggregate, the consideration here is not the recovery of nuggets of 8 mesh (2.36 mm) or larger, but the rather difficult problem of recovering fine gold in the size range between minus 8 mesh (2.36 mm) and 400 mesh (.038 mm).

It was recognized in the early 1800's that there was a progressive or a linear distribution of the finer gold particles within a sluicing system; the particle size decreased down the length of a single channel sluice box. It was also recognized that considerable losses were occurring in the minus 12 mesh (1.70 mm) fractions. Subsequent studies indicate that a volume of water suitable to wash a gravel aggregate through a length of single channel sluice box of uniform width will reach mechanical energy equilibrium within two to three feet of the head of the box. There then becomes a relation between the size of a rock aggregate and the size of a gold particle that will be transported by a given uniform hydraulic energy. Tourtelot (1968) used the concept of hydraulic equivalence, which states, "Hydraulic equivalence is the concept that describes the relation between grains of a mineral of a

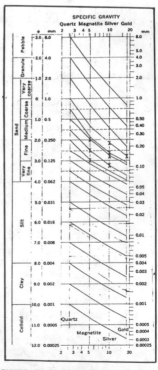

FIGURE 1: Hydraulic-equivalent grain sizes of quartz, magnetite, silver, and gold (From Tourtelot, 1968).

given size and specific gravity and the smaller grains of heavier minerals that are deposited simultaneously under given hydraulic conditions", to show (figure 1) that a sphere of quartz 8 mm (5/16 inch, 2.5 mesh) in diameter would be deposited in an aqueous medium contemporaneously with a sphere of gold 0.80 mm (approximately 20 mesh) in diameter. Conversely, the hydraulic action of water within a sluice box capable of carrying an 8.0 mm particle of quartz to tail would also transport a 0.80 mm gold particle to tail. This hydraulic equivalence of grains becomes an important consideration when the minus 40 mesh (.425 mm) gold constitutes 50 to 80 percent of the product as it does in some geographic locations (Walsh, 1985).

SHAPE:

Particle shape also plays an important role in entrapment. The shape is commonly referred to in terms of the Corey Shape Factor, which is simply the sphericity of the gold particle. The Corey Shape Factor is derived from the formula $CSF = C/\sqrt{A \times B}$, where A equals the length, B equals the width and C equals the thickness of the particle. A sphere or a cube has a Corey Shape Factor of 1, simply because the three dimensional axes are the same length in all directions. Any other form has a Corey Shape Factor of less than 1.

Figure 2 shows two groups of nuggets of approximately equal weight. The flat nuggets in the group on the right have a much smaller Corey Shape Factor than the more spherical nuggets on the left.

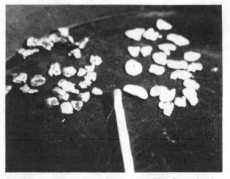

If a small rounded nugget and a nugget of similar weight, but much flatter, were both deposited in a column of stream gravel under agitation by running water, the rounded nugget would sink into the gravels, while the flat nugget would be carried farther along with

FIGURE 2: Variations in Corey Shape Factor. Note flat gold grains on right and more spherical grains on left.

the stream flow. This analogy helps explain the effect of shape on the transportation and entrapment of fine placer gold within a sluicing system.

Field tests have shown that, unless appropriate measures are taken, a large percentage of gold particles in the minus 20 mesh (0.85 mm) fraction, which exhibit small Corey Shape Factors, are commonly carried along with the aggregate out to tail in a single channel sluicing system.

SPECIFIC GRAVITY:

The third parameter which controls the collection of fine gold within a sluicing system is specific gravity. Natural gold has a

specific gravity which ranges between 15 and 19.3. This variation in
specific gravity of gold is produced most often by associated alloyed
metals. The major alloying metal with gold is silver, which has a
specific gravity of 10.1 to 11.1. Natural placer gold may have a
gold/silver ratio as high as three to one (3/1). As a result, with
the increase in alloyed silver, there will be a corresponding decrease
in specific gravity of the particle.

Effects of Porosity on Specific Gravity: In addition to the
alloyed metals, which will decrease the specific gravity of gold,
there are other factors to consider. Among them is the porous nature
of placer gold found in certain geographic areas. It should be
pointed out that other valuable placer minerals, such as magnetite,
cassiterite and ilmenite, are compact and non-porous, whereas placer
gold can be naturally honeycombed with cavities. This is enhanced
when alloyed elements such as silver and copper are leached from the
surface of the gold particle, thereby adding to the surface porosity
of the gold. If these pores are filled with clay or other low specific gravity materials, there is a corresponding decrease in the
overall specific gravity of the particle. In addition if the small
particles of gold have a significant amount of quartz or encrustations
such as iron oxide, these will also lend to a lower apparent specific
gravity.

Note the 120 mesh (0.125 mm) brown-grey residue in figure 3A
which was recovered from an Alaskan placer mining operation. In addition to the dark brown color, this material had a specific gravity
ranging about 9. It would not amalgamate and was difficult to separate, by tabling, from the other concentrate. However, because of the
rather large volume, some of the material was sent to a laboratory
for analysis, where it was found to be over 970 fine gold. It was
thought at first that there might be an oxide coating on the gold to
prevent amalgamation and produce variations in color. However, exami-

nation under a microscope (figure 3B) revealed that the gold
particles exhibited a highly pitted
spongy character which did not
reflect light as did the more typical gold and therefore was responsible for the brown-grey color.
Also, the rather porous nature of
the particles contributed to the
lower specific gravity observed.
In addition, it is thought that the
porosity also contributed to the
resistance to amalgamation.

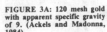

FIGURE 3A: 120 mesh gold
with apparent specific gravity
of 9. (Ackels and Madonna,
1984).

FIGURE 3B: 120 mesh gold
under high powered magni-
fication showing porosity.
(Ackels and Madonna, 1984).

Effects of Hydrophobicity On Specific Gravity: It has been
suggested that gold has a natural hydrophobicity (lack of wetability)
which causes floatability. According to a review by Walsh (1985),
there has been some controversy in the scientific literature regarding
the hydrophobic nature of gold. It appears that gold particles are

normally hydrophobic and hence floatable due to contamination by organic matter. In his summary treatment of hydrophobicity, Walsh suggested that if significant amounts of alloyed metals such as silver are present, contamination is less prevalent and the gold particle will be less hydrophobic. On the other hand, silver leached from the surface of the particle produces porosity and increases the hydrophobicity of the gold. Gold's resistance to wetting leaves organic matter or oxygen free to coat its surface and fill associated surficial cavities, thereby contributing to the gold particle's apparent low specific gravity. It also lends to the gold's floatability and is a particular problem with fine gold recovery should the particles become even momentarily exposed to the air during the sluicing action.

PROCESSING EQUIPMENT

CLASSIFICATION:

Because the entrapment of gold particles is much more efficient when the ratio between the maximum size particle and the minimum size particle is very low, sizing the material prior to concentration is desirable. Sizing can occur in two areas, the pre-sluice system and the inner-sluice system.

Pre-Sluice Sizing: Pre-sluice sizing can be conducted utilizing several types of sizing equipment. Recently, in the Alaskan (USA) and Yukon (Canada) placer mining industry, three pieces of equipment have found favor. First, the standard grizzly which consists of stationary bars or gridwork through which the finer material passes. Secondly, a shaking screen type sizing system which sizes the material over three banks of screens with progressively smaller openings. And, finally, a trommel screen which is an elongate barrel with orifices through which the fine grained material passes into the gold recovery plant.

Inner Sluice Sizing: Punch plate has been used as the common classification system for inner sluice sizing. Its positioning will be discussed in detail during treatment of the various sluice designs.

To better understand the technological advances in gold recovery systems that have occurred in the past 100 years, it is necessary to examine a variety of placer mining operations in Canada and Alaska (see location map, figure 4).

FIGURE 4: Location of mining sites.

SLUICE BOXES

A sluice box is an elongated three-sided rectangular wooden or metal trough, equipped with riffles and matting, through which gravel is washed to recover gold and/or other valuable heavy minerals.

RIFFLES:

Riffles are obstructions, placed in the bottom of a sluice box, which create a turbulence during hydraulic transportation of valuable minerals and gravel passing over them. The interaction of the riffles with the suspended gravel aggregate results in a retarded flow of material, which permits the higher specific gravity minerals to settle. Proper shape of the riffles produces agitation within the riffle pockets, which returns collected low specific gravity materials to the hydraulic flow. In addition, the proper form, combined with the proper spacing, provides a crude classification of the material retained in the riffle pockets.

Some of the riffle systems that were used in the early days included the transverse riffles, the longitudinal riffles and the block riffles. More recently, however, placer miners have settled on angle iron transverse (Hungarian) riffles as shown in the sluice of figure 8A, and raised expanded metal riffles as shown in the sluice of figure 12B.

MATTING:

Matting is the fabric material that separates the riffles from the sluice floor. The major function of the matting is to provide a secure area for fine grained ore minerals, should excessive scouring of the riffle pockets occur. In addition, as Poling and Wang (1983) have pointed out, it should be durable, easy to clean, and prevent material flow beneath the riffles.

Early placer miners used burlap, corduroy and cocoa matting beneath the riffles to trap the finer gold. However, more recently, miners have begun to use ozite, astroturf or Nomad carpet.

EARLY WOODEN SLUICE BOXES:

In the interior of Canada and Alaska the stream gravels are perpetually frozen. The early miners, of the late 1800's, were called drift miners because they sank shafts through the frozen gravels to bedrock, then began drifting out laterally to either locate or follow the pay. The pay gravel was hauled to the surface and washed through small wooden sluice boxes similar to that shown in fig. 5.

FIGURE 5: Single channel wooden sluice box typically used by Alaskan and Canadian gold miners prior to 1930.

This particular sluice box is roughly 16 feet long and one foot wide. Although block and longitudinal riffles were common, this sluice box employed wooden transverse riffles. Wooden sluice boxes of this style, which could process between one and two cubic yards an hour, were common in the smaller placer mining operations of the Klondike in Canada (see location map, figure 4).

In larger operations, the sluice boxes could be extended to lengths of sixty feet or more. The purpose of the longer sluice box was fourfold. First, there was a limited amount of water that was channeled by a flume, ditch or pipe into the head of the sluice box. To make full use of the water, the sluice boxes extended the length of the mining section of the claim. Secondly, the miners shovelled in from multiple piles located along the length of the sluice box. Third, reaching the area appropriate for tailings disposal dictated the length of the sluice box. Finally, the capture of fine gold was also a consideration. Sluices of this style could process between 4 and 6 cubic yards of minus 6-inch material per hour. It required a high mechanical energy for the water to transport the larger gravel through. As a result, significant fine gold losses in the minus 12 mesh (1.70 mm) fraction were experienced.

STEEL SINGLE CHANNEL SLUICE BOXES:

With the development of the diesel engine and corresponding improved lightweight earth moving equipment in the 1930s, there was a change from the old style wooden sluice boxes to the more durable steel boxes, similar to that used at a mine located near Fairbanks, Alaska, shown in figures 6A and 6B. This sluice is 2 feet wide and 30 feet long, with 26 feet of transverse riffles followed by 4 feet of punch plate which provides an undercurrent at the end. Boxes of this style can process between 25 and 40 cubic yards of unsized gravel per hour. However, to wash the larger boulders through, a significant amount of mechanical energy is required. As a result significant losses in gold finer than 20 mesh (0.85 mm) occur.

Figures 7A and 7B show a more advanced single channel sluice box currently used at a placer mine near Fairbanks, Alaska. This box is 60 feet long and 54 inches wide,

FIGURE 6A: Single channel steel sluice box introduced into the mining industry in the 1930's.

FIGURE 6B: Sluice box as shown in FIGURE 6A in operation. Note the high water energy.

FIGURE 7A: Modern steel single channel sluice with hungarian riffles and white fiberglass undercurrent punch plate at the tail end.

FIGURE 7B: Sluice box as shown in FIGURE 7A in operation. Note the high water energy necessary to transport larger boulders.

with fifteen feet of transverse angle iron (Hungarian) riffles, one
foot of steel punch plate followed by 24 feet of ribbed fiberglass
punch plate underlain by Nomad matting. In this case, the fiberglass
punch plate provides an undercurrent at the tail end of the sluice.
There is no sizing of material; it is pushed directly onto the sluice
plate where the water force is capable of moving 3 foot diameter
stones. Boxes of this type are capable of processing up to 150 cubic
yards per hour. Mechanical energy of this magnitude within a single
channel sluice box contributes to significant fine gold losses in the
minus 20 mesh (0.85 mm) range.

MULTIPLE CHANNEL SLUICE BOXES:

Since 1975, more advanced sluicing systems, using more sophisticated pre-sluice and in-sluice sizing, have been developed.

Triple Channel Sluices: A Fairbanks, Alaska, mining company
(figure 4) uses a triple channel sluicing system (figure 8B), which
employs a three-deck screening system that permits only the 1/2 inch
minus ore to be deposited on the slick plate leading to the recovery
sluice. The 1/2 inch minus material that passes through the bottom
sizing screen and onto the slick plate encounters a 1/4 inch punch

plate shown in figure 8A. The 1/4
inch minus material passes through
the punch plate where it is directed to the two satellite sluice
runs. Simultaneously, the 1/4 inch
plus to 1/2 inch minus material is
directed down the central sluice.
A punch plate which extends over
the first two feet of the central
sluice provides an undercurrent, in
contrast to the previously studied
single channel sluice box where the
fiberglass ridged punch plate was
positioned in the last 24 feet of
the channel to provide an undercurrent.

FIGURE 8A: Steel triple
channel sluice box with
angle iron (hungarian)
riffles and undercurrent
punch plate at entry of
center sluice run.

FIGURE 8B: Sluice box as
shown in FIGURE 8A in
operation. Note the
decreased water energy of
the satellite sluices
compared to the center
sluice.

As shown in figure 8B, there is a considerable amount of mechanical energy in the central sluice box, when compared to that existing
in the satellite boxes. As the water moves down the box, there is a
diminishing effect in the mechanical energy. However, the mechanical
energy diminishes only to a point of equilibrium which will be maintained throughout the remaining length. As a result, a particle of
gold of the proper size, shape and specific gravity which can be
suspended and carried by a mechanical energy of this particular
magnitude will be transported to tail. As a result, additional length
would not contribute significantly to the capture of these finer gold
particles. To overcome these gold losses, the finer material is
directed to the satellite sluice boxes where both the decrease in
mechanical energy and particle size interrelate in an attempt to

satisfy the hydraulic equivalency of grains and as a consequence recover a higher percentage of fine gold.

Five Channel Sluice: A brief summary of an open cut mining operation, located at Central, in the Circle Mining District of Alaska (see figure 4), which uses a five channel sluice box, may clarify the steps in ore processing. In this example, the gold has been deposited in a river channel which overlies a silica and clay rich alluvial fan. The dark material shown in the background of figure 9 is the gold-bearing alluvial channel, whereas the light colored material in the foreground is the uneconomic alluvial fan.

FIGURE 9: Alaska placer mining operation. Note the dark pay channel in the background which overlies light colored uneconomic channel exposed in the foreground.

FIGURE 10: Material pushed to a stock pile is transferred by loader to a three deck sizing screen where the minus 1/2 inch material is separated for sluicing.

A Caterpiller D-7 earthmover pushes the ore approximately 150 yards to the stockpile where a Caterpiller 988 loader then transfers the gravel to a triple deck shaking screen which leads to the sluicing system. The coarse plus 1/2 inch material is washed and dropped out the back end of the shaking screen into tailings (figure 10), while the 1/2 inch minus material falls through the shaking screen onto the slick plate leading to the five channel sluicing system.

As shown in figures 11A and 11B, the 5-channel sluice box has raised expanded metal riffles with 1/2 inch wide by 1+1/2 inch long openings. The riffles overlie astroturf carpeting. Just prior to entering the center channel of the sluice box, the 1/2 inch minus material encounters a wash bar which adds additional washing to the material in an effort to free any fine gold particles that might be attached by clay. After passing the spray bar, the material passes across a 1/4 inch punch plate. The 1/4 inch minus material that

FIGURE 11A: 1/4 inch sizing and undercurrent punch plate located at head of central sluice run in the five channel sluice box of FIGURE 12A.

FIGURE 11B: The head of the secondary (left) and tertiary (right) sluice runs in the five channel sluice box of FIGURE 12A. Note the 1/8 inch sizing punch plate in the secondary sluice run and the raised expanded metal riffles overlying astroturf carpet in the tertiary sluice run.

passes through the punch plate is directed to two adjacent satellite sluice boxes. The remaining 1/2 inch minus material then encounters a two foot length of 3/8 inch punch plate that permits the finer material to drop through and into an undercurrent, thereby producing more efficient gold recovery. At the end of the punch plate, the coarse material recollects and is washed across the expanded metal riffle system for the remaining length of the central sluice. Simultaneously, the 1/4 inch minus material that has been directed into the first set of satellite sluices encounters a 1/8 inch punch plate. The fine 1/8 inch minus material passes into the second set of satellite sluices, while the coarse 1/4 inch minus to 1/8 inch plus material is washed across the expanded metal riffle system in the first set of satellite sluice boxes. Finally, the minus 1/8 inch material is washed down the two outside channels.

All this is an attempt to progressively decrease the mechanical energy and particle size, and therefore satisfy the hydraulic equivalence, thereby permitting the extraction of finer and finer gold particles. This is the effect of classification on the extraction of fine gold. As shown in figure 12B, the water in the center sluice box has a high mechanical energy, compared to that of the first set of satellite sluices, and they, in turn, have a much greater mechanical energy than the water flowing through the outside set of sluices.

FIGURE 12A: Three deck shacker screen feeds steel five channel sluice box with raised expanded metal riffles underlain by astroturf.

FIGURE 12B: Sluice box as shown in FIGURE 12A in operation. Note the decrease in water energy in the two sets of satellite sluice runs.

SLUICE BOXES WITH PULSATING RIFFLES:

The final sluicing system to be considered is located at Atlin, British Columbia, Canada. This particular sluicing system is a pulsating riffle system, manufactured by RMS Westpark of Canada (figures 13A and 13B). In this particular plant, the material is sized using a trommel. The one and three-quarter inch minus material that passes through the primary trommel screen drops onto a slick plate which channels the material through a nugget trap and onto two 44-inch wide sluice boxes. The sluices are equipped with hydraulically driven pulsating riffles with 1/8 inch rubber on the bottom which provides mobility. The movement from top to bottom is 7/16 of an inch, with a

FIGURE 13A: Trommel screen feeds two sluice boxes with pulsating riffles.

FIGURE 13B: Sluice boxes as shown in FIGURE 13A in operation. Note the high water energy.

pulse rate of 42 cycles per minute. The use of hydraulically driven pulsating riffles has eliminated clay packing previously experienced using a conventional triple sluice fitted with standard transverse (Hungarian) and/or expanded metal riffle systems. As a result, the sluice bed remains soft and open to receive fine gold particles. Sizes of gold captured in these pulsating sluice boxes are in the minus 8 mesh (2.36 mm) range. However, losses in the minus 30 mesh (0.60 mm) fractions were still experienced.

SUMMARY

In summary, it is recognized that over the past 100 years there have been significant advances in technology regarding sluice box recovery systems. Wooden sluice boxes, with wooden longitudinal and transverse riffles, first gave way to metal single channel sluice boxes with steel transverse riffles and punch plate undercurrents at the tail end of the sluice. These single channel units were replaced in the 1970's by multiple channel sluice boxes, which use diminishing mechanical energies and particle sizes in an effort to obtain the hydraulic equivalence necessary to extract the very fine particles of gold with very low Corey Shape Factors. Through such use of classification it is clear that the technological advances in sluicing systems alone have led to fine gold recovery not achieved using the more traditional style single channel sluice boxes. As a result of these advances, it is now possible to mine ground that could not be mined economically using the earlier designed sluicing systems. However, it is recognized, through testing, that even with these advanced technologies which employ undercurrents, sizing, and decreases in mechanical energy, there still exists significant fine gold loss, especially in the range of minus 30 mesh (0.60 mm).

Researchers have discovered that in some geographic locations the minus 40 mesh (0.425 mm) gold constitutes as much as 50-80% of the product (Walsh 1985). Under these conditions the mechanical energy of the water within even a modern sluice box system will be sufficient to meet the requirements of collecting and transporting a particle of gold of the proper size, shape and specific gravity out the end along with the gravel aggregate as tailings. This suggests that other forms of mechanical concentrators should be considered when the gravel aggregate contains large percentages of gold smaller than 30 mesh (0.60 mm).

REFERENCES

Ackels, D., 1985, "Some Aspects of Gold Recovery with IHC Jigs," Proceedings of the Seventh Annual Conference on Alaskan Placer Mining, Madonna, J.A., ed., pp. 86-99.

Poling, G. W. and Hamilton, J. F., 1986, "Fine Gold Recovery of Selected Sluice Box Configurations," University of British Columbia.

Walsh, D. E., 1985, "Evaluation of the 4-inch Compound Water Cyclone as a Fine Gold concentrator Using Radiotracer Techniques," University of Alaska, School of Mineral Engineering.

Tourtelot, Harry A., 1968, "Hydraulic Equivalence of Grains of Quartz and Heavier Minerals, and Implications for the Study of Placers," U.S. Geological Survey Professional Paper 594-F.

Chapter 28

INNOVATIVE GOLD DUAL DREDGE ALLOWS SIMULTANEOUS
OVERBURDEN REMOVAL AND MINERAL RECOVERY

Martin H. Col and Wayne F. Crockett

R.A. Hanson Company, Inc.
Spokane, WA U.S.A.

ABSTRACT

Gold, as a precious metal, has been sought for several thousand
years. During this time economics have generally determined whether
or not a deposit is minable. Today, few would spend $500.00 or
$1,000.00 to produce one ounce of gold, yet most gold deposits
currently being mined would have such costs if they were mined using
the methods of the 18th century. Innovation has been the key to
success in the world of mining. We are constantly having to look for
new ways to mine deposits of lower grade, at greater depths, or with
higher stripping ratios.

The mining of placer gold deposits has always been subject to the
volatile swings of the world price and the constant rise in production
costs. The rise in costs is compounded by the drop in overall ore
grades. Occasionally, a major technological innovation will dramatically reduce the costs of placer mining to allow previously marginal
or unminable deposits to be mined at a profit. The continuous bucketline dredge was the last such major innovation in the gold placer
mining.

The continuous bucketline dredge has allowed large volumes of
placer deposits to be picked up and processed on a continuous basis.
Unfortunately, as the overall grade of the deposit declines, it
becomes impossible to process enough ore to pay for the operating
costs. There are many gold deposits around the world waiting for the
next technological innovation that will allow them to be mined.

One such deposit has been the Grey River deposit near Greymouth in
New Zealand. The deposit contains gold reserves of 8992 kg, yet the
deposit has been sitting idle for many years because the average value

is $0.127g/m^3$. Recently, an innovative dredge was designed to mine
this deposit at a very low cost. This dredge, with two excavating
systems, independently controllable discharge, and complete on-board
computer control system, is currently under construction and will be
operational in mid-1988. (Patents have been applied for in the
United States and will be applied for worldwide.) The operating costs
are projected to be less than $100.00/ounce of gold produced.

This paper will examine the design, how it was conceived, how it
will operate, and the impacts on the Grey River and other low-grade
placer deposits.

INTRODUCTION

For more than 100 years, there have been no major changes in
placer mining dredge design or operation anywhere in the world. The
last major innovation was the introduction of the continuous bucketline dredge in 1882 in New Zealand. Dredges are still designed and
built much the same as they were in the early to mid-1900's, and many
of the dredges put into operation at that time are still operating.
Current dredges use either conventional bucketlines or cutting heads
that pick up ore, put the material through the dredge recovery system,
and discharge the tailings at the stern of the dredge, to remove the
desired minerals in the process.

Undoubtedly, there have been significant improvements in dredges
in the last few years. Perhaps the most significant is the addition
of microprocessors and computers to the actual operation of dredges.
These control systems have allowed the dredges to operate closer
to their optimum design ratings a higher percentage of the time.
Improvements in cutterhead design and operation have also increased
dredge performance, particularly in more cohesive materials. Mineral
recovery has improved over the years as well, and the circular jig
has been the most significant improvement to date in that area.

The project on the Grey River affords several distinct challenges:

1. Very low average grade ($127mg/m^3$).
2. More than 90% of the gold is located in the lower 25 - 30% of
    the deposit, leaving an average overburden depth of 15m (49 ft).
3. Strict environmental controls, particularly concerning
    reclamation and final contouring.
4. The deposit is located in an active river channel and flood
    plain.

Grey River - History

There is a long history of bucketline dredges operating in the
Grey River area. More than 20 dredges have operated in the area,
mostly in the tributaries of the Grey River. Kanieri Gold Dredging,
Ltd., (KGD) of New Zealand began dredging operations in the area in

1936 and ceased operations in the nearby Taramakau River in 1981.
KGD had planned to build a new dredge on the Grey River, but due to
the low grade in the area and financial difficulties with its parent
company (a U.S. oil company), the property stood idle for several
years and finally was offered for sale.

The R.A. Hanson Company, Inc. (RAHCO), of Spokane, Washington,
U.S.A., recently acquired KGD, including mining and prospecting
licenses covering 2904 hectares. In mid-1987, RAHCO entered into a
joint venture agreement with Australian-based Giant Resources, to
form Grey River Gold Mining, Ltd. (GRG), to mine the deposit. In
addition to being a mining partner, RAHCO is designing and constructing the innovative dredge required to make the project successful.
While the dredge is being constructed, GRG will be conducting
drilling and evaluation of additional adjacent ground to prove up
additional reserves, under Giant Resources supervision.

DEPOSIT DESCRIPTION

The Grey River Gold project is located in the Grey River Valley
on the South Island of New Zealand (see Figure 1). The River Valley
and associated alluvial gold deposits are situated in the Grey-
Inangahua Depression, which is a middle Tertiary, linear tectonic
trough more than 80 kilometers in length. The west side of the
trough is occupied by wide floodplains, terraces, and the present
braided course of the Grey River, which is fed by numerous tributaries. The alluvial gold deposits were formed by the fluvial
working and reworking of gold-bearing glacial morain detritus, which
was derived from gold-bearing rocks of the Greenland Series.

The alluvial gold deposits are underlain by the Tertiary Blue
Bottom formation in the lower course of the river and Pleistocene
Old Man Gravels in the upper course. The Blue Bottom formation
consists of glaucontic marine clay, silt, and minor sand, while the
Old Man Gravels consist of loosely consolidated well-rounded conglomerate. Both deposits form the bedrock of the alluvial gold
deposits and are soft enough to allow for excellent clean-up of
gold concentrated on the bedrock surface.

Figure 1

The gold is distributed throughout the alluvial section, but the
greatest concentrations are in the lower portion of the deposit. It
is estimated that 93.5% of the total gold value is contained in the
lower 28.6% of the deposit. Reserves currently under mining license
are estimated to be 9991 kg (310,824 ounces) (see Table 1).

TABLE 1

	Under Mining License	Under Prospecting License
Plan Area (ha)	420	1,185
Average Depth (m)	28	24
Volume (m^3 x 10^6)	118.3	284,000
Grade (mg.m^3)	127	114
Gold Content		
Raw (kg)	15,024	32,422
Fine (kg @ 950 fines)	14,273	30,801
Recoverable Gold		
(@ 950 fines x 70% recovery)		
kg	9,991	21,561
ounces	310,824	670,758*

*Based on 50% of the area being minable at 90% of the grade of the
area currently under mining license.

RAHCO DUAL DREDGE CONCEPT

Kanieri Gold Dredging, Ltd., had many components on site that
would be suitable to build a standard 20 ft^3 bucketline dredge which
would have a capacity in excess of 4 million m^3 per year. Careful
analysis of a standard bucketline dredge had shown that 4 million m^3
per year was an economically viable size if the overburden could be
removed ahead of the bucketline.

In order to exploit such a low-grade deposit profitably, the
overburden has to be stripped at a cost of less than $1.00 per cubic
meter, as this is the approximate average value of the entire deposit.
Standard means of stripping the overburden were considered, but
methods such as dragline, or even the use of a cutterhead gravel pump
dredge, were considered marginally profitable.

Several different concepts were explored and computer models
developed to examine both the operational and economic feasibility of
each concept. It was determined that, if the overburden removal
system could be incorporated into the normal operation of a standard
bucketline dredge, the incremental cost of it would be minimal. As
this appeared to be the most cost-effective and theoretically possible, this concept was pursued.

A successful overburden removal system would need to have sufficient capacity so that the bucketline would operate only in the pay
zone. Since the upper two-thirds of the deposit contained very little
value, a successful overburden removal system would have to be able
to remove at least 8 million m^3 per year.

The suction cutterhead gravel pump eventually emerged as both the
lowest cost and the lowest risk overburden removal system. Since both
the suction cutterhead gravel pump and the bucketline were proven
systems with known operating parameters and predictable costs, there
was very little risk associated with the combination (Figure 2).

The overburden removal system consists of an underwater cutterhead/
suction gravel pump which takes off the top 18.3m (60 ft) of very low-
grade overburden. The 4.9m (16 ft) diameter rotary cutter (Figure 3)
feeds a 660mm (26 in.) underwater gravel pump mounted to a support
structure on the front of the dredge forward of the bucketline. The
cutter is designed to operate bidirectionally, independently and/or
simultaneously with the bucketline. This allows either system to
operate by itself in the event one is down for maintenance. The
capacity of the cutter suction dredge is 1530 m^3/hr (2000 yd^3/hr).
The overburden is pumped through a pipeline to a large, rotating,
dewatering trommel, which loads the material onto the dry stacking
conveyor to be deposited as tailings 61m (200 ft) behind the dredge.
The bucketline has a capacity of 800m^3/hr (1046 yd^3/hr) for a total
combined capacity of 2330m^3/hr (3046 yd^3/hr). The net result is a
tremendous increase in production and lower cost per cubic meter of
mined material. This combined operation has the effect of increasing
the overall capacity of the dredge by 300%, to 12,000,000m^3
(15,695,322 yd^3) per year.

The 12,000,000m^3 capacity is based on a very conservative 5150
hours of operation per year. This number is cited as it was used in
the financial analysis. Actual hours of operation are expected to be
in excess of 7000 hours per year. Several years ago RAHCO supplied
four computer-controlled dredges for potash mining in Jordan, and
those dredges have averaged 94% availability. Additionally, a dredge
owned and operated by R.A. Hanson Company had an availability in
excess of 90% while operating in the harsh conditions of Alaska.

In all design discussions, the driving criteria have been to keep
the bucketline operating with maximum bucketfill, a minimum of non-
pay zone material in the buckets, and minimum downtime. Several
design elements ensure accomplishment of these goals.

1. Design capacity of overburden removal system is more than
    two times the capacity of the bucketline. This ensures
    that downtime on the overburden removal system does not
    slow down the bucketline system, and that the bucketline
    system is always operating in the gold-rich pay zone.

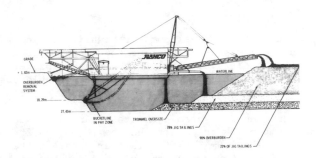

Figure 2

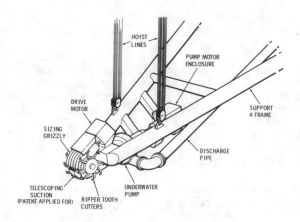

Figure 3

2. The dredge layout provides the ability to service overburden removal system while the bucketline system continues to operate.

3. The layout also places the overburden removal system in an area well forward of the bucketline system. This maintains a bench approximately 16.1m to 19.5m (53 - 64 ft) wide. The material that slumps or caves in from the overburden face will not fall to the bucketline system.

4. A computer control system will ensure consistent, maximum bucket fill.

TREATMENT PLANT/RECOVERY SYSTEM

Perhaps the most critical function on any dredge is the mineral recovery system. The total cost of owning and operating a dredge is nearly identical whether 1% of the total gold is recovered or 100% of the total gold is recovered. RAHCO has taken several steps to maximize gold recovery on the Grey River Gold Dredge:

1. The mineral recovery jig surface area to jig input volume ratio will be very high.

2. The system will use high efficiency jigs of circular and trapezoidal design.

3. The large radius of the jigs will provide an increased transit time across the jig bed.

4. The material being discharged as waste will be repeatedly sampled and analyzed for gold content. If sufficient values are found, that portion of the waste stream will be reprocessed prior to discharge.

5. A computer-controlled leveling system that will keep the entire dredge (and therefore the mineral recovery system) level independent of the position of the discharge conveyor, bucketline, or suction cutterhead. The system will operate by pumping water in and out of ballast tanks to keep the dredge level. The leveling system will keep the entire dredge level to within a small fraction of a degree both fore and aft and port to starboard.

6. Security will be increased by the automated mineral recovery system which will require very little maintenance.

The overburden material may have some small values and therefore this material will pass through a flow splitter for partial dewatering prior to entering the dewatering trommel with the large oversize passing to the dry stacking conveyor to be placed as tailings. The undersize will be pumped to the primary jigs.

The bucketline excavator digging in the pay zone will discharge
this material through a moving grizzly onto a chute. This chute
will feed one 3.5m (11.5 ft) diameter trommel screen approximately
11.12m (36.5 ft) in length. The undersize from the trommel will pass
through a dewatering tank and then onto the primary (rougher) jigs.
The oversize greater than 25.4mm (1.0 in.) will be discharged directly
into the pond, while the oversize less than 25.4mm (1.0 in.) will pass
over a nugget trap prior to discharge into the pond. (See the
generalized flow diagram in Figure 4)

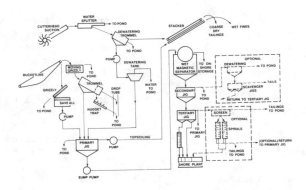

Figure 4

The trommel will serve two purposes: scrubbing and screening.
To aid in maximizing the recovery rate, the incline of the trommel
will be adjustable. Thus, the retention time in the trommel will
be accurately controlled. The screens in the trommel will be made of
high density plastic and have rectangular holes to increase the
efficiency. The total effective screen area will be nearly twice
that of most trommel screens.

The trommel undersize material will pass through a wet magnetic
separator to remove excess magnetite before it is fed through a
splitting system to ensure even distribution on two 10.97m (36 ft)
diameter circular, primary jigs. Each circular jig has 18 trapezoidal cells, with each cell having an area of $4.27m^2$ (46 ft^2), for
a total primary jig area of $153m^2$ (1656 ft^2).

RAHCO has, in keeping with its use of modern, high performance
equipment, opted to use these circular jigs (and cells from circular
jigs in the case of the cleaner and scavenger jigs) for their high

efficiency, high recovery, and ease of feed distribution. The jigs, also designed and manufactured by RAHCO, have several innovative features. As is common on all modern jigs, they will have a fast upstroke and a slow downstroke which increases recovery. The primary jigs, with 18 cells each, have several distinct advantages:

1. All 18 cells are driven by a single motor driving an adjustable cam. Thus the speed and general stroke cycle are adjusted for all cells simultaneously. Also, the drive system is very simple providing excellent reliability and minimal maintenance. However, the total stroke distance can be independently adjusted for each cell if this can be shown to increase recovery.

2. The jig cells have a smooth transition from the trapezoidal section to the circular section on each cell. This reduces build up of gold "inventory" in the cells.

3. The cells will have a very low water requirement.

4. Large radius, 5.5m (18 ft) provides longer transit for material as it moves across the jig surface.

5. The RAHCO jigs will include an individual cell monitoring system. This system will detect when a cell is beginning to decrease in efficiency and will schedule maintenance for that cell. This will reduce the possibility of sudden failure of a cell diaphragm.

The primary jig hutch product will be collected in a sump and be pumped up to the wet magnetic separator and then onto the three-cell, secondary (cleaner) jig.

A portion of the primary jig overflow will be pumped up the stacker through a pipe that is situated beside the conveyor from the overburden system and deposited on top of the dewatered suction cutter tailings as fines for land reclamation. This should provide 0.5m - 1.0m of topsoil. The remainder of the primary jig overflow material will be discharged to the pond well beyond the stern of the dredge.

The overflow of the secondary jig cells will flow to the pond. However, if sampling of the overflow shows sufficient values, some of the overflow may be directed to a scavenger jig with the hutch product to be routed back to the primary sump pump. The hutch product from the secondary jig will go to the tertiary jig. The resultant hutch product from the tertiary jig will be taken to shore for final treatment. The overflow from the tertiary jig flows across a sizing screen with the fines going to a set of spiral separators. The screen oversize will go to the pond with constant sampling to determine if any values are present.

The heavy product from the spirals will be taken to shore for
final treatment. The lighter fraction from the spirals will be discharged to the pond with constant sampling to determine if significant
values are present. If values are found, the lighter fraction will
be pumped back to the primary jigs.

Again, the treatment system is designed to recover the maximum
amount of gold that is economically justified. The treatment
system will be installed on the dredge in such a manner as to be
easily modified to match the actual character of the ore body and
the operational economics.

Reclamation

Concern for the protection of the environment is a concern in
all parts of the world, but New Zealand is known for its very strict
environmental protection policies.

The land must be reclaimed to equal or better than original
condition. In most cases this would be difficult and expensive,
but that is not the case with the RAHCO Dual Dredge.

The RAHCO Dual Dredge with its unique built-in reclamation system
will leave the ground in a better, more production condition. First,
the swell normally associated with dredging will increase the
elevation slightly and improve the drainage. Second, since the
dredge will have a computer-controlled, slewable discharge conveyor,
the dredge will automatically place material according to a predetermined contour. Third, some of the fines from the bucketline
will be wet-placed on top of the material placed by the dry stacking
conveyor. This will result in a uniform profile with good drainage
and sufficient topsoil, yet the reclamation costs will be minimal.

Another benefit provided by the computer-controlled, slewable
discharge conveyor will be the ability to build stop banks and
river control structures while the dredge continues its productive
mining.

AUTOMATIC CONTROL SYSTEM

Computer technology has revolutionized the method of operating
large, complex machinery. A computer and programmable logic
controllers (PLC) will be used to operate the dredge controlling the
two excavation systems, an extensive mineral recovery system, and a
slewable and luffable discharge conveyor. The system will sense,
and either directly control or suggest to the operator, the most
appropriate control action for all functions on the dredge.

Modern dredges (such as those designed and manufactured by RAHCO
for the Arab Potash Company in Amman, Jordan, and the Qinghai Potash
project in China) are completely controlled by on-board computers.

The PLC receives singnals from sensors located at critical locations which feed back position, speed, fill and other information necessary to control the dredge. The PLC is programmed to optimize all dredge functions.

Additional instruments and sensors monitor the dredge's mechanical systems and supply information to the PLC. Data collected includes slurry density and flow rate for control of pump speed, cutter rpm, bucketline swing speed, and depth of cut. Sensors mounted on the bucketline ladder and suction head monitor position and cutting force. In a matter of milliseconds, the controller continuously scans the sensors and notes any change in the sensors' output. This data is then processed by the computer and control signals are sent out to initiate operational functions. The constant scanning ensures the consistent optimum operation of the dredge.

The precise position of the dredge in the pond is determined by a microwave triangulation system using two or more shore-mounted remote transponders as a baseline. An electronic compass provides orientation information. Movement of the dredge is accomplished by microprocessor control of the headline and anchor winches to position the dredge in the proper location and orientation as required by the predetermined dredging pattern stored in the microprocessor.

All critical process and guidance control functions are continuously monitored and displayed to the operator by a PLC-based supervisory control program. This system reads the key process and control parameters from the PLC and displays the information graphically on a video display at the operator's station. The operator has full control of key functions and may alter predetermined decision points as required.

The controller also stores historical data about the operation of the dredge that can be recalled at a later date by the controller or by operations personnel interested in analyzing the operation of the dredge. The controller will be able to compare existing conditions with those encountered earlier. This feature is extremely helpful. It enables the PLC and the operations personnel to repeat or to avoid certain dredge operations. Management will be able to analyze the performance of the dredge in any area of the mine. In short, the PLC has the ability to increase dredge production by providing more uptime and improving recovery through increased efficiency.

Some of the key items to be senses and/or controlled are:

-Dredge position
-Dredge orientation
-Pond water level
-Dredge cross level (both side to side and fore and aft)
-Cutterhead position

-Cutterhead drive torque
-Gravel pump flow density
-Gravel pump flow velocity
-Bucketline position
-Bucket fill
-Bucketline speed
-Bucketline drive torque
-Material flow from each jig cell
-Overburden dry stacking conveyor (both absolute and relative
 positions in 3D space)
-Depth of tailings in pond
-Depth of tailings beneath dry stacking conveyor

In addition, all I/O or control points such as pumps and motors
will be sensed and logged for the maintenance record, and the data
collected will be used to predict failure. This will allow repair
or replacement to occur during scheduled maintenance, thus reducing
unscheduled maintenance. For instance, critical pumps will be
monitored for vibration, a key indicator of impending failure.
Similarly, all critical bearings will be monitored for temperature.

ECONOMICS

Operating Costs

The RAHCO Dual Dredge is engineered to maximize gold recovery and
pay zone production while minimizing net operating costs. The three
factors that most influence net operating costs per ounce of gold
produced are ore grade, total expenses, and total production. Since
we cannot influence ore grade, we must concentrate our efforts on
reducing total expenses and increasing total production.

In order to maximize production, the dredge needs to have a
minimum amount of downtime. The downtime on this dredge will be
minimized by the following:

1. Careful component selection including;

    a) Oversized, antifriction bearings
    b) Totally enclosed gearboxes

2. Recirculating, filtered lubrication system including
    temperature sensors and an on-board oil analysis
    system;

3. Computer-controlled operation to minimize over-
    stressing of components;

4. Easy access for both preventative maintenance and
    repair. This includes several overhead cranes, as
    well as easily accessed pumps, motors, gearboxes,

TABLE 2
AVERAGE ANNUAL OPERATING COSTS (US$)
GREY RIVER PROJECT

	Total Cost ($000/yr)	Unit Cost ($/m^3)
Pay Zone (4 million m^3/yr)		
Labor (36 people*)	533	0.133
Maintenance	400	0.100
Power ($0.0301/kW hr)	377	0.094
Reclamation	150	0.038
River control	150	0.038
Infrastructure	120	0.030
Exploration	150	0.038
Contingency (15%)	311	0.078
Subtotal	$2,341	$0.587
Management Fee**	200	0.050
Subtotal Pay Zone	$2,541	$0.637
Overburden (8 million m^3/yr)		
Labor (4 people*)	74	0.009
Maintenance	245	0.031
Power ($0.0301/kW hr)	190	0.024
Subtotal	$509	$0.064
Total (12 million m^3/yr)	$3,050	$0.254

Operating costs	Average Annual Production	Average Cost $/unit
kg	1,013	$3,010.86
	31,529	96.74

*Four (4) workers will be required to maintain the overburden removal system. Thirty-six (36) workers will be required to operate and maintain the basic dredge, as well as the rest of the project:

4 operators x 4 shifts = 16; 3 for office support and purchasing; 1 dredgemaster; 1 assistant dredgemaster; 1 heavy equipment operator; 1 project supervisor; 1 engineer' 5 for dredge maintenance; 3 for project development; and 4 for general project support and exploration. **Management fee is 7% of total operating costs for both the pay zone and overburden.

Chapter 29

POTENTIAL FOR PLACER GOLD OFFSHORE
LUNENBURG COUNTY, NOVA SCOTIA

by G.F.T. Lay and M.C. Rockwell

Technical University of Nova Scotia
Dept. of Mining and Metallurgical Engineering

ABSTRACT

Promising placer gold exploration sites offshore Lunenburg County
are drowned pre- and post-glacial river channels and auriferous
glacial tills. Gold values up to 5.67 g/m^3 have been encountered in
the vicinity of the Ovens Peninsula offshore Lunenburg County.
Moderate gold values associated with notable amounts of heavy
minerals were reported from various other exploration sites in this
area.

This paper highlights the major geological and geomorphological
features for placer gold depositions in southern Nova Scotia. The
past to present gold exploration acitivities nearshore and offshore
Lunenburg County are summarized and potential areas for placer gold
identified.

GEOLOGIC AND GEOMORPHOLOGIC SETTING

The gold in Nova Scotia has been mainly confined to the southern
part of the province, hosted in metasedimentary rocks of the Cambro-
Ordivician Meguma Group. There are over 200 occurences of load,
paleoplacer and placer gold compiled by Ponsford and Lyttle (1984)
and adapted by Miller and Fowler (1987), displayed in Figure 1.

The Meguma Group underlays approximately sixty percent of the
southern mainland and is comprised of Goldenville quartzites and
overlaying Halifax slates, Figure 2. Devonian granites containing
gold-veined quartz stringers have intruded into the Meguma at various
places as depicted in the structural map of the province, Figure 3.
Numerous folds trending northeast-southwest and faults trending

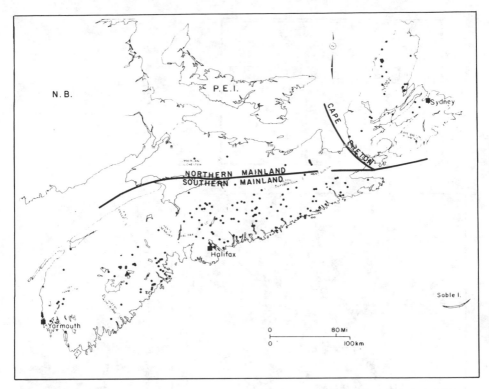

Figure 1.   Gold Occurrences of Nova Scotia (Miller and Fowler, 1987).

northwest-southeast are apparent throughout the Meguma complex.
Stevenson (1959) suggests that longer folds (up to 160 km) were formed
in a series of domes.   These dome structures have been the focal point
of past gold exploration activities on mainland and offshore Nova
Scotia.

Rocks of the Meguma Group and intruding granite extend offshore
onto the Scotian Shelf for approximately 40 km, bounded to the west by
Georges Basin and to the east by the Laurentian Channel, Figure 4,
corresponding to an offshore area of about 26,550 km^2 (King and
MacLean, 1976).   The existence of auriferous granite intrusions
suggested by Hood (1966) could lead to the assumption of at least 50
favorable sites for gold mineralization comparable in size to the
former gold producing areas onshore Nova Scotia (Miller and Fowler,
1987).

The folding and faulting of the Meguma and intrusions of quartz-
bearing granites was followed by erosion, possible faulting and
inundation of mainland southern Nova Scotia by Windsor seas in
Carboniferous times.   The uplift and prolonged erosion created a

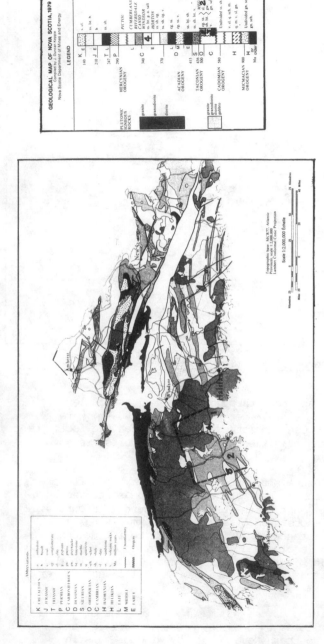

Figure 2.   Geological Map of Nova Scotia (N.S. Department of Mines and Energy, 1979).

────────   Lunenburg County boundary line
1,2,3,4   Various bedrock types

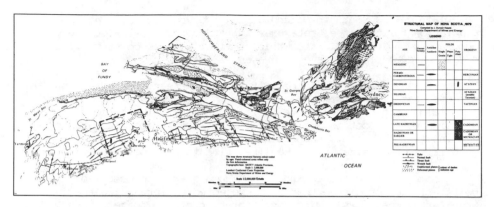

Figure 3.   Structural Map of Nova Scotia (N.S. Department of Mines and Energy, 1979).

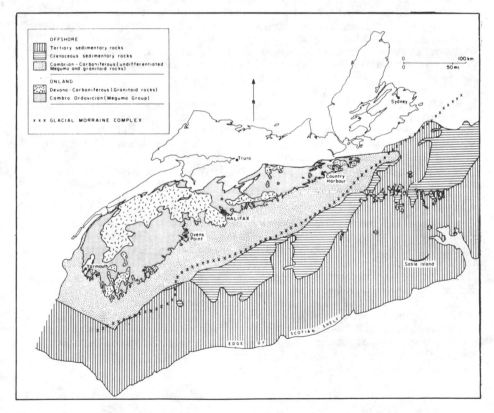

Figure 4.   Offshore bedrock geology of Nova Scotia (King and McLean, 1976).

peneplain on the Southern Upland during the Cretaceous. The last glaciation followed the Cretaceous uplift approximately 100,000 years ago, and by 50,000 B.C., the whole province was ice-covered, to the edge of the continental shelf (Hunter, 1968).

The following period of ice melting and retreat created ancient streams that flowed from the northwest in New Brunswick through the peneplain to the southern Nova Scotia coastline. The sediments supplied to the Scotian Shelf during glaciations were estimated by Hopkins (1985) to contain up to 2.27 g/m^3 of gold and unidentified quantities of cassiterite, scheelite, ilmenite, garnets and zircon.

In some areas of the Upland, the Meguma quartzite occupied narrow belts, flanked by slates that were subjected to intense weathering when left exposed, following disentegration of the softer slates. Ice movement through granites and quartzites quarried large blocks and produced boulders, but created little overburden, due to the hardness of these rocks. By contrast, the slate covered areas produced large amounts of crushed material which when mixed with clay, cobbles and boulders, formed drift deposits in the shape of drumlins (Hunter, 1968). Auriferous materials were thus incorporated into an easily erodable depositional feature, which frequents the southern coast of Nova Scotia, particularly Lunenburg County. Grant (1963) concluded that the composition of examined glacial materials was relatively similar to the underlying bedrock.

Drumlins orientation from northwest to southeast interferred with normal drainage patterns and frequently resulted in the formation of lakes and the slowing of river waters. This slowing effect conceivably led to silting of potential placer development sites and a shift upstream to sites with steeper gradients and smaller cross-sectional areas. This tended to diminish the potential for creating nearshore placer deposits via river laiden sediments (Hunter, 1968).

Erosional features like drumlins dot the southern Nova Scotia coastline. Truncated drumlins at Mahone Bay are evidence of wave action (Hunter, 1968). Bowen (1975) has estimated the coastal erosion of drumlins to range from 0.05 to 2.3 m^3 per year. King (1967) stated that sand does not move substantially below 15 m depths, although waves can set particles in motion at depths up to 35 m. Piper et al. (1985) confirmed that mineral grains set free from drumlins and the exposed bedrock by the coastal erosion will travel only a short distance from their origin, to the nearshore zone. Most coastal beaches are eroded in winter and then rebuilt during the summer months, due to higher winter storm waves moving over the coastal slope of Nova Scotia.

STUDY AREAS

Considering the evidence regarding the source rock, river drainage and fluvial transport, coastal erosion and nearshore transport phenomena, and landward sediment transport during transgressions, a number of sites exhibiting potential for placer deposition in Lunenburg County were investigated. The paper groups the sites into three main study areas, as displayed in Figure 5, the index map, to facilitate the presentation of data. Specifically, Green Bay, False LaHave and Mosher Harbour are found on Area 6, Mosher Bay, Hartling Bay, Kings Bay, Rose Point, Rose Bay, Cross Island, the Ovens Peninsula and Lunenburg Bay represent Area 7. Mahone Bay is encompassed by Area 8.

PREVIOUS WORK

Matachewan Canadian Gold Limited in 1968 collected 148 km of seismic sub-bottom profile lines to locate potential sample sites, then drilled 122 sample holes ranging in depth from 0.8 to 7 m adjacent to the Ovens Peninsula, at Mosher Island, Rose Point, Rose Bay, and in Lunenburg Bay, enclosed in Study Area 7 of Figure 5. They concentrated on sites containing medium to coarse gravels and sands in scour basins and channels, and ancient stream channels. Narrow tidal estuaries were sampled since the currents may have preferentially scoured the higher level sediments and concentrated the heavier gold particles (Hunter, 1968). A $3.5KH_z$ pinger system was employed to delineate surficial sediment depth to bedrock, while the drill samples provided evidence of surficial material type and thickness.

Estimates indicate that a minimum of 4.6×10^6 m^3 of material was deposited within 350 m of the Ovens Peninsula and contained gold values between a few cents and $4.19/m^3$ (1968 U.S. dollars) (Hunter, 1968). Matachewan followed-up the preliminary investigations in 1969 with a dredging programme employing a 22 m by 9 m wide steel structure barge, mounted with a suction dredge. The dredge was capable of accessing below normal depths and removing boulders from the working face. It also had an on-board gold processing system and was a combination of several previous Ocean Engineering designs (Libby, 1968). The exploration target at Ovens point was a buried and submerged channel lying 340 m offshore in water depths of 15 m. This target displayed the most promising values from drill samples. Upon anchoring, it was discovered that most of the previous surveyed alluvium had disappeared, probably removed by the severe winter and spring storms and deposited in deeper water. The remaining alluvium was dredged, and a team of divers equipped with hand held 7.5 cm airlift pump and hose vacuumed out the small crevices and depressions of the channel. Gold values ranging up to 5.67 g/m^3 were recovered (Libby, 1969). The claims were released in 1970 due to technical difficulties and low gold prices.

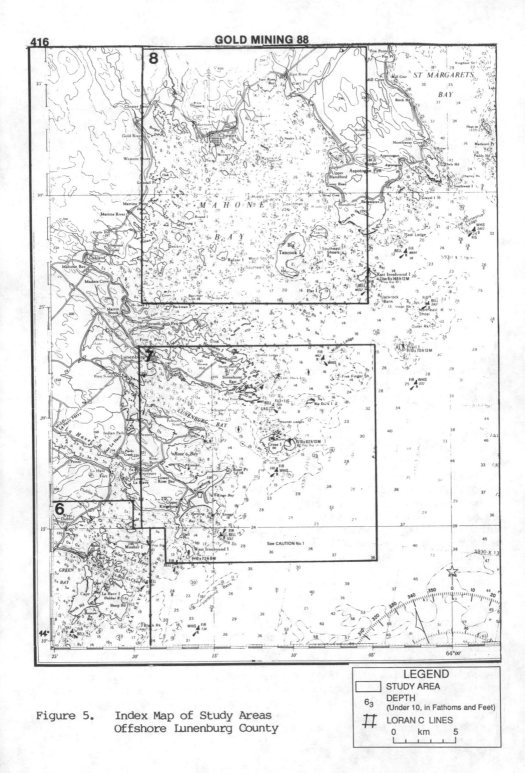

Figure 5. Index Map of Study Areas
Offshore Lunenburg County

LEGEND

☐ STUDY AREA

6_3 DEPTH
(Under 10, in Fathoms and Feet)

LORAN C LINES

0   km   5

RECENT EXPLORATION ACTIVITY

Mer Resources of Dartmouth, Nova Scotia conducted a province wide program of beach sampling, followed by an offshore drilling program during the summer and fall of 1985. They collected beach grab samples and vibracore drill samples from the Lunenburg study Areas 6 and 7 of Figure 5. The drilling program employed a sonic sampler from W.I.N.K. of Toronto with an air lift system designed by Mer Resources (MacPherson, 1985). The vibracore drill was mounted on a platform aft of a 12 m fishing vessel. The sample locations are depicted on study area maps, in Figures 6 and 7.

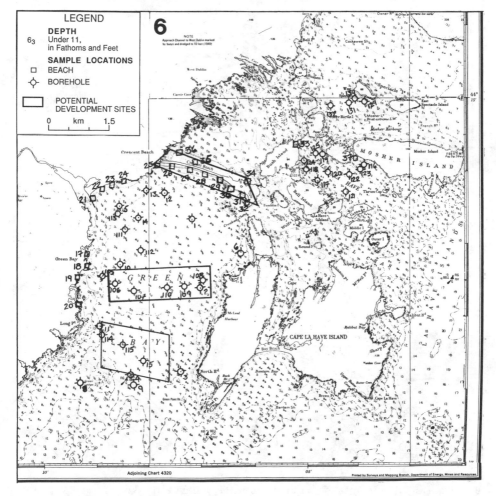

Figure 6.   Study Area showing Green Bay, False LaHave, and Mosher Harbour, in Lunenburg County; adapted from MacPherson (1985).

Referring to Figure 6, on the previous page, the beach samples
locations #16 to #32 from Green Bay, #33 to #36 from Dublin Bay, #37
from False LaHave, #38 to #41 from Hartling Bay and #42 from Rose Bay
displayed values of less than 0.032 g/m^3 gold, except for #39 which
was 0.063 g/m^3 and samples #28, #30 and 31 which averaged 0.084
g/m^3.   Vibracore samples produced better gold results than beach
samples collected from the same study area. The vibracore samples
#16 to #125 from False LaHave, #105 to #115 from Green Bay and #126
to #132 from Mosher Harbor contained up to 0.095 g/m^3 gold.   The
#108, #109 and #115 from Green Bay and #132 from Mosher Harbour
contained 0.126 g/m^3 gold (MacPherson, 1985).

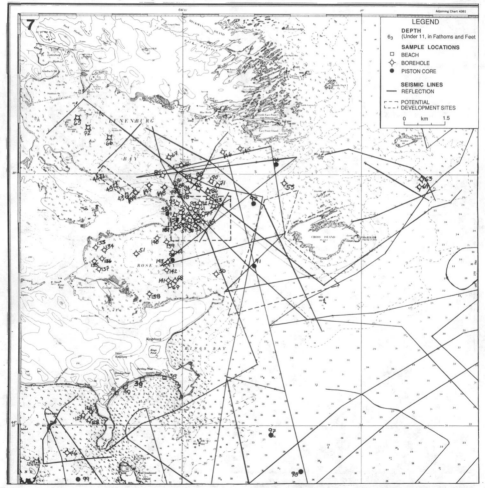

Figure 7.   Study Area showing Mosher Bay, Hartting Bay, Rose Bay,
             Lunenburg Bay, Cunard Cove, and the Ovens Peninsula in
             Lunenburg County; adapted from Libby (1968), and
             MacPherson (1985).

The beach samples collected by Mer Resources from locations at Cunard Cove, Rose Bay and Lunenburg Harbour, Figure 7, attained relatively higher gold content. The beach samples from Cunard Cove and Rose Bay displayed values of up to 3.75 g/m^3 gold, and those from Lunenburg Harbour of up to 1.57 g/m^3 gold. More notable gold assay results were from vibracore samples collected from Cunard Cove and Rose Bay. Gold values ranging from 0.084 g/m^3 to 8.53 g/m^3 were reported, as following: #146, 0.084 g/m^3; #147, 1.07 g/m^3; #148, 1.16 g/m^3; #149, 1.87 g/m^3; and #153, 8.53 g/m^3 gold (MacPherson, 1985).

Gedoro Mining Limited of Halifax, Nova Scotia, recovered 35 diver assisted samples and 20 km of side scan sonar lines during the summer of 1987 from Lunenburg Bay. They also collected approximately 45 km of side scan sonar lines from a 32 m converted steel fishing trawler during the summer of 1988, from Little Duck Reef to Five Finger Shoal, then south past Cross Island and West to Rose Bay. Gedoro devised their own Loran C based lattice system to isolate sample locations interpreted from side scan evidence. Currently, they are retrieving a 38,000 kg airlift sample from four to eight predetermined sample locations. The results are expected to be available later this year   (Private communication, 1988).

Considerable data describing the bedrock, surficial geology and geomorphology of Mahone Bay basin, Figure 8, were reported by Barnes (1976), Letson (1981), and Piper et al., (1985). They collected borehole (gravity core), box core, piston core, and grab samples.

A Dietz-Lafond snapper and a VanVeen grab were used by Barnes (1976) to collect 140 samples from eastern half of Mahone Bay for grain size distribution and textural analyses. Barnes also collected 425 km of MS26B sounder profiles and sparker lines. Grab sampling was designed to determine the type and areal extent of surficial materials. Sounding provided a check on bathemetry, and an account of bottom texture and depth to bedrock and till exposures. Seismic reflection profiling permitted determination of the total sediment till thickness and the depth and type of bedrock.

Letson (1981) collected 222 grab and borehole samples and 70 km of echo sounder lines from southwestern Mahone Bay. Echograms were employed to interpret bottom textures after the technique developed by King (1967). Grain size analysis were performed and textural facies plots were developed. Gravity core, box core and piston core samples were analysed and interpreted to determine acoustic bedrock units and lithofacies characteristics (Piper et al., 1985).

Grab samples collected by Piper et al. (1985) from Mahone Bay contained an average of 10% by weight resistates (tourmaline, zircon, apatite, rutile and sphene) compared to those collected from the Green Bay area, which exhibited an average of 3.2% resistates. Gold content was not determined. The Rose Bay and Cunard Cove study areas indicated an average of 4% resistates.

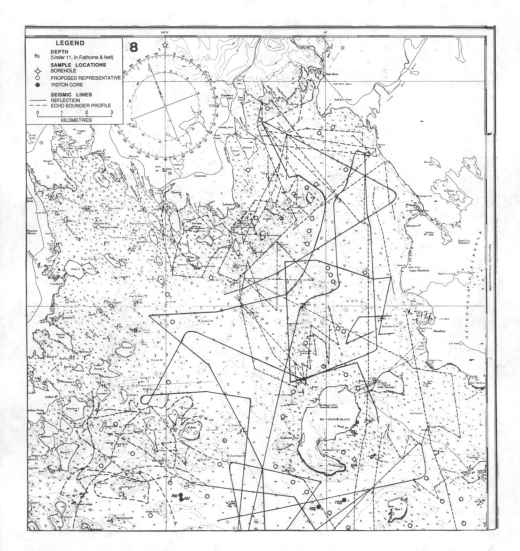

Figure 8.   Study Area of Mahone Bay in Lunenburg County, adapted
             from Piper et al., (1985).

Resource Planning Consultants Ltd. of Halifax plans to perform an
integrated heavy mineral analysis program on representative sand and
gravel samples from Mahone Bay. The company is currently collecting
several kilometers of side scan sonar lines ranging from Big Tancook
Island in Lunenburg County to Prospect Bay in Halifax County
(Private communication, 1988).

CONCLUSION

The placer exploration to date has been limited to isolated areas, with insufficient sampling for gold and heavy minerals, to provide a high level of confidence in the potential of placer gold deposits offshore Lunenburg County. The placer exploration program conducted recently by Mer Resourses Ltd. reported isolated sample locations with relative potential for gold.

Three sites at Green Bay, outlined in Figure 6, have a gold content ranging from 0.084 to 0.126 g/m^3 and a content of 3.2% by weight of heavy minerals. The two sites outlined in Figure 7, have a moderately higher gold content. The site enclosing Cunard Cove and Rose Bay displays a gold content ranging from 0.084 to 8.53 g/m^3 and a heavy minerals content averaging 4.0% by weight. The other site, at Lunenburg Harbour, has a gold content of up to 1.57 g/m^3 and an unknown heavy minerals content.

Matachewan Canadian Gold Ltd. in the late 1960's estimated a minimum of 4.6 x 10^6 m^3 of auriferous sediments off the Ovens Peninsula with a gold content of up to 5.67 g/m^3. The area off Ovens Point to Cross Island, and the area to the north and south of Cross Island are currently under evaluation for gold content by Gedoro Mining Ltd.

Several locations in Mahone Bay near eroding coastal drumlins and drowned river channels could prove potential for placer gold deposits. An average of 10% by weight of heavy minerals content was reported, which could warrant consideration as a by-product to a dredging operation.

There is still an extensive unexplored offshore and nearshore area for placer gold deposits. The history of gold liberated by glacial communition of the auriferous rocks of the Meguma Group and the subsequent sea level rise, indicate favorable environments for potential placer deposits offshore Lunenburg County.

ACKNOWLEDGEMENTS

The authors wish to express their gratitude to William MacPherson of Mer Resources Limited, Gordon Fader, Bob Miller and David Doucet of the Bedford Institute of Oceanography in Halifax for assistance. Appreciation is also extended to N.S.E.R.C. for publication support.

REFERENCES

Barnes, N.E., 1976, "The Areal Geology and Holocene History of the Eastern Half of Mahone Bay, Nova Scotia," M.Sc. Thesis, Dalhousie University, Halifax, N.S., 124 pp.

Bowen, A.J., 1976, "Maintenance of Beaches," Technical Report, Institute of Environmental Studies, Dalhousie University, Halifax, N.S., 582 pp.

Gedoro Mining Limited, 1988, Private communication.

Grant, D.R., 1963, "Pebble Lithology of the Tills of Southeast, Nova Scotia," M.Sc. Thesis, Dalhousie University, Halifax, N.S., 104 pp.

Hopkins, R., 1985, "Placer Gold Potential Country Harbour, Nova Scotia," COGLA-EMR, Ocean Mining Division, Canada, 94 pp.

Hunter, N.J., 1968, "Geological Background for Placer Exploration on the Scotian Shelf," plus Maps, N.S.D.M., A.F. - Offshore, 21-W-07, 254 pp. and 8 plates.

King, L.M., 1967, "Use of a conventional echosounder and textural analysis in delineating sediment facies of Scotia Shelf," Can. J. Earth Sci., V.4, pp. 691-708.

King, L.M., and MacLean, B., 1976, "Geology of the Scotian Shelf," Geological Survey of Canada, Paper 7331, 31 pp.

Letson, J.R., 1981, "Sedimentology in Southwestern Mahone Bay," M.Sc. Thesis, Dalhousie University, Halifax, N.S., 189 pp.

Libby, F., 1968, "Ovens Gold Dredging," Progress Report, N.S.D.M., Ovens Gold, A.F. - Offshore, 21-W-10, 23 pp.

Libby, F., 1969, "Searching for Alluvial Gold Deposits Off Nova Scotia," Ocean Industry, Jan., pp. 43-46.

MacPherson, W.A., 1985, "Potential for Placer Gold on the Scotian Shelf, Nova Scotia", N.S. Dept. of Mines and Energy, Report 86 056, 168 pp.

Miller, C.K., and Fowler, J.H., 1987, "Development Potential for Offshore Placer and Aggragates Resources of Nova Scotia, Canada," Marine Mining J., V.6, pp. 121-139.

Piper, D.J.W., Mudie, C.F., Letson, J.R., Barnes, N.E., and Iuliucci, R.J., 1985, "The Marine Geology of the Inner Scotian Shelf Off the South Shore of Nova Scotia," G.S.C. Paper 85-19, 65 pp.

Ponsford, M., and Lyttle, N.A., 1984, "Metallic Mineral Occurences Map and Data Compilation, Eastern, Central and Western Nova Scotia,", N.S. Dept. of Mines and Energy, Open File Reports 559, 600, 601.

Resource Planning Consultants Limited, 1988, Private communication.

Stevenson, I.M., 1959, "Shubenacadie and Kennetcook Map Areas, Colchester, Hants and Halifax Counties, Nova Scotia," Geol. Soc. Can. Mem., 302, 88 pp.

8

Environmental

Chapter 30

DRAINAGE OF TAILINGS IMPOUNDMENTS
APPLICATION OF PRINCIPLES AND EXPERIENCE

by Clint L. Strachan and John D. Nelson

Senior Geotechnical Engineer
Water, Waste and Land, Inc.
Fort Collins, Colorado

Professor and Department Head,
Civil Engineering Department, Colorado State University
Fort Collins, Colorado

ABSTRACT

Drainage of tailings impoundments is a key issue for
impoundment stability, water control, seepage control, and
reclamation. Drainage control is an established operating practice
where climate and site conditions permit, such as in South Africa
and various mines in North America. In these areas, drainage
control is primarily used to maximize tailings densities and to
provide adequate impoundment stability. Drainage practices during
operation also become an important issue when tailings impoundments
are reclaimed. Time required for cover placement and post-
reclamation seepage control depends greatly on the ability for
tailings to drain and consolidate. This paper discusses drainage
issues associated with gold tailings impoundments in terms of
mechanics of saturated and unsaturated flow and experience with
design and reclamation of tailings impoundments.

INTRODUCTION

Due to the slurry handling and discharge of mill tailings,
management of the liquid portion of the slurry and tailings
drainage are key components of tailings impoundment operation and
reclamation. In impoundments, the tailings slurry is separated
into liquid and solid components by settlement at various flow
velocities.

Drainage of the tailings in this process affects the
following:

a.   Static and seismic stability of the tailings impoundment,

b.   Recovery of the tailings liquid (for reuse in the mill),

c.   Control of interstitial pore fluids and their migration
      from the tailings (as seepage), and

d.   Settlement and consolidation of the tailings solids.

These issues have an impact on how the impoundment is
operated, what the downstream groundwater and surface water impacts
will be, and how the impoundment is reclaimed after mine closure.

THEORETICAL CONSIDERATIONS

Analysis of tailings impoundment drainage requires careful
consideration of both saturated and unsaturated flow under varying
material characteristics. Due to the method of tailings discharge,
the characteristics of the settled tailings vary spatially and
directionally. The upper surface of the tailings typically moves
upward with continued tailings discharge, and layered systems may
be created. Drainage from the tailings in an impoundment may occur
under either saturated or unsaturated conditions.

A thorough discussion of drainage analysis techniques is
provided in Nelson and Davis (1987). A brief review of drainage
analysis techniques used for saturated and unsaturated flow in
tailings impoundments is provided below.

Seepage in the Saturated Zone

The flow of water in saturated soils is described by the
equation

$$K_x \frac{\partial^2 h}{\partial x^2} + K_y \frac{\partial^2 h}{\partial y^2} + K_z \frac{\partial^2 h}{\partial y^2} = 0 \qquad (1)$$

where h is the total head,

K_x, K_y, K_z are the hydraulic conductivities in the x, y and
z directions.

A number of computer modelling techniques have been developed
for solutions of this equation. Typically, these techniques
utilize either the finite element method or finite difference
techniques for solution.

These numerical methods are used to predict the location of the phreatic surface in the area of interest in the tailings impoundment. The location of the phreatic surface is influenced primarily by the relative hydraulic conductivities of the tailings and foundation materials. These methods are also used to locate drains and determine ways to reduce the phreatic surface to acceptable levels.

One factor that needs to be considered in seepage from tailings impoundments is the effect of infiltration from the beach during tailings discharge. In many impoundments, tailings are typically discharged at one location for a period of time after which the spigots are moved and discharge is carried out at a different location, while the previous location is allowed to drain.

When discharge is initiated at a new location, the seepage into the beach results in a saturated (or nearly saturated) zone that moves downward. During the period of time that the wetting front of this zone is moving downward, but has not yet contacted the original phreatic surface, there is no influence on the phreatic level in the impoundment. The time required for the wetting front to move downward represents the time required to fill the void spaces in the tailings. Once the wetting front contacts the phreatic surface, the additional amount of water required to raise the phreatic surface is relatively small. Consequently the phreatic surface rises very rapidly after that point. In order to remedy this rapid rise in the phreatic surface, spigots should be moved to a different location.

Seepage in the Unsaturated Zone

McWhorter and Nelson (1979) developed procedures for making engineering estimates of seepage from impoundments located above a partially saturated zone. The objective of that paper was to provide engineering solutions that do not require impractical quantities of data or the use of numerical methods and computer applications. Only those aspects of the flow phenomenon that have first order effects on seepage rates were included in the analyses.

The mechanics of seepage and methods of analyses are discussed briefly below. For more detailed discussion the reader is referred to McWhorter and Nelson (1978, 1979, 1980) and McWhorter (1985).

Seepage from beneath an impoundment can be characterized by four distinct stages. The first stage involves the downward movement of the wetting front from the bottom of the impoundment. After the wetting front has reached a lower boundary or good water table, development of a saturated mound begins. This is followed by lateral spreading of the saturated mound, and following

dissipation of the source of water for seepage the mound
dissipates.

Mechanics of Seepage

Seepage in the unsaturated zone is also governed by Darcy's
law.   The continuity equation must also take into account the
storage of water in the unsaturated pore spaces.

In an unsaturated soil the pressure head, h, will be negative
and will be a function of the water content.   Also, K_e will vary as
a function of the water content, or it may be expressed as a
function of the capillary pore pressure head, h.   In order to solve
the appropriate equations, relationships are needed between
volumetric water content and hydraulic conductivity as functions of
the capillary head of the pore water in the soil.   Typical
functional relationships for these two parameters are shown in
Figures 1 and 2.

An important parameter shown on those Figures is h_d, the
displacement pressure.   This capillary pressure represents the
negative pressure that must be applied to a soil during drainage in
order to initiate desaturation of the sample.

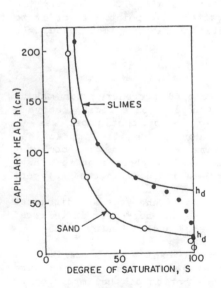

FIGURE 1- Typical Relationship Between Capillary Head and Saturation

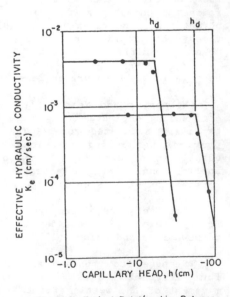

Figure 2- Typical Relationship Between
Hydraulic Conductivity (K_e) and
Capillary Head

Lateral Seepage

For the case of lateral seepage migration (such as through a sloping liner), these techniques may still be used. Seepage will be controlled by Equation 2, and analyzed in a manner similar to that for Stage I. In this case, however, the flow is horizontal and no difference in elevation needs to be considered. Consequently, seepage takes place due to a gradient imposed only by the negative pore water pressure at the wetting front and whatever positive pore water pressure exists along the face of the liner. Storage of water in the pore spaces must be taken into account as well. If the water pressure along the liner can be kept low, and if the storage in the unsaturated soil adjacent to the liner is large, such as in coarse materials, the movement of the wetting front may be small.

Equilibrium Water Content Profiles

In predicting the response of tailings impoundments to reclamation schemes and the ability to predict whether or not drainage equilibrium has been reached, it is important to know the water content profile that will exist after long-term periods. Investigations have been carried out on a number of inactive uranium mill tailings impoundments in the United States.

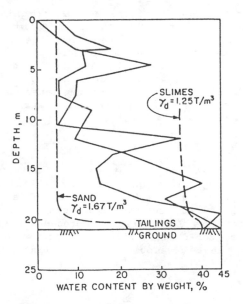

FIGURE 3 – Water Content Profiles in Tailings, Rifle, Colorado

Water content profiles at a site near Rifle, Colorado, USA, are shown in Figure 3. At first appearances the data appears to be erratic and the high water contents associated with certain zones appear to indicate that the tailings have not yet drained.

Also on Figure 3 are the relationships measured in the laboratory between degree of saturation and capillary pressure head for both sand tailings and slimes tailings taken from the same impoundment. The two curves shown by dashed lines represent the extremes measured for a relatively pure sand and a relatively pure slimes sample. Similar curves for samples having different percentages of sands and slimes fall intermediate between these two curves depending upon the gradation of the sample. It is seen that the extreme points of observed water contents in the impoundments fall relatively close to the laboratory measured curves.

The pressure head within the impoundment will be hydrostatic if the sample has drained and the agreement between the field values and laboratory values indicate that the tailings have drained at this point in time. The erratic nature of the water content in the impoundment reflects the existence of lenses or layers of slimes zones intermediate between the sands. Intermediate peaks are believed to represent samples having different mixtures of sand and slimes.

The tailings shown in Figure 3 are approximately 20 to 25 years old. The high water contents in the slimes zones at first led many investigators to conclude that the tailings had not drained and would continue to do so for years to come. It is evident, however, that full drainage has been completed. It is to be expected, however, that upon covering the impoundment, the applied load will cause excess pore water pressures to be developed, accompanied by further drainage and, of course, additional settlement of the tailings.

Mechanics of Consolidation

Consolidation and settlement may be analyzed by classical consolidation theory (Terzaghi and Peck, 1967). However, for materials at high void ratios (such as tailings slimes), classical consolidation theory has serious drawbacks. Recent theoretical developments have provided analytical procedures for materials at high void ratios (Schiffman, 1980; Schiffman et al., 1984). The application of these developments to tailings is discussed in Caldwell, et al. (1984).

DRAINAGE ISSUES

As described above, drainage of tailings affects several aspects of tailings impoundment operation, including stability, water recovery, seepage, and settlement. Drainage issues involved in the design of tailings impoundments are discussed in Nelson and McPhail (1987). The techniques reviewed in the previous section are used to assess drainage and determine acceptable performance. Several methods discussed below can be used to help achieve acceptable performance.

Tailings Water Management

Operation of the tailings impoundment should utilize a water management scheme. Use of this scheme depends on the climate at the site and the percent solids of the tailings slurry. Water management is helpful in promoting drainage by minimizing the pool volume on the impoundment as well as moving the pool from place to place, and keeping ponded water away from tailings embankments. These techniques, developed since the early 1900's in South Africa, are summarized in Smith et al. (1987). However, minimizing pond size may be problematical for materials that have airborne transport problems. For tailings that lack cohesion, large ponds are used as a control measure for blowing tailings.

Drains

Tailings drainage can be augmented with drains, whether under or inside the impoundment. Drains can be used as a continuous drainage blanket or constructed in specific areas of concern (such as near or beneath embankments). Drains can also be constructed with selected tailings materials, created by cycloning or controlled spigotting.

One type of drain that has been used to accelerate consolidation is the use of wick drains or band drains. In this technique a drain consisting of a central core wrapped with geofabric filter material is pushed into the tailings. Spacing generally ranges from 8 ft to 12 ft. The drains serve to decrease the drainage path and hence facilitate consolidation.

Drains have been shown to be effective in accelerating consolidation. Nelson and Wrench (1988) reported on a project in which wick drains were effective in reducing the time for consolidation of soft clay soils for several years (without drains) to less than six months (with drains).

Few applications for tailings have actually been tried. However, the success that has been observed for other soils indicate that they would be effective in tailings if properly

designed.   At the present the authors are considering this method
for use in tailings reclamation schemes to shorten reclamation
time.

Dewatering

Another method of tailings drainage is dewatering the tailings
prior to discharge in the impoundment.   Dewatering consists of
pressure or vacuum filtration or centrifugal separation, depending
on the characteristics of the tailings.   Dewatering has been used
as a method to recover cyanide solutions for increased precious-
metal recovery, or at sites where a conventional tailings
impoundment would not be feasible (such as high seismic regions or
at sites with limited impoundment space).   In specific situations,
dewatered tailings result in some cost savings or ease in licensing
over conventional discharge methods.

Integrated Approach

Achieving acceptable drainage is also the result of
integrating the design and operation into a system that includes
water management, appropriately placed drains, and other
techniques.   Design and construction of drains in selected areas of
the impoundment will work only if tailings discharge is managed so
that the drains will intercept water.

                          RECLAMATION ISSUES

Our experience in reclamation of tailings impoundments has
shown that the operational procedures have a large effect on the
cost of reclamation.   Although tailings impoundments are primarily
designed and operated to contain tailings and perhaps recover
process water for reuse, there are some basic drainage strategies
that can significantly reduce the cost of reclamation.
Discharge Points and Ponds

The authors have been involved with reclaiming several
tailings impoundments where the discharge points were limited or
fixed, such that the pond was located in one area through the life
of the impoundment.   This pond area was subsequently underlain by
the tailings slimes.   This area was not a problem during operation,
but requires special consideration during reclamation.

Upon reclamation this area typically exhibits significant
settlement as the tailings dry out, and is a difficult area to
cover with fill due to the low shear strength of the slimes.   These
areas are also the low spots in the impoundment, and storm runoff
accumulates there.

These areas also exhibit low shear strength, and require special consideration during cover placement. Placing random fill in moderate lifts (450 to 600 mm thick) over slime areas, and spreading these materials with a small dozer has been more successful than either very thin lifts or very thick lifts. Also working in different areas of the impoundment every few days allows excess pore pressures in specific areas to dissipate, and has alleviated trafficability problems in some impoundments.

Consolidation

The most significant problem from the pond being confined to one location is the time required for settlement to occur. Since these areas are underlain by slimes with little or no drainage layers of sands, the time required for consolidation and drainage of the slimes is relatively long. This is due to the long path that interstitial fluids must take to drain from the slimes, and the fact that consolidation time varies as the square of the drainage path length.

During reclamation, a cover of relatively uniform slope is typically placed over the tailings and is designed to drain in a prescribed direction. The cover is gently sloping to minimize erosion and also prevent ponding of water to minimize infiltration. For areas of slowly consolidating tailings, reclamation usually must wait for consolidation to occur. If the time for consolidation is unacceptable, methods of shortening this period can be used.

Remedial Methods

Methods of shortening the consolidation period in these areas include surcharging with fill or installing drains. As mentioned above, wick drains have been used to shorten the period for consolidation of soft soils in many mining and non-mining applications. A case history where wicks reduced the consolidation time from several years to six months is described in Nelson and Wrench (1988).

Operational Remedies

Many reclamation problems can be multiplied by operating the tailings impoundment such that the pond is moved around occasionally during the life of the impoundment. Even though an area may contain predominantly slimes, a few interbedded sand layers (created by having the pond in a different area) would reduce the time required for consolidation of the slimes to a fraction of the time required for consolidation of the slimes without any sand layers.

CONCLUSIONS

The topics discussed in this paper have shown that drainage of tailings is a key factor in successful operation and stability of tailings impoundments.   Drainage also plays a key role in controlling tailings fluids and seepage from both water reuse and metals and process reagent recovery standpoints.   Good drainage practices exercised during operation make seepage control and tailings drainge easier than after mine shutdown.   Tailings drainage also has an effect on the unit weight of the settled tailings, which affects the tonnage of tailings that can be contained in the impoundment.

Tailings drainage also affects how easily the impoundment can be reclaimed, and what additional measures may be involved to get the tailings consolidated and covered.   Tailings that have been discharged in a manner that provides for internal drainage and relatively rapid consolidation make the reclamation effort after mine closure much less costly.

REFERENCES

Caldwell, J. A., Ferguson, K., Shiffman, R. L., and van Zyl, D., 1984, "Application of Finite Strain Consolidation Theory for Engineering Design and Environmental Planning of Mine Tailings Impoundments", Sedimentation Consolidation Models, Prediction and Variation, R. N. Yong and F. C. Townsend, Eds., ASCE, pp. 581-606.

McWhorter, D. B., 1985, "Seepage in the Unsaturated Zone", Seepage and Leakage from Dams and Impoundments, R. L. Volpe and W. E. Kelly, Eds., ASCE, pp. 200-219.

McWhorter, D. B. and Nelson, J. D., 1978, "Seepage in the Partially Saturated Zone Beneath Tailings Impoundments," Reprint No. 78-AG-306, Soc. of Mining Engrg., AIME Fall Meeting, Lake Buena Vista, Florida, September.

McWhorter, D. B. and Nelson, J. D., 1979, "Unsaturated Flow Beneath Tailings Impoundments," J. Geotechnical Engineering Division, ASCE, Vol. 105, No. GT11, November, pp. 1317-1334.

McWhorter, D. B. and Nelson, J. D., 1980, "Seepage in the Partially Saturated Zone Beneath Tailings Impoundments," Mining Engineering, J. AIME, April 1980, pp. 432-439.

Nelson, John D., and Davis, L. A., 1987, "Drainage and Water Movement in Tailings Impoundments", State-of-the-Art Address, International Conference on Mining and Industrial Waste Management, South African Institution of Civil Engineers, Johannesburg, S.A.

Nelson, John D., and McPhail, Gordon I., 1987, "Design and Construction of Gold Mill Tailings Impoundments", State-of-the-Art Paper, First Intl. Conf. on Gold Mining, SME of AIME, Vancouver, B.C.

Nelson, J. D. and Wrench, P. B., 1988, "Construction of Road Embankments Over Very Soft Soils Using Band Drains and Preloading", Second International Conference on Case Histories in Geotechnical Engineering, June.

Schiffman, R. L., 1980, "Finite and Infinitesimal Strain consolidation", J. Geotechnical Engineering Division, ASCE, Vol. 106, No. GT2, pp. 203-207.

Schiffman, R. L., Pane, V., and Gibson, R. E., 1984, "Nonlinear Finite Strain Sedimentation and Consolidation", Sedimentation Consolidation Models, Prediction and Variation, R. N. Yong and F. C. Townsend, Eds., ASCE, pp. 1-29.

Smith, M. E., Bentel, D. L., and Robbertze, J. B., 1987, "Management Guidelines for the Construction of Gold Tailings Dams," Proceedings, International Conference on Mining and Industrial Waste Management, South African Institution of Civil Engineers, Johannesburg, S. A., pp. 31-38.

Terzaghi, K., and Peck, R. B., 1967, Soil Mechanics in Engineering Practice, 2nd Ed., John Wiley and Sons.

9

Case Examples

Chapter 31

A SLOPE STABILITY STUDY FOR THE
ANNIE CREEK AND FOLEY RIDGE MINES

Charles A. Kliche and Anthony D. Hammond

Assistant Professor, Mining Engineering Department
South Dakota School of Mines and Technology
Rapid City, South Dakota

Mine Planning Engineer
Wharf Resources
Deadwood, South Dakota

ABSTRACT

A slope stability study for the Annie Creek and Foley
Ridge Mines was undertaken to provide Wharf Resources,
Inc. with a technical assessment of the stability of the
slope for an overall pit slope angle of 60°. A rock
analysis of the most critical and persistent combination
of discontinuities was carried out using probabilistic
methods.

A detailed discontinuity mapping and assessment of the
rock mass provided the orientation data used in this
study. The orientation data were reduced and analyzed to
identify the discontinuity sets. A statistical analysis
provided information about the distribution of the data.

A deterministic wedge stability analysis identified
the critical combination of discontinuities; those with a
low factor of safety were further studied using
probabilistic kinematic and kinetic stability analyses.
The results from this approach were used to obtain the
reliability or probability of failure of the slopes.

INTRODUCTION

The Annie Creek and Foley Ridge Mines are located 6
miles west of Lead, in Lawrence County, South Dakota
(Figure 1). Since 1983, Wharf Resources, Inc. has been
working the Annie Creek mine by open pit methods. In
1986, the adjacent Foley Ridge deposit was developed and
is currently in operation.

441

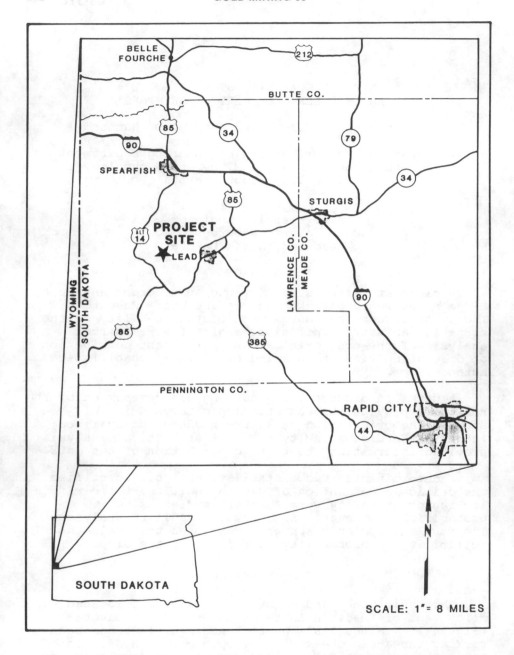

Figure 1.   Location Map, Annie Creek Mine Project, Wharf Resources

The Annie Creek pit (Figure 2) was designed and initially permitted with a 45° overall slope angle. With higher gold prices, an area of lower grade under the present highwall became attractive. Wharf Resources decided to investigate the viability of a 60° overall slope angle in this area.

Figure 2. The Annie Creek Mine

The Annie Creek Gold Mine is a sediment-hosted hydrothermal deposit in the Deadwood Formation. The ore was deposited along vertical fractures as replacement deposits. The most important deposition is found in the dolomitic members of the Deadwood Formation. The sedimentary rocks of the upper contact of the Deadwood Formation have been altered by metamorphism associated with the intrusion of Tertiary age porphyry sills.

From a structural standpoint, the NE trending fractures are the most strongly developed and most distinctive fracture system. These fractures are essentially vertical. A detailed line mapping of discontinuities identifed the dominant orientation of the

so called "Vertical Structures." The mean attitude is
N25°E with the dip ranging from 73° SE to 84° NW (cluster
Annie 3 in Figures 3 & 4). A conjugate system of joints
strikes N 70° E and dips 77° SE to 85° NW (Annie 5 and
Annie 6 in Figure 3), but is not as strongly developed as
the major system. In the southern part of the east wall a
bedding plane daylights at the face of the benches. It
has a mean strike of N10°E and a mean dip of 28° NW (Annie
7 in Figures 3 & 5).

These discontinuity sets are also present in the
porphyry domain of the Annie Creek Mine with little
variation in the orientation. The roughness on the joint
surface varies from slickensided to rough in the
sedimentary rocks, whereas in the porphyry the joint
surface has defined ridges. The aperture of the
discontinuities varies from narrow (6 to 20 mm) to
extremely narrow (less than 2 mm). They are clean in the
majority of cases, but when infilling is present its
nature is non-cohesive and its consistency is weak.

COMPUTER CODES USED IN THE STABILITY STUDY

Intensive usage of computer codes was done in all
stages of the project. Most of the computer codes were
developed in previous works, but some supporting programs
were developed for this project. Among the computer codes
used are NEWSNAP, originally developed by the U.S. Bureau
of Mines and modified to treat the orientation data
according to the engineering convention of dip direction
and dip.

A copy of the computer codes developed for the
research project "Risk Analysis for Rock Slopes in Open
Pit Mines" funded by the U.S. Bureau of Mines and carried
out by the Department of Civil Engineering of MIT was
acquired by the Mining Engineering Department of South
Dakota School of Mines and Technology. Figure 6 shows a
flow chart of the computer codes used in the different
phases of the Project.

DETERMINISTIC WEDGE ANALYSIS

A deterministic wedge analysis narrowed the vast scope
of possible wedge formation to focus on the critical
cases. The wedge stability analysis examined the
kinematic and kinetic feasibility of wedge formation from
the identified sets of discontinuities. The kinematic
analysis, which is purely geometric, examines which
movements can occur. In the kinetic analysis, the driving

ANNIE CREEK MINE, LEVELS 6320-6060, SED. DOMAIN

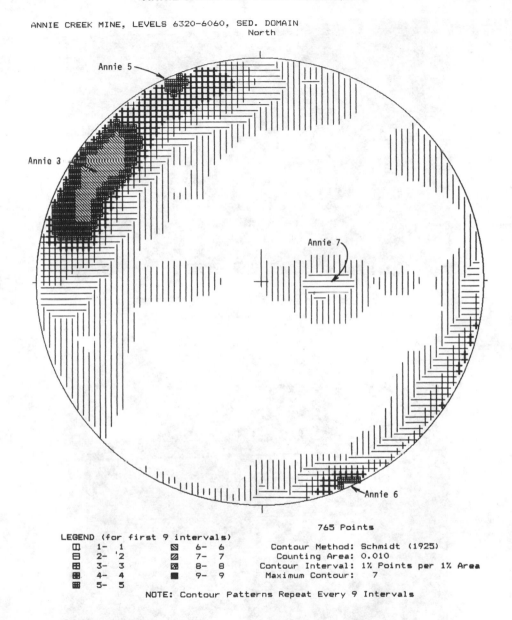

765 Points

LEGEND (for first 9 intervals)

	1- 1		6- 6
	2- 2		7- 7
	3- 3		8- 8
	4- 4		9- 9
	5- 5		

Contour Method: Schmidt (1925)
Counting Area: 0.010
Contour Interval: 1% Points per 1% Area
Maximum Contour: 7

NOTE: Contour Patterns Repeat Every 9 Intervals

Figure 3. Stereonet plot of the Annie Creek sedimentary domain.

Figure 4.   The predominant set of joints at the Annie Creek Mine (Annie 3).

Figure 5.   Shallow dipping planes at the Annie Creek Mine (Annie 7).

The zones are multiple quartz vein systems, consisting of at least two generations of gold bearing quartz, a later stage of more massive coarse grained white quartz superimposed on a multitude of thinner sheeted darker quartz veins. Arsenopyrite tends to accompany the early set, forming disseminated haloes in the walls and vein septa, while coarse grained base metal sulphides and coarse gold appears to favour the later. Wall rock alteration is not extensive consisting mainly of pale green diopside with rare calcite.

The quartz veins and even pegmatite dikes which locally cut them have been slightly squeezed and boudinaged, providing evidence of post mineral elevated rock temperatures and ductile flow. This probably causes annealing and exsolution of gold from sulphides.

RESERVES

Currently, reserves are calculated at 3.9 million tons grading .23 ounces of gold per ton. At an overall extraction of 75 percent there are sufficient mineable reserves for an eight year mine life.

Ore reserve calculations were based upon a model which treats each zone as a thin seam of 1.83 metres in thickness. Computer generated polygons are used to predict tonnages and grade.

Further diamond drilling down dip and economic evaluation of deeper mine workings will determine ultimate mine life.

CONSTRUCTION

On January 19, 1987, Pioneer announced the decision of its Board of Directors to place the Puffy Lake gold deposit into production. Initial plans were to build a 500 tonne per day mill and produce 40,000 ounces of gold per year. Three diamond drill rigs continued to drill infill and step-out holes and the result was the announcement in May, 1987 that reserves had been expanded and the mill capacity would be increased to 1,000 tons per day. Annual gold production would be 72,000 ounces.

Clearing of a powerline right-of-way to the substation at Sherridon, began in February. In May the mill site was selected and clearing began at once. Throughout June and early July, the site was leveled and backfilled to grade.

forces (weight, water, external driving forces) are
compared to the resisting forces (shearing resistance of
joints and external resisting forces) and a limit
equilibrium analysis is conducted to compute a factor of
safety. The factor of safety is computed as:

$$F.S. = \frac{\text{Resisting Forces}}{\text{Driving Forces}}$$

The stability analysis for each wedge is conducted for
both peak and residual joint shear resistance.

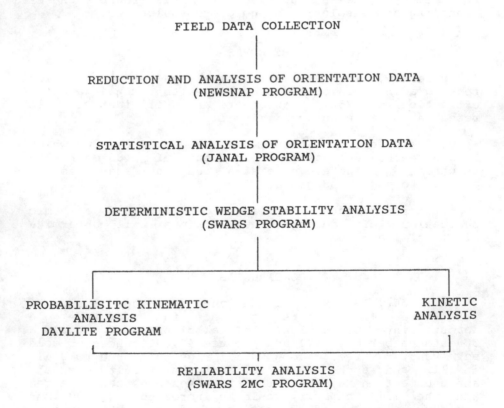

FIELD DATA COLLECTION

REDUCTION AND ANALYSIS OF ORIENTATION DATA
(NEWSNAP PROGRAM)

STATISTICAL ANALYSIS OF ORIENTATION DATA
(JANAL PROGRAM)

DETERMINISTIC WEDGE STABILITY ANALYSIS
(SWARS PROGRAM)

PROBABILISITC KINEMATIC   KINETIC
ANALYSIS   ANALYSIS
DAYLITE PROGRAM

RELIABILITY ANALYSIS
(SWARS 2MC PROGRAM)

Figure 6.   Flow chart of phases of the study and computer
            programs used

A sensitivity analysis was conducted to test the effect of the following parameters in the calculation of the factor of safety: cohesion, friction angle, dip of slope, vertical limits of the wedges, and presence of water. The friction angle and cohesion values used in this study were taken from a report on the stability of the Annie Creek Mine issued by a consulting firm retained by Wharf Resources.

The following parameters were used in the wedge analysis of the sedimentary and porphyry domains of the Annie Creek and Foley Ridge mines:

- Strike of the slope face = N14°E for Annie Creek
                                               N23°E for Foley Ridge
- Height of the excavation = 79 m (260 ft)
- Length of the excavation = 213 m (700 ft)
- Top slope angle = 5°
- Unit weight of rock = 2300 kg/m^3 (144 pcf) for sediments
                                               2500 kg/m^3 (155 pcf) for porphyry
- Friction angle = 20° to 30°
- Asperity angle = 5° to 10°
- Persistence (% of un-attachment of the joint) = 70%
- Cohesion = 193 KPa (28 psi)
- Water depth = 0 to 30 m (0 to 100 ft)

A surcharge of 575 kPa (83 psi) was considered in the analysis of the NW Foley Mine. This surcharge is equivalent to the effect of a waste dump 30m (100 ft) high placed adjacent to the north east side of the NW Foley Mine.

A factor of safety of 1.3 was considered acceptable for the stability of final pit walls.

The results of the slope stability analysis showed that at the Annie Creek Mine, the bedding plane daylighting in the southern part of the pit controls the stability of the east highwall (Figure 5). For a 60° slope angle, the lowest value of the peak factor of safety is 1.74. The factor of safety without cohesion is 0.88.

At the NW Foley Mine, the analysis was conducted for different orientations of the final highwalls. A 60° slope angle is deemed safe except in the northeast side where the dump will be constructed.

PROBABILISTIC KINEMATIC STABILITY ANALYSIS

The potentially unstable combination of planes were analyzed using a probabilistic approach that treats the orientation of planes as random variables defined by probability density functions (PDF's). Knowing the positions and forms of the PDF's of 2 joint sets in relation to the orientation of the slope, one can calculate the probability that the line of intersection daylights in the slope.

Computer program DAYLITE was used to determine the probability of kinematic instability of a 2 joint wedge. Fisher distributions were used to define the probability density function of the orientation data. The Fisher distribution is an analog of the planar normal distribution on the surface of the sphere.

In the Annie Creek sedimentary domain, the critical joint set Annie 7 combined with four other sets of planes was analyzed for slopes striking N14^{O}E and dipping 60° NW. The highest probability of kinematic instability is 13%, occuring with the combination of planes from the set Annie 5. The orientation of the line of intersection has a plunge of 12.5^O. At this shallow angle, wedges may have relatively small driving forces and consequently negligible probabilities of failure since the friction angle of the joints is 25°. Also, the formation of wedges with narrow central angles makes the sliding of the wedges unlikely because the narrow geometry of these wedges induce high normal stresses in the joint plane and thus high resistance.

The probability of kinematic wedge instability in all three domains is very low. The results obtained from this analysis has an overriding effect on the stability as will be seen in the reliability study.

RELIABILITY ANALYSIS

A reliability analysis for 2-joint wedges was used to determine the reliability, i.e. probability of failure of rock slopes, at the Annie Creek Mine with an overall pit slope angle of 60°. The reliability depends on the probability that a wedge is kinematically and kinetically unstable:

$$Pf = \Sigma [P U_i / \theta_i, \phi_i] \times P [\theta_i, \phi_i]$$

where $P[\theta_i, \phi_i]$ is the probability that the wedges with lines of intersection in the θ_i, ϕ_i direction are kinematically unstable, i.e. daylight, and $P[U_i / \theta_i, \phi_i]$ is the probability that the wedges represented by θ_i, ϕ_i are kinetically unstable.

The probability that the wedge is kinematically unstable, i.e. daylights, was estimated in the Probabilistic Kinematic Analysis. The probability that the wedge is kinetically unstable is estimated by means of the computer program SWARS-2MC. This program performs Monte Carlo simulation based on user specified distributions of both intact rock and joint resistance.

The Factor of Safety against ultimate failure was adopted in this study. This resulted in simplifications in the kinetic analysis as this approach neglects the cohesion, persistence, and asperities along the joints leaving the friction angle of the joints as the only resisting force.

The geometry of the wedge and the friction angle were probabilistically defined by assigning uniform distributions to these parameters. The uniform distribution was chosen due to a lack of data to determine the distribution of the strength parameters.

Two cases were analyzed, one for wedges formed by joints from sets Annie 7 and Annie 3, and another for wedges formed by joints from set Annie 3 alone. The first case represents the most apparent failure mode in the Annie Creek Mine which was alluded to in the deterministic wedge analysis; and the second case represents the possible combination of wedges from the dominant set of discontinuities in the Annie Creek Mine, namely, the vertical structures.

The following information was used in both cases:

- strike of the excavation   = N14°E
- dip   = 60° NW
- range of friction angle   = 25° to 40°
- height of slope   = 79m (260 ft)
- inclination of top slope   = 5°
- unit weight of rock   = 2300 kg/m³ (144 pcf)

Two hundred simulations were performed. The
reliability analysis concluded that the probability of
failure of the slopes striking N14°E and dipping 60° NW at
the Annie Creek Mine is only 3%. This very low
probability of failure is controlled by the low
probability of kinematic instability of the wedges.

CONCLUSIONS

The most apparent failure mode at the Annie Creek mine
is toppling of the steeply dipping joints of the
predominant set of discontinuities, aggravated in the case
where this set sits on the shallow dipping planes that
daylight in the southern part of the east highwall. In
the northern part of the east highwall, there is a
potential for local slumps in the highly fractured and
weak rock mass.

The overall slope angle of 60° for the final pit
limits of the Annie Creek Mine can be undertaken with only
3% risk of failure. This low probability of failure is
dictated by the unlikely formation of kinematically
unstable wedges from the identified sets of joints in the
final orientation of the east highwall.

Further mapping of discontinuities in the NW Foley
Mine should be undertaken when more benches are exposed as
the mining progresses and a refinement of the risk of
failure should be done with the recent data made available
by the rock mechanics testing program carried out in
1988.

REFERENCES

Einstein, H.H. et al., 1979, "Risk Analysis for Rock
    Slopes in Open Pit Mines," U.S. Bureau of Mines,
    National Technical Information Services, Springfield,
    Virginia.

Hammond, A.D., 1987, "A Slope Stability Study for the
    Annie Creek and Foley Ridge Mines, Lead, South Dakota,"
    M.S. Thesis, South Dakota School of Mines and
    Technology, Rapid City, South Dakota.

Jeran, P.W., and Mashey, J.R., 1970, "A Computer Program
    for the Stereographic Analysis of Coal Fractures and
    Cleats," U.S. Bureau of Mines, I.C. 8454.

Klohn-Leonoff, 1986, Internal report for Wharf Resources,
    Inc.

Chapter 32

A COMPARATIVE REVIEW OF QUEENSTAKE RESOURCES LTD.
ALLUVIAL GOLD MINING OPERATIONS

by Gordon C. Gutrath, P.Eng

President and Chief Executive Officer
Queenstake Resources Ltd.

ABSTRACT

Queenstake Resources Ltd. has been exploring and mining alluvial gold deposits in the Yukon Territory and British Columbia since 1980. The five mines in production produced, in 1987, a total of 13,222 ounces. This paper will briefly review Queenstake's exploration program, property descriptions and mining methods, environmental concerns and a comparison of costs. Queenstake has been successful in developing a cash flow base from these operations that provide the capital for the acquisition and exploration of new, and hopefully, larger scale gold producers.

NOTE: The historic alluvial mining unit of ounces per cubic yard and English units of measurement have been used throughout this paper. Dollars are Canadian unless otherwise noted.

INTRODUCTION

Queenstake Resources is a Canadian mining company with its executive head office in Vancouver, British Columbia and regional offices in Whitehorse, Yukon and Reno, Nevada. The Company was listed on the Vancouver Stock Exchange in 1979 and on the Toronto Stock Exchange in 1982.

Queenstake's corporate strategy has been to concentrate its activities in precious metal exploration and development in Western Canada and the United States and in maintaining a cash flow base from alluvial gold mining operations in the Yukon and British Columbia.

453

In 1987, Queenstake had five alluvial mines in production, three in the Yukon and two in British Columbia. The Yukon mines are all located in the Dawson area on creeks peripheral to the Klondike gold fields. Queenstake has operated the only bucketline dredge in Canada for the past seven years but in 1987 the last of the dredgeable reserves were depleted. Queenstake's two other Yukon mines are on Black Hills Creek and Maisy May Creek. These are bulldozer - sluice box operations and have reserves from five to ten years of mining. In 1987, Queenstake acquired the Pine Creek placer property at Atlin, British Columbia and upgraded the existing operation. The Moyie River placer property near the town of Cranbrook in southern B.C. was placed into full scale production late in 1987 after one and a half years of exploration.

It is often difficult to acquire recent operating costs for alluvial mining operations since the majority of alluvial mines are privately owned. Queenstake is a public company so its financial information is readily available.   The tables in this paper give a general description of the properties, a description of the deposits, the mining method and equipment, 1987 production results and the distribution of 1987 operating costs per cubic yard processed.

Alluvial mining has had a rich history and has been the early foundation of numerous major mining companies. Queenstake has demonstrated that alluvial mining in the 1980's can be a profitable enterprise and a method for a small company to develop a significant cash flow at an early stage and at low capital cost.

ALLUVIAL EXPLORATION

Queenstake has concentrated its alluvial exploration in the Yukon and British Columbia, although a number of properties have been, and are being, evaluated in the western United States and Alaska and on a world wide basis.   Over the past seven years some twenty-five properties in the Yukon and another fifteen in B.C. have been explored utilizing overburden drilling and/or backhoe pit sampling through to major bulk sampling programs of 25,000 to 50,000 cubic yards. The typical exploration program is initiated by mapping the surface and outcrop geology followed by drilling and pit sampling. In order to qualify the drill and pit sampling data a bulk sample is mined and processed to determine grade and mining characteristics of the gravels and bedrock.

GENERAL PROPERTY DESCRIPTION AND MINING METHODS

Every alluvial deposit is different even though they may only be a few miles apart, such as Maisy May Creek and Black Hills Creek. At these two deposits the surficial bedrock geology and gold fineness are very similar.

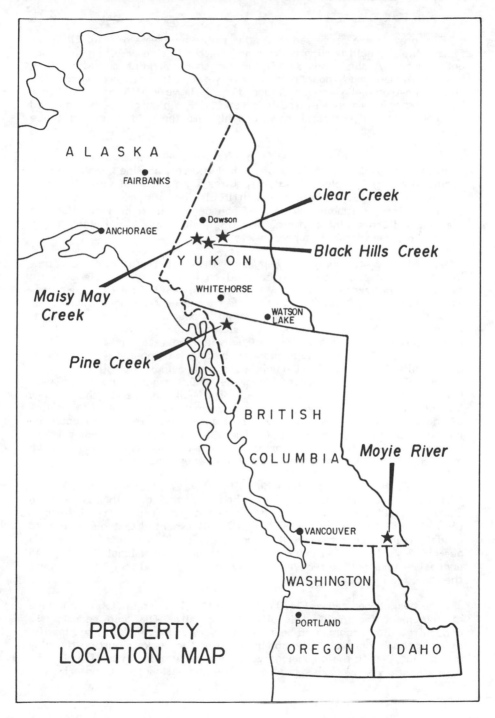

PROPERTY
LOCATION MAP

The gravels are processed in similar sluice box systems (Fig. 2) but at Maisy May the gold is slightly finer and flatter in size resulting in higher gold losses. A sluice box system similar to Black Hills and Maisy May would produce very poor gold recovery at Clear Creek where the bucketline-trommel equipped dredge (Fig. 1) is very efficient. At Clear Creek the gold is concentrated in a very sticky, high clay - decomposed bedrock, that will not break down and liberate the gold in a short sluice box.

Pine Creek and Moyie River (Table 7) are very similar deposits as to depth, nature of gravels and size of gold. However, at Pine Creek the gold occurs in a very tenacious clay that requires a great deal of scrubbing prior to screening. At Moyie there is very little clay and the gravels wash easily without a large scrubbing section. Each deposit must be viewed as site specific and the exploration program, final mining plan, and plant design must take these factors into account.

As with any mine, each year that the property is operated, new information is gained that can be used to improve it's efficiency. An example is the transition at Maisy May and Black Hills Creek from a dump-sluice box recovery unit (Fig. 2) to a skid mounted trommel screen-sluice recovery system (Fig. 3). The original production units were simple, inexpensive to build, dump plate-sluice boxes giving very little classification of the gravels and an uneven feed rate. In 1987, a more efficient trommel screen sluice plant replaced the dump plate-sluice box at Black Hills Creek. A 1,400 cubic yard bulk sample of tailings from a reserve block having a recoverable grade of 0.015 ounces per cubic yard was washed by the trommel sluice. The bulk sample of the tailings indicated that the sluice system was recovering approximately 70% to 75% of the gold, and that with the trommel screen unit recoveries could be increased to 85% and possibly 90%. The better gold recovery was a result of material classification, a better scrubing-washing action and a much more uniform flow of material to the sluice recovery tables.

In 1988, the Maisy May sluice box (Fig. 2) was replaced with a 150 cubic yard per hour trommel screen - sluice box system (Fig. 3) that is fed by a Caterpillar 300 backhoe. Previous recovered grades at Maisy May are in the range of 0.014 ounces per cubic yard. It is expected that the recovered grade will increase between 0.002 ounces per cubic yard and 0.003 ounces per cubic yard. At 150 cubic yards per hour an additional 0.30 to 0.45 ounces per hour will be recovered or approximately 500 to 700 ounces over the short mining season of 120 days.

The bucketline dredge (Fig. 1) at Clear Creek was an outstanding piece of equipment for mining this difficult deposit. The gold deposit was concentrated on decomposed bedrock in a very sticky, difficult to classify, clay-rich cobble gravel. The combination of the buckets cutting the clay and mixing it with the cleaner gravels, prior to washing in the trommel by high pressure water nozzles, produced a 'milling action' as the material was

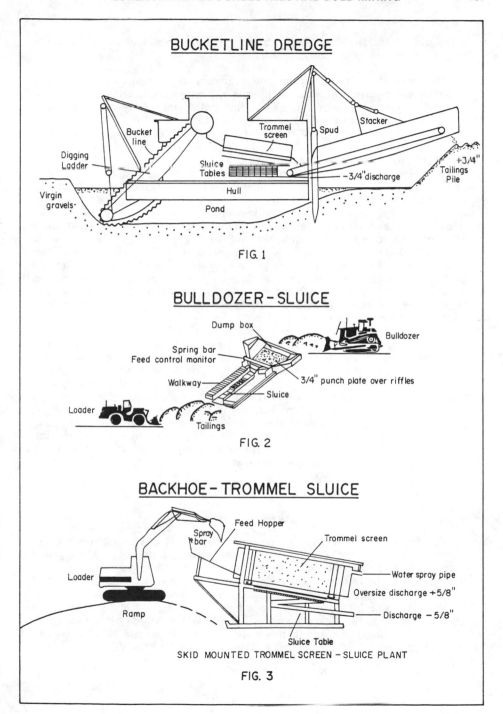

BUCKETLINE DREDGE

FIG. 1

BULLDOZER-SLUICE

FIG. 2

BACKHOE-TROMMEL SLUICE

SKID MOUNTED TROMMEL SCREEN – SLUICE PLANT

FIG. 3

being classified. It was a rare occurrence to see gold bearing clay balls escape the scrubbing action of the trommel.

The next step in improving gold recovery at Queenstake's operation would be the addition of jigs. Gold from alluvial deposits in B.C. and the Yukon has generally been defined as "fine gold" when it is in a size range between -10 to +40 mesh (-1.5 mm to +0.35 mm) with -40 mesh being referred to as "flour gold" (Wang and Poling, 1981). Jigs are more efficient for the recovery of fine gold with a reported 93% recovery of +40 mesh gold for a jig system as compared to 82% recovery for a riffled sluice system (Wang, 1979). However, recent work completed at the University of British Columbia using sample material from a typical deposit in the Klondike area of the Yukon has demonstrated much better fine gold recovery in a sluice-riffle unit. Using expanded metal riffles as opposed to "Hungarian" angle-iron riffles in the sluice, gold recoveries ranged from 85% to 100% for +40 mesh gold. For -40 to +100 mesh gold the recoveries ranged from 73% to 97%. It was concluded that sluice boxes used in the Yukon can be an effective method of recovering gold as fine as 150 mesh. (Poling and Hamilton, 1987). Queenstake's properties in the Yukon and B.C. have only a small percentage of gold less than 100 mesh so the trommel-screen-sluice recovery plant is very effective in recovering 85% to 90% of the gold in the deposit.

A jig plant is more efficient at fine gold recovery and has the advantage of a continuous clean up of the gold and heavy mineral (black sand) concentrates. The disadvantage of the jigs is the additional weight of the jig units, the higher capital cost and a more complex plumbing system that will be susceptible to freezing, with more equipment maintenance and lower availability than a sluice box system. Queenstake has used a jig plant to process a bulk sample from "white channel" gravel deposits is the Klondike where there was a high percentage of gold in the -100 mesh size fraction. Queenstake has not incorporated it into its present operations because marginal improved gold recovery would not offset the additional capital and operating costs.

ENVIRONMENTAL

For the past few years enviromental regulations governing the alluvial mining industry in the Yukon were not stringently enforced. During this period new environmental guidelines were being developed jointly by the Klondike Placer Miners Association, representing the local miners and Federal and Territorial government agencies. These new guidelines were in place for the 1988 mining seasons and they reflect the importance of alluvial mining to the Yukon economy. In 1987 the Yukon placer industry, comprised of some 200 small mining operations, produced 100,000 ounces of gold or approximately $60 million in revenue for the Yukon economy.

The guidelines also take into account that the majority of the alluvual mining operations are not in conflict with other resource users or an important wildlife or fish resource.

Water quality standards are liberal compared with any other Canadian or U.S. jurisdiction. The guidelines are based on a stream classification that indicates the sensitivity of the streams to water quality and potential disruption of fish habitat. Streams that are presently being mined, have been mined historically or have a natural low sensitivity, have a water quality standard for mine discharge water of not more than 5.0 millilitres per litre settleable solids and there is no turbidity standard. These standards apply to the majority of streams presently being mined in the Yukon. For moderately sensitive streams the water quality standards will range from 1.0 millilitre to 0.2 millilitre per litre settleable solids and for very sensitive steams, such as a salmon spawning stream there is to be zero sediment discharge to the stream.   The reclamation requirements emphasize stream stabilization suitable for the development of fish habitat. Tailings piles are to be contoured to minimize erosion and to promote natural revegetation. Reseeding is not required.

In British Columbia, environmental guidelines relative to the alluvial mining industry are much more stringent than those found in the Yukon. Queenstake's Pine Creek mine at Atlin is in a historic mining camp that has seen continuous alluvial mining activity since the late 1800's. Water quality and reclamation standards are more liberally interpreted because the mining activity is not in conflict with any other resource user and it is the only industry, other than tourism, in the local area.

At the Moyie River operation in southern B.C. environmental standards are enforced to the "letter of the law". The Moyie River is a popular recreation fishing stream and as a result there can be no sediment discharge to the river, and reclamation, including reseeding, immediately follows mining.

PRODUCTION RESULTS

Production data is outlined in Table 7 for the 1987 mining season. For all properties, a total of 1,891,715 cubic yards were stripped and 847,476 cubic yards were processed producing 13,222 ounces of gold. The average grade mined was 0.016 ounces per cubic yard and the cost of processing was $5.44 per cubic yard. Average gold price realized by Queenstake in 1987 was Cdn$604 per ounce producing a value per yard of $9.66 and an operating profit margin per cubic yard of 43% or $4.22 per yard.

GENERAL DESCRIPTION OF YUKON PROPERTIES

	Maisy May	Black Hills	Clear Creek
Location	Yukon 60 air miles SW of Dawson City	Yukon 55 air miles SW of Dawson City	Yukon 60 miles east of Dawson City
Access	104 miles of gravel road from Dawson City	80 miles of gravel road from Dawson City	84 miles of gravel road from Dawson City
Elevation (feet above sea level)	2,400	2,400	3,200
Vegetation	muskeg and stunted black spruce	muskeg and stunted black spruce	muskeg and stunted black spruce
Water Quantity (cubic feet per second)	30	40	24
Rainfall (inches per year)	15	15	15
Ground Condition	permafrost	permafrost	permafrost
Valley Width (feet)	200 - 600	400 - 1500	200 - 500
Valley Gradient (percent)	1.5 - 2	1 - 2.5	1.5 - 3

Mining Season
  Start up date
  Stripping
  Mining
  Shut down date
	April 20
	April 25 - Oct. 15
	June 10 - Sept. 30
	Oct. 15
		April 15
		April 15 - Oct. 15
		June 10 - Sept. 30
		Oct. 15
			May 10
			May 15 - Oct. 20
			June 10 - Oct. 20
			Oct. 25

Environmental Sensitivity	not sensitive arctic grayling stream	not sensitive designated placer stream	medium sensitive arctic grayling and coho salmon rearing in lower reach

Table 1

GENERAL DESCRIPTION OF BRITISH COLUMBIA PROPERTIES

	Pine Creek	Moyie River
Location	Northern B.C., 5 miles from Atlin, B.C. and 100 miles S of Whitehorse, YT	Southern B.C. 14 miles NW of Cranbrook
Access	5 miles of gravel road from Atlin	10 miles of paved road and 5 miles of gravel road from Cranbrook
Elevation (feet above sea level)	2,600	5,000
Vegetation	sparse pine (not commercial)	spruce, tamarack, fir (sub-commercial)
Water Quantity (cubic feet per second)	50	40 (can run underground in summer)
Rainfall (inches per year)	10	20
Ground Condition	thawed	thawed
Valley Width (feet)	2,000	400 - 700
Valley Gradient (percent)	2	1.5 - 3
Mining Season Start up date Stripping Mining Shut down date	April 1 April 15 - Nov. 15 June 1 - Oct. 31 November 20	January 10 Jan. 10 - Dec. 20 May 15 - Oct. 31 December 20
Environmental Sensitivity	not sensitive historic and active placer camp	very sensitive poor fishery but important recreational area

Table 2

MINING METHODS AND EQUIPMENT AT YUKON ALLUVIAL PROPERTIES

	Maisy May Creek, Yukon	Black Hills Creek, Yukon	Clear Creek, Yukon
Mining Method	Bulldozer/sluice box/loader A bulldozer pushes up a ramp into the sluice box, the material is washed and the tailings are stacked with the 980C loader. (Fig. 2)	Bulldozer/sluice box/loader A bulldozer pushes up a ramp into the sluice box, the material is washed and the tailings are stacked with the 980C loader. (Fig. 2)	Bucket ladder dredge The dredge floats in its own pond, the bucketline elevates the material which is washed inside dredge and tailings are stacked with the conveyor. (Fig. 3)
Stripping Method	Overburden - rip and stack Overburden - bulldozer and water ditches, pump and monitors Gravel - stack with dozer	Overburden - rip and stack Overburden - bulldozer and water ditches, pump and monitors Gravel - stack with dozer	Overburden - bulldozer and water ditches Gravel - stack with dozer
Mining Equipment	2 - D9L Caterpillar tractors (19 yard) 1 - 980C Caterpillar loader (5 yard) 1 - 100C Hough loader (5 yard)	2 - D9H Caterpillar tractors (16 yard) 1 - D355A Komatsu tractor (16 yard) 1 - D8K Caterpillar tractor (10 yard) 1 - 980C Caterpillar loader (5 yard) 1 - 980B Caterpillar loader (4 yard) 1 - 225 Caterpillar backhoe (1 3/4 yard)	1 - Bucketline dredge (3.5 cubic feet) 1 - D9H Caterpillar tractor (16 yard) 1 - D8H Caterpillar tractor (10 yard)
Total Site Personnel	11	11	6

Table 3

MINING METHODS AND EQUIPMENT AT BRITISH COLUMBIA
ALLUVIAL PROPERTIES

	Pine Creek, B.C.	Moyie River, B.C.
Mining Method	Truck - Trommel/Sluice	Scraper - Trommel/Sluice
	35 ton trucks are loaded with a backhoe, material is hauled to the wash plant, a 966 loader feeds the wash plant, and a 988B loader removes the tailings.	Scrapers load and haul the material to the wash plant, a backhoe feeds the plant, a conveyor stacks the tailings, and a bulldozer levels and contours the pile.
Stripping Method	Bulldozer pushes to backhoe which loads a truck for waste pile or backfill.	Scrapers load and haul the material out of the pit to contoured waste piles.
Mining Equipment	1 - D10L Caterpillar tractor (28 yard) 1 - D9H Caterpillar tractor (16 yard) 2 - 988B Caterpillar loaders (7 yard) 1 - 966C Caterpillar loader (3.5 yard) 3 - 769B Caterpillar trucks (35 ton) 1 - 245 Caterpillar backhoe (3.75 yard) 1 - 235 Caterpillar backhoe (2.75 yard)	3 - 627B Caterpillar scrapers (20 yard) 2 - D9H Caterpillar tractors (16 yard) 1 - D8H Caterpillar tractor (10 yard) 1 - 245 Caterpillar backhoe (3.75 yard) 1 - 235 Caterpillar backhoe (2.75
Total Site Personnel	25	10

Table 4

DESCRIPTION OF GOLD RECOVERY PLANTS

Property	Processing Plant	Sluice Area (square feet)	Capacity (yards per hour)
Maisy May, Yukon	28 feet of wash section, 16 feet wide, followed by 30 feet of sluice, 4 feet wide. Total length 58 feet. The dump box has hinged 3/4 inch punchplate covering 1 inch riffles. The sluice has three, 6 inch drops. At each drop there is 4 feet of punchplate over 2 inch riffles and then 6 feet of open 2 inch riffles.	216	175
Black Hills, Yukon	20 feet of wash section, 16 feet wide, followed by 30 feet of sluice, 5 feet wide. Total length 50 feet. The dump box has hinged 3/4 inch punchplate covering 1 inch riffles. The sluice has 3, 6 inch drops, at each drop there is 4 feet of punchplate over 2 inch riffles and then 6 feet of open 2 inch riffles.	294	200
Clear Creek, Yukon	Continuous 3.5 cubic foot bucketline feeds a 4 foot diameter by 20 foot long trommel. There is a 6 foot wide sluice run with one inch riffles directly below the trommel, then the material is split and runs over ten, 3 foot by 15 foot sluice runs.	930	80
Pine Creek, British Columbia	Two 5.5 foot diameter trommels with 30 foot scrubber section and 10 feet of 2 inch screen. The material is screened again with punchplate to -3/4 inch. The gold is trapped in and recovered by expanded metal riffles.	300 per plant each	75 per plant each
Moyie River, British Columbia	A 6.5 foot diameter by 36 foot long trommel is used to wash and screen the material to -5/8 inch. A continously turning belt-sluice run with 1.5 inch riffles is used to recover the gold. The belt is 16 feet long and makes a half revolution every 40 minutes.	90	100

Table 5

DESCRIPTION OF DEPOSITS

	Maisy May	Black Hills	Clear Creek	Pine Creek	Moyie River
Depth of overburden stripped (feet)	20	15	4	35	40
Nature of overburden	frozen silt and peat moss	frozen silt and peat moss	frozen silt and peat moss	glacial till	glacial till
Depth of upper gravels stripped (feet)	3	5	2	-	-
Nature of gravels	loose cobble gravels	loose cobble gravels	loose cobble gravels	oxidized compact boulder gravel	compact boulder gravel
Average depth of bedrock mined (feet)	1.5	1.5	1.5	1	1
Average depth of gravel processed (feet)	4	4	8	15	10
Type of bedrock	decomposed soft micaceous shist	decomposed soft micaceous shist	decomposed soft micaceous shist	blocky to sheared serpentine	fracture siliceous argillite

Table 6

1987 PRODUCTION RESULTS

	Maisy May	Black Hills	Clear Creek	Pine Creek	Moyie River	Totals
Overburden Stripped (cubic yards)	412,964	400,000	0	0	0	812,964
Gravel Stripped (cubic yards)	90,000	90,000	50,000	664,100	184,651	1,078,751
Gravel Processed (cubic yards)	170,000	255,000	191,415	166,025	65,036	847,476
Total Material Handled (cubic yards)	672,964	745,000	241,415	830,125	249,687	2,739,191
Fine Gold Produced (troy ounces)	2,334	1,764	679	7,002	1,444	13,222
Grade (fine ounces per cubic yard)	0.014	0.007	0.004	0.042	0.022	0.016
Operating Costs (Canadian dollars)	$ 915,130	$ 1,100,215	$ 443,107	$1,498,780	$ 650,350	$4,607,582
Cost per yard Processed ($ per cubic yard)	$ 5.38	$ 4.31	$ 2.31	$ 9.03	$ 10.00	$ 5.44
Cost per yard Handled ($ per cubic yard)	$ 1.36	$ 1.48	$ 1.84	$ 1.81	$ 2.60	$ 1.68

Table 7

CONCLUSION

These small alluvial mines are not low cost per ounce gold producers. In 1987 direct operating costs per ounce were Cdn$348 or approximately U.S.$285 per ounce. However, these mines have a very low capital investment compared with hardrock mines and all the capital equipment is mobile allowing a great deal of operational flexibility.

Queenstake's corporate objective of developing a cash flow base from small alluvial gold mining has been successfully reached. The challenge now will be to maintain and expand production levels.

REFERENCES

Wang, W. and Poling, G.W., October, 1981 - "Methods for Recoverying Fine Placer gold", paper for Sixth Annual District C.I.M.M Meeting, Victoria, B.C.

Poling, G. W. and Hamilton, J. F., 1987 - "Pilot Scale Study of Fine Gold Recovery of Selected Sluicebox Configurations" Department of Mining and Mineral Process Engineering, University of British Columbia, Vancouver, B.C.

Sidon, T., Minister of Fisheries and Oceans, April 1988, Yukon Fisheries Protection Authorization Applicable to Placer Mines in the Yukon Territory, Government of Canada.

Chapter 33

RECENT DEVELOPMENTS IN YORBEAU'S ASTORIA PROPERTY
ROUYN-NORANDA AREA, QUEBEC.

Daniel Kelly, P.Eng.

Yorbeau Resources Inc.

INTRODUCTION

The Astoria property is located in the prolifically mineralized Abitibi greenstone belt and is one of the most advanced projects presently under development by Yorbeau Resources Inc.

This property, which hosts the Astoria gold deposit, is located in the southern part of Rouyn township, Abitibi area, Quebec, 4 km south of the city of Rouyn-Noranda (Figure 1).

The Astoria deposit has been the subject of underground development for over a half century, beginning with Granada Rouyn Mining Co. in the 1930's and followed by Astoria Rouyn Mines Ltd. in the 1940's. Yorbeau Resources Inc. acquired the property in 1984 and has, since then, conducted numerous exploration programs on this gold deposit.

The diluted reserves of the Astoria deposit are estimated at 1,3 million metric tons grading 6,14 grams of gold per metric ton (1.4 million st at 0.18 oz/st). This reserve figure is based on the results of surface and underground exploration carried out from January 1986 to June 1987 and includes ore in the proven, probable and possible categories.

This paper is divided into two sections. The first part includes a summary of the geological environment including a description of the gold bearing zones and a short discussion on wall rock alteration. The second part of this paper reviews the underground work and the objective of the ongoing exploration program.

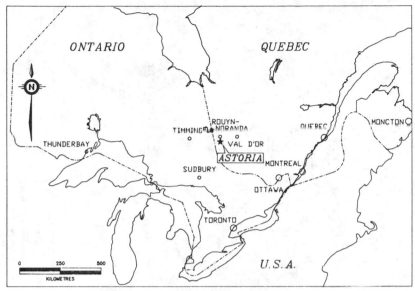

Figure 1
General location map

PART 1: GEOLOGY OF THE ASTORIA DEPOSIT

Regional geology

The Astoria property lies within the Superior structural province
of the Canadian Shield close to the southern margin of the Abitibi
greenstone belt. The Abitibi belt, extending over 600 km from east to
west and 200 km from north to south, is the largest and most prolifically mineralized belt of Archean rocks in the Precambrian shields of
the world. Figure 2 shows the location of the Astoria property within
this greenstone belt.

The Abitibi belt is composed of volcanic[1] and sedimentary supra-
crustal rocks containing numerous mafic to felsic volcanic cycles
(Goodwin and Ridler, 1970).   These volcanic piles typically have
basaltic and associated gabbroic intrusions at their base and exhibit
differentiation through an andesitic phase to an upper rock sequence
of dacitic to rhyolitic composition. Chemical and clastic sedimentary rocks are interbedded throughout these differentiated volcanic

---

1. The "meta" prefix is here implied for all the statigraphic units of
the Abitibi belt, as all rocks have undergone metamorphic recrystala-
zation generally within the greenshist facies.

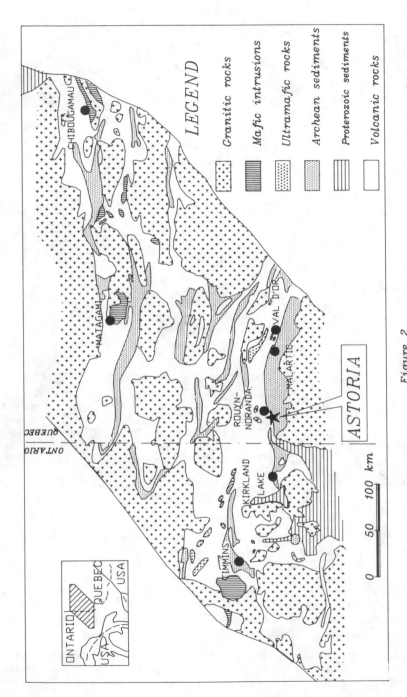

Figure 2
Simplified geology of the Abitibi greenstone belt
(modified from Goodwin and Ridler, 1970)

piles.

The Astoria property covers an area of 2 km^2 along the Cadillac-Larder Lake fault zone, a major structural feature which extends over 200 km in an east-west direction from Kirkland Lake to Val d'Or. A large number of gold producers are located along the Cadillac-Larder Lake fault zone, the most important of which have been illustrated in Figure 3.

This spacial relationship between the Cadillac-Larder Lake fault zone and gold mineralization has long been known and used by geologists as an important guide for the exploration of gold deposits. Although the precise location of the fault is not always clear, it appears that mineralization can be found in a wide variety of geological environments on either side of the fault, as well as some distance from it.

In addition to this "major break - gold association", the various directions of diabase dyking and fault zones appears to have a spatial association with gold mineralization. Although the later diabase dykes clearly postdate the metamorphism, deformation and mineralization, they may, in some cases, have occupied deep-seated fractures

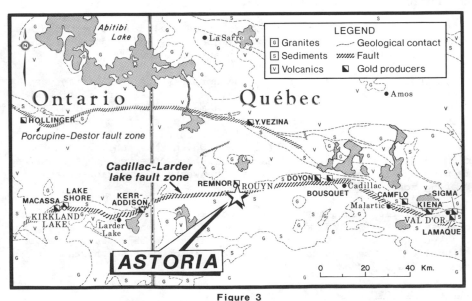

Figure 3
Gold producers along the Cadillac-Larder lake fault zone

along older zones of weakness which have focussed the mineralizing
fluids into structurally prepared host rocks. These dykes or fault
intersections would therefore appear to be particularly favorable
sites to explore for gold. The Astoria deposit, for instance, is one
such example. It should be noted though, that the presence of dykes
and fault systems serve only as a general guide to a prospective area
rather than pin pointing sites of possible mineralization.

Local geology

On the Astoria property (Figure 4), the area to the north of the
Cadillac-Larder Lake fault zone is underlain predominantly by
volcanics of the Blake River Group. The rocks consist of a suite of
andesite and basalt with subordinate amounts of intermediate to felsic
tuffs and felsic flows.

Stratigraphically below the Blake River volcanics, encompassing the
Cadillac-Larder Lake fault zone, is a group of sediments (conglomerate, greywacke and siltstone) with interbedded and interfingering
tuffs and lenses of talc-chlorite schist and magnesian carbonate
schist (altered ultramafic rocks, possibly of volcanic origin). This
group has been named the Granada Group by Goulet (1978) and is
considered by Gauthier (1988) to be a part of the Cadillac-Larder Lake
fault structural complex. Yorbeau geologists, however, have
interpreted the actual placement of this fault to be on the south
contact of this structural complex.

To the south of the Cadillac-Larder Lake fault zone is a broad area
underlain by sediments (dominantly greywacke and conglomerate, with
minor siltstone) belonging to the Timiskaming Group.

Intruded into the sediments to the northwest of the Astoria deposit
is a felsic intrusion of unknown size. The composition of this intrusion, which is probably caused by a magmatic differentiation, ranges
from granitic in the centre to gabbroic on the contacts. It is
presently interpreted to be a synvolcanic intrusion, following a N50°E
direction. This felsic intrusion, which has only been noted in
diamond drill hole intersections has not been observed by the surface
mapping. It probably displaces the western extension of the gold
bearing zones to the north.

The Astoria gold zones are transected by two major Proterozoïc
diabase dykes. The most important dyke, which has an average
thickness of 100 metres, was injected into a N55°E bearing regional
structure. The second dyke, which has an average thickness of 30
metres, was injected into a N-S structural feature.

The Astoria deposit

The Astoria deposit is located within, or near, the contacts of a
talc-chlorite schist unit (Figures 5 and 6), consisting of five gold

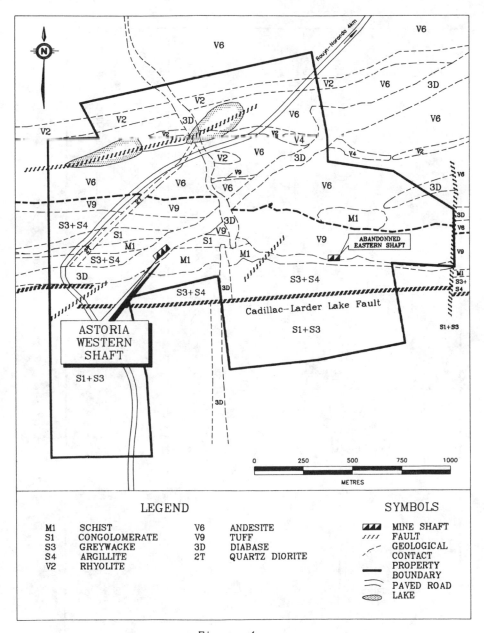

Figure 4
Geology of the Astoria property

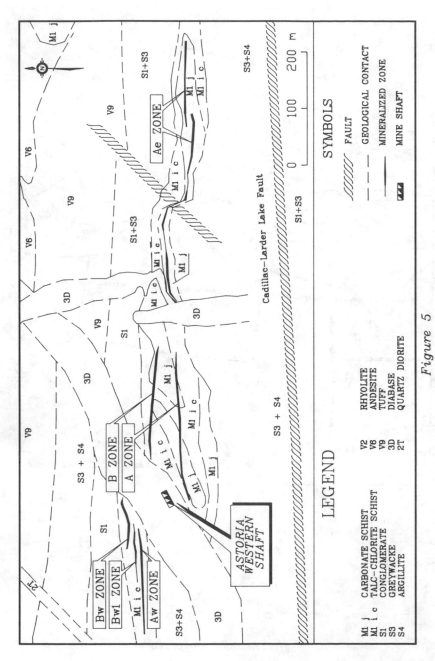

Figure 5

Distribution of the Astoria gold bearing zones

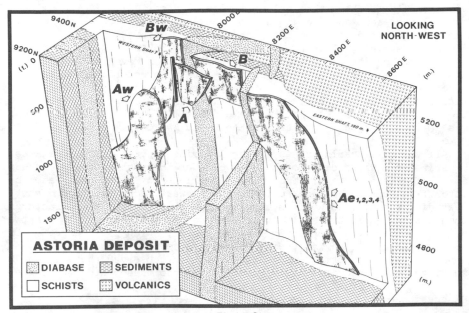

Figure 6
Astoria gold bearing zones schematic view

bearing zones: A, B, Aw, Bw and Ae. According to the terminology
used at the Kerr-Addison Mine (Kishida, 1984), which is geologically
similar to the Astoria deposit, the mineralization within these five
zones has been classified into two distinct types: "Carbonated Ore"
and "Flow Ore".

The "Carbonated Ore" type is observed to be associated with
carbonated alteration zones, located within, or near, the north
contact of the talc-chlorite schist unit. The gold typically occurs
as fine grains and occasionally forms coarse flakes, in fractures,
either within quartz-carbonate-tourmaline stockworks or silicified
zones. These carbonated alteration zones are characterized by the
following mineralogy: iron-magnesium carbonate (magnesite, siderite
and ankerite), quartz, tourmaline, chlorite, sericite and fuchsite.
We have also observed minor albite and disseminated sulphides
(arsenopyrite, gersdorsfite, pyrite, chalcopyrite), however these
minerals are not always present within the altered rocks. The
"Carbonated Ore" type is present in the B, Bw and Ae zones.

The "Flow Ore" mineralization is characterized by a strong
sulphide association and has been observed near the south contact of

the talc-chlorite schist unit, occuring either at this rock contact or
within the sedimentary rocks; although gold has been observed
occasionally within the talc-chlorite schist unit. The gold is
typically present as fine grains in fractures or as inclusions within
the disseminated arsenopyrite crystals. While arsenopyrite is the
dominant sulphide, other sulphides are present and include:
pyrrhotite, pyrite and chalcopyrite. The gold however is exclusively
associated with the arsenopyrite. The characteristic alteration
observed in these zones (besides the presence of sulphides) is albitization. Quartz-sulphide injections, which are often observed in
these zones are folded in the same direction as the main foliation.
The "Flow Ore" type is present in the A and Aw zones.

All the gold bearing zones of the Astoria deposit follow east-west
shear zones, which dip north from 80° for the western zones, to 65°
for the eastern zones. The mineralization is often displaced by
later fractures which have northeast, northwest, north-south and east-
west directions. The horizontal displacement of these later fractures
ranges from centimetre scale to several metres and generally exhibit a
dextral movement. The vertical displacement, which is interpreted to
be more important, varies from a few metres to some tens of metres and
generally shows a normal movement. We have observed that in the B
zone, the gold is occasionally remobilized into these later fractures.

Discussion

While the mineralization of the Astoria deposit can be broken down
into two distinct ore types, each with its attendant alteration style,
carbonatization for the "Carbonated Ore" and albitization with
sulphides for the "Flow Ore", the alteration patterns for the gold
bearing zones are considerably more complex than that presented in the
previous section. The following paragraphs will help to illustrate
the complexity of the alteration of the rocks hosting the gold mineralization.

While carbonate is the dominant characteristic of the "Carbonated
Ore", carbonatization has also affected the wall rocks of the other
mineralized zones of the Astoria deposit. The pervasive alteration
has been defined as having greater than 5% iron-magnesium carbonate
content in the form of magnesite, siderite or ankerite, whereas this
content usually ranges from 40% to 80% in the "Carbonated Ore" zones.
Lower than a 5% iron-magnesium carbonate content, which is also a
visible criteria, the rocks are not classified as part of the
carbonated alteration zone.

For example, in the A gold bearing zone, which is a "Flow Ore"
type, carbonatization is generally well developed. This zone is also
injected by barren quartz stockworks which are similar to the gold
bearing quartz stockworks observed in the Ae zones, a "Carbonated Ore"
type. The carbonatization and the quartz stockwork injections
gradually disappear at the western end of this zone. The second

example of "Flow Ore", the Aw zone, has similarly been affected by carbonatization. Here, at depth, at the eastern extension of the Aw zone carbonate has been noted in the drill holes. A vertical longitudinal section (Figure 7) shows the limit of this carbonate alteration.

In the Bw and B gold bearing zones, both of which are "Carbonated Ore" types, underground mapping has reported small lenses of about 4 metres in width of "Flow Ore" mineralization (sulphides in albitized zones). The distribution of these lenses however, is quite erratic. We have also noted that the carbonatization of the rocks disappears westward. In the Bw gold zone the carbonatization has been displaced to the southwest both in the horizontal and vertical directions and is now referred to as the Bw_1 zone. At depth the Bw zone becomes a silificied and chloritized shear zone.

To conclude this discussion, it has become apparent that the western limit of the carbonated alteration is an important factor for the distribution of gold within the sulphide zones ("Flow Ore" type). Immediately to the west of this boundary, the higher grade mineraliza-tion of the Aw zone is developed, averaging 7.3 g/mt of gold (0.21 oz/st). The gold content of the A zone ("Flow Ore" type), located within the carbonated alteration influenced area, however, is lower grade, averaging only 4.0 g/mt of gold (0.12 oz/st). Based on the results to date, it seems that this carbonate overprint of the sulphide zone has affected the distribution of the gold, with the better grade lying outside of the affected areas. This alteration boundary may define the eastern limits of the higher grade sulphide mineralization.

PART II: ASTORIA EXPLORATION PROGRAM TO DATE

Previous work

Historically, the property on which the Astoria deposit is located was held, in part, or in its entirety by the Granada Rouyn Mining Co. and the Astoria Rouyn Mines Ltd. After substantial diamond drilling between 1928-1934, a shaft was sunk in 1935. Extensive diamond drilling, surface and underground exploration, sinking of a second shaft and additional drifting was carried out in the following 11 years, with both companies focussing their efforts on the sulphide gold bearing zones (A and Aw). In 1947, the operations at Astoria were suspended, although available records do not indicate the reason for this suspension. It is thought that the relative low grade and small size of the deposit developed above the 120 metre level (400 feet vertical), combined with the wartime labour shortage and availability of funds prohibited further development. An additional factor, no doubt, was the relatively stable price of gold; at that time only $35./oz.

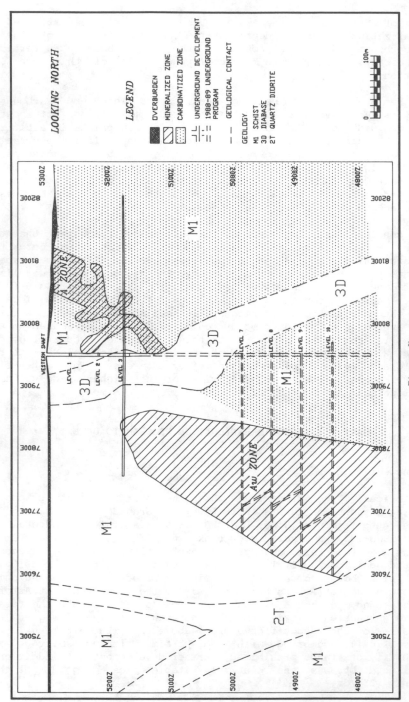

Figure 7

Longitudinal view combining the A & Aw zones

In 1984, Yorbeau Resources Inc. became the operator and all data, from 1926 to the present, was systematically compiled. An initial plan of action based on this information was developed and agreed upon by both management and mine geologists. The initial drilling on the project began in the summer, using a 30 metre grid pattern of diamond drilling from the surface, beginning from the known mineralized areas towards areas of unknown mineral potential.

The results of this drilling program has indicated additional defined gold mineralization, for the B, Bw and Ae zones ("Carbonated Ore" type), above the 150 metre level (500 feet vertical). The results were not as good for the A and Aw zones ("Flow Ore" type), which were previously developed from underground workings in the 1930's and 1940's. Reserves, which initially were established at 540,000 metric tons (600,000 st) have been upgraded to 1.1 million metric tons (1.2 million st) by 1985. This is almost a doubling of reserves within that time period. The grade has also been increased by 50%, from 4.11 g/mt to 6.17 g/mt of gold (0.12 to 0.18 oz/st). This concerted exploration effort led the company to the second step of its development program.

The underground and surface exploration program carried out between January 1986 and June 1987 was focussed on the B, Bw and Ae mineralized zones ("Carbonated Ore" type), and included a bulk sample taken from the first three levels, above the 122 metre level (400 feet vertical). A 10 000 mt (11,030 st) bulk sample was sent to the McBean mill in Dobie, Ontario. This mill was designed for a standard industrial cyanide leaching circuit. This sample was extracted by drifting and raising in the Bw and B zones mainly from low grade mineralization prior to October 1986. It returned 2,91 g/mt of gold (0.085 oz/st) and the gold recovery reached 88,9%. The conclusions drawn from this test is that a gravity circuit would have been necessary to increase the gold recovery; most of the coarse gold particles being trapped in the crushing circuit (Lachance, 1987). It is anticipated that recoveries could reach 95% given the appropriate mill design.

The results from the underground and surface diamond drilling completed during this period, were very encouraging. The discovery of higher grade gold mineralization in the Aw zone, below the 150 metre level (500 feet), heightened our interest for this type of sulphide gold bearing zones ("Flow Ore" type). No reserve was estimated from this zone in the 1985 study. Probable and possible diluted reserves of 280.101 mt grading 7,26 g/mt of gold (308,759 st grading 0.21 oz/st) are now estimated for the Aw zone. Diamond drilling, with a success ratio in excess of 85%, indicates an additional 500 000 mt of potential gold bearing mineralization with this zone still open to the west and at depth. The success ratio of the diamond drilling in the "Carbonated Ore" type of mineralization (B, Bw and Ae zones) is averaging 60%. Note that this "Flow Ore" type of mineralization has accounted for most of the gold production of the Kerr-Addison Mine. If we compare the Astoria deposit to the Kerr-Addison Mine model, the

Aw zone represents the best potential for further increasing of the
reserves.

The diluted reserves of the Astoria deposit (as a result of the
1986-87 exploration program) are now estimated at 1,3 million metric
tons grading 6.14 grams per metric ton of gold (1.4 million st grading
0.18 oz/st) in the proven, probable and possible categories. Our
recent experience from the underground work has suggested that this
is a conservative reserve estimate as we have used the polygon method
instead of the section method previously employed to prepare the 1985
reserve estimate.   Roche Ltée, a mining and geological consulting
firm supervised the 1986-87 exploration program and prepared the
reserve estimate (Lachance, 1987).

Present and future programs

The objective of the underground exploration program currently in
progress, is to assemble the necessary parameters required to make a
production decision.   To realize this objective, the reserves in the
proven and probable categories have to be increased and upgraded.
According to the compilation from previous exploration programs, a
confirmation of the Aw zone reserves could lead to a production
decision.   To this end Belmoral Mines Ltd. has recently committed 10
million dollars for this work and in turn will obtain a 50% interest
in the Astoria property.

This ongoing exploration program includes enlarging the existing
shaft which will increase the hoisting capacity from 750 to 1000 tons
per day and the deepening of this shaft to the 512 metre level (1680
feet). It also includes the exploration of the Aw zone on the 7th,
8th, 9th and 10th levels (respectively 314, 360, 407 and 457 m depth)
by drifting and raising in the mineralization. It is planned to take
a bulk sample in order to test the Aw zone. 6000 metres of underground diamond drilling is planned to outline the gold bearing zones. A
22 month schedule is foreseen for this work which began in March 1988.
The 1988-89 underground program for the Aw gold bearing zone is shown
by dashed lines on Figure 7.

Yorbeau is optimistic that this underground exploration program
will lead to a production decision. In the near future Yorbeau, with
the help of Belmoral, could take the final steps to becoming a
producing gold mining company. A thorough understanding of the gold
mineralizing processes, in combination with a well planned exploration
program will have greatly contributed to making this possible.

REFERENCES

Gagnon, Y., 1986, MAKING NEW MINES OUT OF OLD ONES, Unpublished text of the conference presented on March 10th, 1986 at the Annual Prospector and Developers Association Convention, Royal York Hotel, Toronto, Ontario, Canada, Yorbeau Resources Inc. files, 15 p., 14 slides.

Gauthier, N., Rocheleau, M., St-Julien, P., 1988, THE CADILLAC-LARDER LAKE FAULT ZONE, ADITIDI DELT, CANADA. AN EXAMPLE OF AN ARCHEAN ACCRETIONARY PRISM HOSTING GOLD DEPOSITS, GOLD' 88, Extended Abstracts, Poster program, Vol. 1, Geological society of Australia, Challis House, 10 Martin Place, Sydney, Australia, p. 13-16.

Goodwin, A.M., and Ridler, R.H., 1970, THE ABITIBI OROGENIC BELT, In A.J. Baer (ed.), Geol. Surv. Can., paper 70-40, p. 1-30.

Goulet, N., 1978, STATIGRAPHY AND STRUCTURAL RELATIONSHIPS ACROSS THE CADILLAC-LARDER LAKE FAULT, ROUYN-BEAUCHASTEL AREA, QUEBEC, Ph.D. thesis, Queen's University, Kingston, Ontario, Canada, DP-602, M.E.R.Q., 141 p.

Kelly, D., Mémoire M.Sc.A en préparation, "GITOLOGIE DU GITE ASTORIA, CANTON DE ROUYN, QUEBEC, Ecole Polytechnique de Montréal, Montréal, Québec.

Kishida, A., 1984, HYDROTHERMAL ALTERATION ZONING AND GOLD CONCENTRA-TION AT THE KERR-ADDISON MINE, ONTARIO, CANADA, Unpublished Ph.D. thesis, University of Western Ontario, London, Ontario, 231 p.

Lachance, J.P., 1987, 1986-1987 EXPLORATION PROGRAM AND ORE RESERVE ESTIMATE, ASTORIA PROPERTY, YORBEAU RESOURCES INC, Roche Ltée., Groupe conseil, Internal report, 53 p. 6 appendicies and 22 maps.

Chapter 34

THE PUFFY LAKE GOLD PROJECT

James R. Tompkins, P. Eng.

Vice President Mining
Pioneer Metals Corporation

ABSTRACT

This paper details the successful progression of a promising exploration property into a producing gold mine in only 35 weeks on schedule and on budget. While under construction, an extensive exploration program continued to add to the mine life through the increase in reserves, resulting in a decision to increase the mill capacity from 500 tonnes per day to 1000 tons per day while maintaining the original construction schedule. Details are provided on the unique geology at the Puffy Lake Gold Mine, development from the exploration stage through construction and ultimately to production, with a detailed flow sheet. In addition, of note is the successful change over from contract mining to the hiring of Pioneer personnel who are successfully carrying on both stope mining and development. Operating costs per ounce of gold production are projected at $185 U.S. and alternate methods of mining are currently being examined which could further reduce operating costs.

INTRODUCTION

The Puffy Lake Gold Mine is 100% owned by Pioneer Metals Corporation subject to a 20% net profits interest after exploration and capital payback, to Granges Exploration Ltd. During 1987, a process plant with 1,000 tons per day capacity was constructed and the mine was developed for production.

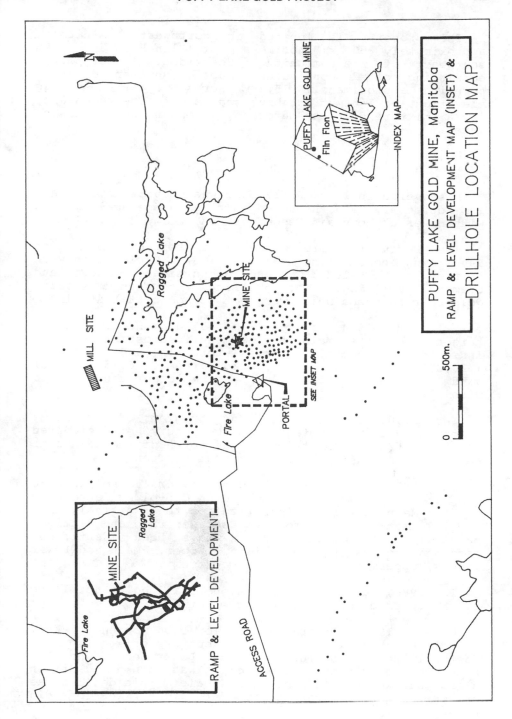

PUFFY LAKE GOLD MINE, Manitoba
RAMP & LEVEL DEVELOPMENT MAP (INSET) &
DRILLHOLE LOCATION MAP

The property is located in northwestern Manitoba, approximately 60 km northeast of Flin Flon and 12 km southeast of the community of Sherridon. The property, consisting of 2 contiguous groups of unpatented claims covering approximately 25,288 acres, is at an average elevation of 240 metres above sea level. The topography is a mixture of rocky outcrops, swamp and small shallow lakes.

HISTORY

Discovered in 1979, Puffy Lake was acquired by Pioneer in September, 1986 through a share exchange amalgamation with Maverick Mountain Resources.

The Puffy Lake area was earlier prospected by Hudson Bay Mining and Smelting Ltd. (HBM & S). Exploration work carried out included ground geophysical surveys and diamond drilling. The drilling intersected several gold-bearing zones which were not economic at that time.

Prior to 1984, a total of 59 diamond drill holes targeted on the Main Zone had outlined estimated drill indicated geological reserves totalling 480,800 tonnes (530,000 tons), with an average grade of 6.15 grams of gold per tonne (0.18 ounces of gold per ton) located within the Bed Claim block.

The 1984 winter drilling program of 29 holes discovered a new, potentially economic zone, designated the "Swamp Zone", on strike to the north of the Main Zone and at the same projected structural horizon.

In 1985, 29 additional diamond drill holes were completed. This confirmed the down dip-plunge extension of the Main Zone gold mineralization to a vertical depth of 183 metres (600 feet). Four separate and distinct parallel gold-bearing lenses have been identified; referred to as the Sherridon, Upper, Main and Lower lenses, uppermost to lowest respectively.

In 1986, a total of 75 diamond drill holes totalling 9,145 metres of drilling were completed.

During 1986, underground exploration was undertaken. A decline was collared and a crosscut was driven from the footwall rocks to intersect the Lower Zone. When the zone was encountered, the decline was turned and driven along strike and slightly down dip. This decline was driven to prove the continuity of the zone. The zone did prove to be continuous and the decline was then spiralled down to intersect the zone at a lower elevation. Development totalled 650 metres of decline.

In early 1987, a raise was driven from the lower portion of the decline up dip to prove continuity in that direction and a test stope was mined. Additional diamond drilling from surface was undertaken, extending the known limits of the orebody along strike to the north and down dip.

To date, 380 diamond drill holes have been drilled on the property. The deepest hole to date intersects the Lower Zone at a vertical depth of 560 metres. The orebody is open down dip beyond this known point.

GEOLOGY

The Puffy Lake deposit is within the Proterozoic Churchill province, specifically in a unique high grade metamorphic terrain, the Kisseynew gneiss belt. This belt which includes metamorphic equivalents of various sediments, volcanics and deformed plutonic rocks is uniquely distinguished by high metamorphic grade and high tectonic strain. The strain is of such character that otherwise coarse-grained gneisses have suffered widespread grain size reduction with development of mylonite and ductile shear zones.

Within the Puffy Lake area all lithologies have developed pronounced planar and linear fabrics with original bedding having been transposed by repeated isoclinal folding and shearing. Competent units such as minor granitic intrusions, mafic sills and flows are flattened, dismembered, enveloped in schists and stretched into a consistently oriented regional lineation.

Foliation parallel gold quartz veins have been emplaced preferentially within the more schistose, pelitic and mafic layered gneisses with ore shoots aligned down dip, following the regional lineation. This regional fabric is more important as it controls the continuity of the mineralization which could be very extensive, particularly down plunge.

Within the property, three main rock assemblages are distinguished. A lower homogeneous light grey to white, lineated granitic gneiss, which forms a distinctive "footwall" for drilling; a central well layered, inhomogeneous, generally mafic schist and gneiss package (Nokimos group) which hosts all the known gold mineralization and an upper unit of more competent felsic and intermediate gneisses. This unit, the Sherridon group, is bounded at the base by a stretched polymictic conglomerate that likely represents a regional unconformity and forms a distinctive marker throughout the area.

Gold mineralization is present in at least four parallel zones of variable extent, averaging about 2 metres thick and occurring tens of metres apart. These have been named from bottom to top as Lower Zone, Main Zone, Upper Zone and Sherridon Zone. The latter being so named as it occurs directly below the Sherridon marker conglomerate. Both the Upper and Main Zones outcrop west of an extensive swamp where they have been trenched and sampled. See Figure 1.

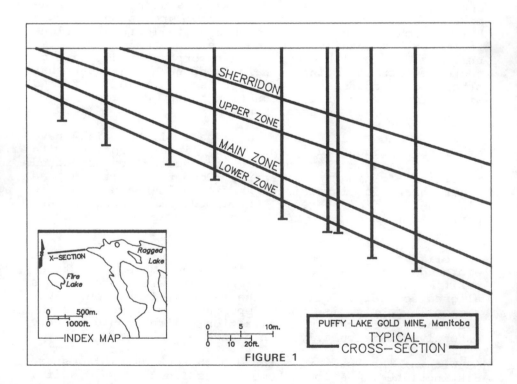

FIGURE 1

The zones are multiple quartz vein systems, consisting of at least two generations of gold bearing quartz, a later stage of more massive coarse grained white quartz superimposed on a multitude of thinner sheeted darker quartz veins. Arsenopyrite tends to accompany the early set, forming disseminated haloes in the walls and vein septa, while coarse grained base metal sulphides and coarse gold appears to favour the latter. Wall rock alteration is not extensive consisting mainly of pale green diopside with rare calcite.

The quartz veins and even pegmatite dikes which locally cut them have been slightly squeezed and boudinaged, providing evidence of post mineral elevated rock temperatures and ductile flow. This probably causes annealing and exsolution of gold from sulphides.

RESERVES

Currently, reserves are calculated at 3.9 million tons grading .23 ounces of gold per ton. At an overall extraction of 75 percent there are sufficient mineable reserves for an eight year mine life.

Ore reserve calculations were based upon a model which treats each zone as a thin seam of 1.83 metres in thickness. Computer generated polygons are used to predict tonnages and grade.

Further diamond drilling down dip and economic evaluation of deeper mine workings will determine ultimate mine life.

CONSTRUCTION

On January 19, 1987, Pioneer announced the decision of its Board of Directors to place the Puffy Lake gold deposit into production. Initial plans were to build a 500 tonne per day mill and produce 40,000 ounces of gold per year. Three diamond drill rigs continued to drill infill and step-out holes and the result was the announcement in May, 1987 that reserves had been expanded and the mill capacity would be increased to 1,000 tons per day. Annual gold production would be 72,000 ounces.

Clearing of a powerline right-of-way to the substation at Sherridon, began in February. In May the mill site was selected and clearing began at once. Throughout June and early July, the site was leveled and backfilled to grade.

All major crushing and grinding equipment had been purchased by July. During the last week of July, the first concrete was poured, and by the end of September, the mill, shop and office complex was closed in and mills, crushers, thickener and tanks were in place.

Piping and electrical systems installations continued through October and most of November. Structural steel for the crusher building and conveyors connecting the crusher to the fine ore bin were constructed despite extreme weather conditions.

By early December, Proton Systems Ltd. was prepared to start commissioning of the mill, and on December 15, 1987 the first dore was poured. From commencement of construction to pouring of dore, a mere 35 weeks had elapsed.

Construction crews took a well deserved Christmas break and returned in January to finish cladding of the fine ore bin and to do interior finishing in the office and mine dry.

MINE DEVELOPMENT

Throughout the surface construction period, the underground workings were being extended to provide stoping areas for full scale production. The decline from surface was slashed to a size which would accommodate 26 tonne capacity trucks and scoops ranging in size from 3 1/2 to 8 cubic yard capacity. Declining proceeded in a northerly direction toward the Main Zone and to allow cross-cutting to the Sherridon and Upper Zones. At regular intervals, levels were driven laterally from the decline to provide access and haulage routes to the stopes. Most of the development was done in the ore zone and by mill start-up approximately 30,000 tons of ore were stockpiled on surface. The first production stope mining started in January 1988 and by May crews were mining in five stopes daily.

Pioneer began staffing of the mine in late April and by July, 1988 had taken over all underground mining and development from the contractor, Canadian Mine Development.

FLOWSHEET

Metallurgical testwork confirmed that a high percentage of the gold was occurring as coarse free gold and that an overall recovery of 93 percent could be achieved through combined gravity and flotation followed by leaching of both concentrates.

Following primary reduction by a Traylor Blake 32 x 36 jaw crusher and a 5 1/2 foot Symons Cone crusher, the product minus 12 mm size is conveyed to a 1,500 tonne capacity fine ore bin.

Three variable speed feeders retrieve ore from the fine ore bin which is conveyed to the feed chute on the rod mill. Rod mill discharge passes over a duplex jig for the removal of coarse gold. Ore in slurry passing over the jig is fed to the cyclones for sizing. Cyclone underflow is fed to the ball mill for further size reduction and cyclone overflow is fed to conditioning and thickener tanks for storage prior to leaching. There are four leach tanks in series. Ore is first conditioned in a preareation tank and then fed to the first leach tank where cyanide is added. The leach process is continuous with total residence time of approximately 24 hours. Pregnant leach solution is recovered by a drum filter. Fresh water sprays wash the filter cake to ensure rinsing of all pregnant solution and cyanide. Pregnant solution is stored in a tank and then fed to a Merrill-Crowe circuit to precipitate gold and silver. Precipitate is trapped in a filter press on a batch basis, and then weighed and analyzed for flux addition then charged to a double reverbatory furnace. See Figure 2.

A typical dore bar contains 65 percent gold, 15 percent silver and minor quantities of base metals.

Barren solution is either returned to be used in the process stream or is treated with hydrogen peroxide to neutralize cyanide and then added to the tailings.

Tailings are pumped to a natural basin where the coarse particles are allowed to settle and decant water is reclaimed for mill process water. Surplus water is allowed to pass out of the impoundment area through a weir where regular sampling ensures that water quality objectives are met.

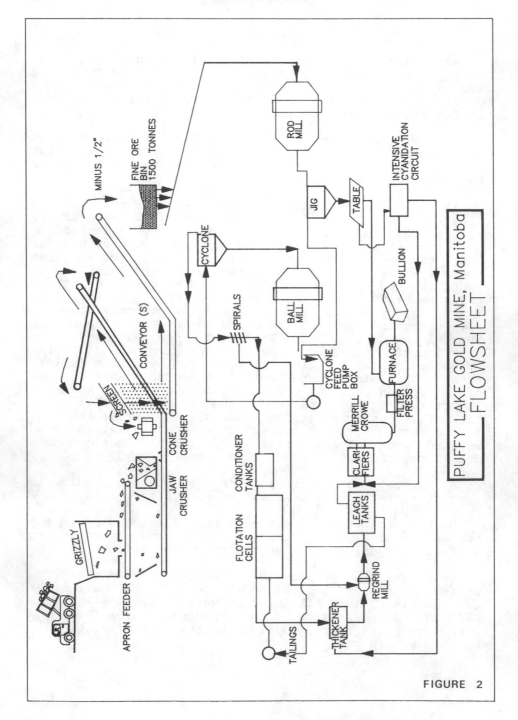

PUFFY LAKE GOLD MINE, Manitoba
— FLOWSHEET

FIGURE 2

MINING

Development and stoping work are done on a seven day per week, two 10 hour shift schedule. The four idle hours per day are utilized by maintenance personnel to service underground equipment.

Development headings are nominally four metres wide and 3.0 metres high with a "shanty" back configuration. Headings are driven within the ore zone, in the footwall, and crosscutting the hangingwall to access overlying zones. All rock is very competent and only minimal ground support is required.

Development equipment includes one two-boom jumbo, a single-boom jumbo, two 5 cubic yard scooptrams and two 24 ton capacity haulage trucks.

Where headings are driven in the ore zone the ore and waste are blasted and mucked separately.

Stope mining is done using jackleg drills and slushers to scrape ore to the haulage drift. Initially a stope cut-out is established to provide muck storage and slusher set-up space at the bottom end of the stope. Stopes are mined up-dip on the 30 degree incline of the orebody and finished stope dimensions are nominally 60 metres long, 20 metres wide and 1.83 metres high. Rib pillars between stopes are 3 metres wide. Back support is accomplished by installing 1.83 metres long rockbolts on a 1.2 x 1.2 metre pattern. See Figure 3.

Each stope is mined by a 2 man crew per shift. The mining cycle consists of drilling, blasting, slushing and rockbolting. Ore is pulled down the 30 degree incline to the muck bay. Target productivity in the stopes is 37.5 tonnes per manshift with bonus incentives offered for additional productivity.

Ore stored in muck bays is next hauled to surface by rubber-tired equipment for delivery to the crushing plant. Haulage equipment consists of three 3.5 cubic yard scooptrams and three 16 tonne capacity haulage trucks.

As the mine workings are extended to greater depth, an inclined skipway hoisting system will be utilized to deliver ore to surface.

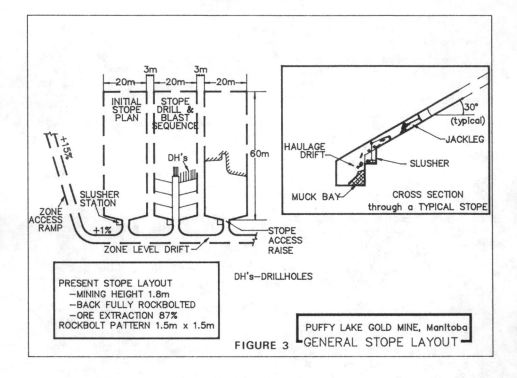

FIGURE 3 GENERAL STOPE LAYOUT

PUFFY LAKE GOLD MINE, Manitoba

PRESENT STOPE LAYOUT
—MINING HEIGHT 1.8m
—BACK FULLY ROCKBOLTED
—ORE EXTRACTION 87%
ROCKBOLT PATTERN 1.5m x 1.5m

DH's—DRILLHOLES

MAINTENANCE

Electricians, millrights and heavy duty mechanics provide the necessary preventive and breakdown maintenance for mine, mill and surface facilities.

A three-bay maintenance shop was incorporated into the mill and office building design where major equipment overhauls are done in addition to fabricating and electrical work. A separate shop was established at the mine portal, a distance of one km from the main shop, to facilitate repairs and preventive maintenance of underground equipment.

WORK FORCE

Puffy Lake Gold Mine currently employees a workforce of 150.

Workers have the opportunity to live in an onsite camp during their scheduled shifts and are provided transportation to nearby communities where they maintain their permanent residence.

The onsite camp has capacity for 150 personnel and rooms will normally be inhabited by shift workers. Shift rotation is one week on, one week off for surface workers and four weeks on, two weeks off for underground workers.

A number of people work a five day week and currently these employees reside in the nearby community of Sherridon with their families.

COSTS

Capital cost of the concentrator plant, office, mine dry, shops, warehouse and lab facilities was $18.1 million. Costs associated with underground development and exploration drilling were approximately $9 million.

Operating costs per ounce of gold production are projected to be $185 (U.S.) for 1989. Pioneer is currently examining alternative mining methods which could further reduce operating costs.

Chapter 35

GOLD DEPOSITION AT GOLD KING, SILVERTON CALDERA, COLORADO

Bernhard C. Koch, Richard W. Hutchinson and Bernhard Free

Ph.D. candidate, Colorado School of Mines
Golden, Colorado

Charles F. Fogarty,Professor, Colorado School of Mines
Golden, Colorado

President, Gold King Mines Corporation
Denver, Colorado

ABSTRACT

The Tertiary Silverton caldera collapsed asymmetrically to the southwest 27.5 m.y. ago. Resurgent volcanic activity is expressed by a quartz-monzonite stock at the southern caldera margin and by the Red Mountain porphyry stock in the northwest of the caldera which were emplaced at 25 m.y. and 11 m.y. b.p., respectively. The apical Eureka graben formed between 27.5-22.5 m.y. ago and transects the Silverton caldera southwestward. In the northeast it terminates at the 22.5 m.y. Lake City caldera margin. This graben as well as caldera faults provide dominant structural control for base and precious metal deposits.

The epithermal gold mineralization of the Gold King mine is proximal to the Red Mountain stock in the north central part of the caldera. Here, west-northwest striking faults related to the collapse of the caldera intersect northeast faults of the Eureka graben. Superimposed upon earlier base metal mineralization at 23 to 22.5 m.y. b.p., the 11 m.y. younger gold metallization was a distinct event at Gold King. Supergene enrichment of the hypogene precious metal assemblage resulted in a metal tenor of up to 3,016 g/t gold and 1,088 g/t silver. The deposit is hosted by quartz-latite tuffs, flows, and flow breccias of the Burns Formation. Veins, up to 5 m wide follow steeply southeast and northwest dipping structures of the Eureka graben. Subsidiary veins up to 2 m wide branch off the main lode and are termed "flat veins" because of their conformable strike but shallow dip of 45° to 50° toward the northwest.

Meteoric water-dominated hydrothermal solutions generated by the Red Mountain stock ascended through intensely fractured, ferruginous

494

Proterozoic formations and overlying Oligocene caldera fill. These solutions were funnelled into the open fracture system causing vein emplacement and argillic, advanced argillic, and phyllic alteration as well as silicification and sericitization of the wall rock. Argillic alteration accompanies elevated gold grades and possibly ore carrying structures at Gold King. Gold was probably leached from Proterozoic basement greenstones and transported as bisulfide complexes. Decay of these complexes and subsequent gold deposition resulted from oxygenation at near neutral pH, triggered by mixing with fresh meteoric water and/or boiling of the hydrothermal solution close to the paleo-surface.

Fluid inclusion data indicate deposition temperatures of 280° $\pm$ 10°C for the vein filling quartz-, electrum (95/5 weight% Au/Ag), hessite-, petzite-, sylvanite-, krennerite-, altaite-, tetrahedrite-, bismuthinite-, $\pm$ pyrite, $\pm$ chalcopyrite, $\pm$ sphalerite, assemblage.

The Paleozoic and Mesozoic rock cover had been mainly eroded when volcanism began, exposing the Proterozoic basement. The gold bearing solutions show predominant meteoric character. It is therefore likely that the underlying basement consisting of Proterozoic olivine and tholeiitic basalt, ferruginous sediments, and iron formation was the source for the gold metallization.

INTRODUCTION

This paper discusses the possible source, transport, and deposition of gold in the northern portion of the Silverton caldera, San Juan Mountains, Colorado. Structural, lithologic, and geochemical factors for the control of ore are considered in an attempt to explain the genesis and distribution of the Gold King lode. Epithermal, bonanza-type precious metal deposits like those of the Gold King mine are coveted because of their economic value, and have been the subject of detailed exploration and research during the past two decades.

Epithermal vein deposits exhibit many common geological features but nevertheless differ individually in their particular geologic setting and characteristics. Generally they are associated with magmatic activity and/or volcanic caldera settings of relatively young geologic age. The Tertiary San Juan volcanic field is located in southwestern Colorado. It hosts 15 calderas of which six are mineralized (Steven and Lipman, 1976). Four of the six mineralized cauldrons constitute the western caldera complex, including the Silverton volcanic center. The remaining two caldera complexes are, Creede to the east and Summitville in the southeast (figure 1), which host rich silver and gold deposits, respectively.

Magmatism in the western complex was accompanied by two major stages of hydrothermal mineralization. An early, base metal rich stage affected the entire western caldera complex, whereas a later gold rich stage was centered more or less on the Silverton caldera only.

As reported by Casadevall et al. (1974), Casadevall and Ohmoto (1977), and Eberl et al. (1987), hydrothermal solutions at

Silverton were predominantly derived from meteoric water and at the
Sunnyside mine became progressively meteoric with time (Casadevall
et al., 1974). The adjacent Sunnyside deposit was initially mined
for base metals only. When mine workings proceded southwest toward
the Gold King mine in the 1970's, increasing gold values were
recovered. The Gold King lode was first mined in 1886. Until 1917
665,500 tons of ore averaging 14.65 g/t gold, 74.32 g/t silver, 0.71
% lead, and 0.52 % copper were produced. No significant ore
production was recorded past 1917. At present the average estimated
ore grade at Gold King is 10.9 g/t gold and 90 g/t silver, for an
ore reserve of 1,459,000 tons (Black, 1987).

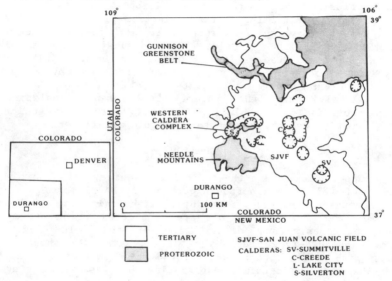

Figure 1: The Tertiary San Juan volcanic field is superimosed
upon the Proterozoic Gunnison greenstone belt. The western caldera
complex hosts the Gold King mine within the Silverton caldera (S)
(modified from Casadevall and Ohmoto, 1977).

REGIONAL GEOLOGY

The San Juan volcanic field is a 2.2 km thick accumulation of
Oligocene pyroclastic rocks and lava flows that rest upon Precambrian
basement and Paleozoic and Mesozoic rock formations (figure 1).
Proterozoic basement crops out to the north of the volcanic field
and constitutes the Gunnison greenstone belt, which contains small
greenstone-hosted gold deposits (Horlacher, 1988). The southern
extension of the belt is covered by volcanic formations of the San
Juan volcanic field. Two windows of these Proterozoic basement rocks
are located just north and east of the Silverton caldera. To the
north quartzites and pelitic schists of the Uncompahgre Formation
are exposed at Ouray. To the east the yet undated granite of
Cataract Gulch mineralogically resembles the 1.35 b.y. Trimble

granite of the Proterozoic Needle Mountains, located just south of the Silverton caldera (Larson and Taylor, 1986). In addition, olivine and tholeiitic basalts of "great but unknown thickness", ferruginous sediments and minor iron formation constitute the Proterozoic Irving Formation in the Needle Mountains (Barker, 1969).

The Proterozoic basement and overlying Paleozoic and Mesozoic rocks were uplifted before explosive volcanism began at the Silverton volcanic center about 28 m.y. ago (Steven and Lipman, 1976). The Silverton caldera collapsed assymmetrically to the southwest at 27.5 m.y.b.p. (Steven and Lipman, 1976). During this event bounding ring faults and west-northwest-trending collapse faults formed. Following this, the Eureka graben subsided from 27.5 to 22.5 m.y. (Steven and Lipman, 1976). This took place in response to doming, centered between the Silverton and the 22.5 m.y. old Lake City caldrons. The graben is an important controlling structure for later mineralization. Its bounding faults and subsidiary fractures became important ore carriers during the following base metal and even younger precious metal mineralization. The base metal stage was preceded and accompanied by a district wide propylitic alteration (Burbank, 1960), whereas the younger precious metal mineralization is linked to local solfataric alteration of argillic and acid-sulfate type.

The following two intrusions are possible heat sources for these mineralizing events. The quartz-monzonite stock at the southern margin of the Silverton caldera is 25 m.y. old (Lipman et al., 1978) and the Red Mountain stock yields a potassium-argon age of 11 m.y. (Gilzean, 1984).

However, location, type of mineralization, and alteration depend upon the geologic setting and the nature of hydrothermal solutions at each individual deposit.

GEOLOGIC SETTING

The Gold King deposit is located in the northern part of the Silverton caldera, it is proximal to the Red Mountain porphyry, the youngest stock of the Silverton volcanic center (Gilzean, 1984). At Gold King, northeast faults of the Eureka graben intersect older, west-northwest striking fractures related to the caldera collapse. The deposit is hosted in rocks of the Burns Formation which is part of the caldera fill and outflow. The caldera fill, a sequence of three volcanic units, becomes less silicic towards the top. The lowermost Eureka member, a lithic rhyolitic ash-flow tuff is covered by relatively homogenous and rigid quartz-latite flows, flow breccias, and tuffs of the Burns Formation, which responded to post-depositional tectonic strain with a more or less consistent fracture system. This resulted in the formation of sufficient open-space sites for later hydrothermal mineralization. Above the Burns Formation a transition zone of quartz-latitic to andesitic tuffites marks the change to the overlying andesitic pyroclastites of the Henson Formation. The chemical, sedimentary, and pyroclastic character of the Henson andesite is manifest by a sequence of laterally and vertically inhomogeneous, interleaved volcanic

breccias, and layers of lapillite, and tuffite. Hence, only an
irregular fracture system developed in these youngest volcanic rocks,
above the Burns Formation.

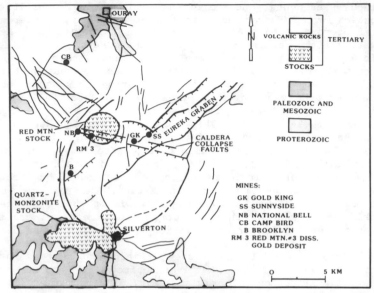

Figure 2:   Location of the Gold King mine in the north central part
of the Silverton caldera, proximal to the Red Mountain porphyry
stock.

THE GOLD KING LODE

The Gold King and Davis lodes consist of two sub-parallel veins
merging downward from the uppermost #1 to the lowermost #7 level
(figure 3). The strong, northeast trending veins are presently known
over a total length of 1.1 km. Their maximum width is 5 m, but
in parts they are represented by fault gauge only. The pinching
and swelling is characteristic of epithermal veins.
      Until 1917, ore was mainly broken from   the   dominant   Gold
King—Davis veins. Subsidiary flat veins branch off the main lode
and yield significant, yet unmined ore reserves. These structures
are a few centimeters to 1 m wide and contain base metal sulfides
as well as precious metal mineralization (figure 3).

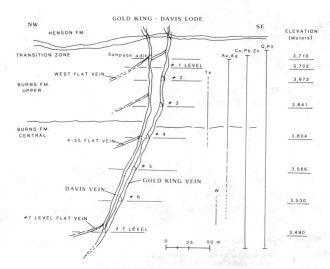

Figure 3:   Schematic cross section of the Gold King–Davis lodes with subsidiary flat veins.   The vertical distribution of precious metals, base metals, and quartz/pyrite gangue is indicated to the right of the vein couple.

The positions of high grade gold shoots are controlled by their specific elevations as well as their locations with respect to the lithologic Burns/Henson boundary (figure 3).   Generally, high grade gold ore is found within the central and upper portion of the Burns formation only.   Although the Henson formation hosts jasperoidal veins and veinlets, it does not yield any significant mineralization above the Gold King lode. The transition zone thus marks the boundary between ore-hosting rock below and barren country rock above (figure 3).   Favorable ground remains north and south, parallel to the Gold King lodes as well as in their northeast extension toward the Sunnyside workings.

ALTERATION

Alteration suites at Gold King are summarized in figure 4.   An earliest, pervasive propylitic alteration of epidote-, chlorite-, calcite, ± pyrite, ± magnetite is recognized throughout the host rock.   According to Burbank (1960) this alteration preceded and accompanied the early district-wide base metal mineralization at 22.5 m.y. (Lipman et al., 1976).

Along faults and surrounding the veins argillic alteration haloes indicate possible gold mineralization.   This alteration was contemporaneous with gold metallization and an argillic (kaolinite) alteration can be distinguished from an acid-sulfate (alunite) type at Gold King (figure 4).   Progressing towards a vein phyllic (quartz-sericite-pyrite) alteration sets in and finally merges with silicification at the vein wall.

Sericite alteration appeared with two different hydrothermal

events. Based upon potassium–argon ages from sericites Eberl et al.,
(1987) report two major hydrothermal events within the Silverton
caldera. The first sericite alteration took place at 21 m.y. (Eberl
et al., 1987) and therefore post–dates the district–wide propylitic
alteration at 22.5 m.y. (Lipman et al. 1976). The second, 13 m.y.
old sericitization (Eberl et al., 1987) pre–dates the argillic
alteration of the Red Mountain district at 11 m.y. (Gilzean, 1984).
Thus, both events of sericite alteration appear to be separate from
the early propylitic and later argillic alteration but probably are
somehow linked to these, respectively. Here, a sufficient explanation
remains to be found.

The latest alteration   at Gold King was supergene in character
and takes the form of opalitic stain upon massive quartz and pyrite
in the lower part of the Gold King and Davis veins. This alteration
oxydized the deposit and caused significant enrichment of the gold
lode.

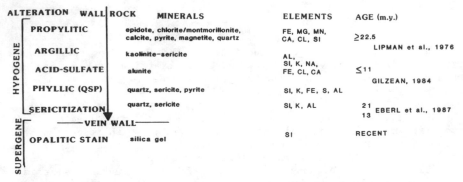

Figure 4: Summary of alteration suites at Gold King.

Elevated gold values appear with argillically altered wall rock
at Gold King, whereas alunite alteration accompanies very low gold
values (figure 5). Comparison of gold assays with alteration suites
of Gold King diamond drill cores GK-1-84 S and GK-12-86 S led to
the recognition of elevated gold values   associated with argillic
alteration.   In contrast, only a very low gold content is recorded
in GK-1-85 U in the presence of acid–sulfate alteration.   The
relatively high sulfur content in all three intercepts can be
explained by sulfides in GK-1-84 S and GK-12-86 S and sulfate in
GK-1-85 U (figure 5).

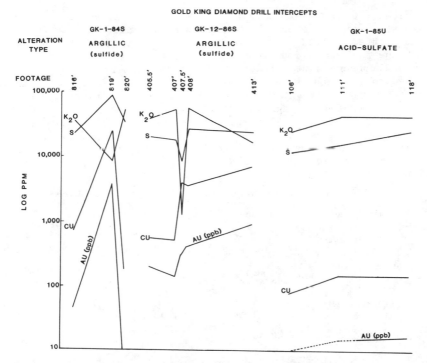

Figure 5: Gold King diamond drill intercepts showing values for gold in ppb, and copper, sulfur and potassium oxide in ppm.

MINERALIZATION

Four phases of hypogene vein filling formed the Gold King deposit (figure 6). Quartz and pyrite first filled large portions of the available open fault-fracture system. Preferentially in the lower part of the deposit massive pyrite is the vein filling gangue mineral together with "bull" quartz.

The second phase is base metal-sulfide bearing. Chalcopyrite, galena, and sphalerite are its main constituents accompanied by pyrite and quartz. This phase is district-wide and probably was linked to the 22.5 m.y. collapse of the Lake City caldera (Slack, 1980).

Superimposed upon phase 2, precious metallization was introduced as a relatively late, single phase. This event is probably linked to the acid-sulfate alteration of the Red Mountain stock at 11 m.y. b.p. (Gilzean, 1984). Still available open spaces in vein centers were filled with free gold and a telluride assemblage of hessite, petzite, sylvanite, krennerite, and altaite. Tetrahedrite-Tennantite and bismuthinite as well as pyrite, some chalcopyrite, and possibly minor sphalerite also accompany this assemblage. Free gold appears as electrum with a relatively high purity, averaging 95% gold and 5% silver at Gold King. Primary fluid inclusions in associated milky quartz indicate temperatures of 270° to 290° for this phase.

The fourth, and last phase of hypogene mineralization at Gold
King precipitated quartz and fluorite only.   This phase probably
represents the cooling period of phase 3.

Following hypogene mineralization, supergene alteration oxydized
and enriched the precious metal assemblage.   Free gold is found along
joints within the 4-55-flat vein.   Oxydational enrichment of the
deposit led to extremely high precious metal grades of up to 3,016
g/t and 1,088 g/t silver.   The supergene alteration also led to the
formation of bornite and covellite.   The opalitic stain in the lower
part of the deposit along joints within the massive quartz/pyrite
"root" of the Gold King-Davis veins is also a result of supergene
alteration.

		MINERALIZATION	ELEMENTS	AGE (m.y.)	ALTERATION
	PHASE 1	quartz, pyrite	SI, FE, S	>22.5	PROPYLITIC PHYLLIC
HYPOGENE	2	chalcopyrite, galena, sphalerite, quartz, pyrite	CU, PB, ZN, S, AG, SI, (low FE)		
	3	electrum, tellurides, tetrahedrite/tennantite, bismuthinite	AU, AG, TE, BI, SB, AS, S, SI, CU	≤11-10	ARGILLIC
	4	fluorite, quartz, huebnerite	F, MN, SI, W	<10	
SUPERGENE	5	electrum, (tellurides), covellite, bornite,	AU, AG, (TE), CU, SI		OPALITIC STAIN

Figure 6:   Summary of five phases of mineralization at Gold King.

GEOCHEMICAL ASPECTS

At Gold King, ore is associated with kaolinite and possibly
sericite alteration. These minerals form from hydrothermal solution
within the pH-ranges of 5.8 to 4 and 5.8 to 6.5 respectively (figure
7).   As pointed out by Romberger (1986), gold is particularly soluble
in hydrothermal solutions of 250° C and pH   6.5 to 8.   Gold was
probably transported in hydrothermal solutions as bisulfide complexes
of $Au(HS)_2^-$ (Romberger, 1986; figure 7). Changes from low log a_{O2}
to higher log a_{O2} cause the decay of the complex by oxygenation.
This can be triggered by mixing with fresh meteoric water and/or
boiling.

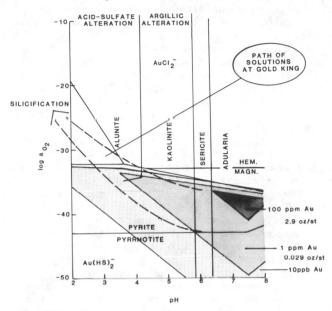

Figure 7: Log a_{O_2}/pH diagram showing gold solubility in hydrothermal solution at 250° C and the above indicated physico-chemical conditions (modified from Romberger, 1986).

The following reaction demonstrates the breakdown of a gold-bisulfide complex by oxygenation:

$$Au(HS)_2^- \;+\; 2\,O_2 \;\Rightarrow\; Au° \;+\; SO_4^{2-} \;+\; 2H^+ \;\Rightarrow\; H_2SO_4$$

$$\text{(aq)} \qquad \text{(aq/g)} \qquad \text{(n)} \qquad \text{(aq)} \qquad \text{(aq)} \qquad \text{(aq)}$$

In this reaction gold is reduced from Au^+ to $Au°$. Once the complex is broken up, gold becomes insoluble and precipitates from hydrothermal solution. Continued release of hydrogen cations and sulfate anions results in the generation of the strong acid H_2SO_4 and decreased pH with a gold solubility possibly four orders of magnitude lower than that of the initial solution. Hydrogen ions released during reactions with wall rock will lower the pH, also causing or re-enforcing gold precipitiation. Decreased pH causes significant precipitation of silica from hydrothermal solution as well (Krauskopf 1979; Holland and Malinin, 1979). Thus, silicification and the growth of quartz veins with euhedral quartz crystals follows alunite alteration. The alteration sequence of kaolinite accompanied by gold precipitation, followed by alunite alteration and finally silicification at Gold King is outlined in figure 7.

DISCUSSION

Magmatic doming caused uplift and subsequent erosion of Paleozoic and Mesozoic formations at the Silverton volcanic center. When volcanism began, probably a minor portion, if any of the Paleozoic and Mesozoic rock cover remained at the apex of the dome. North of Ouray some Paleozoic and Mesozoic rocks are preserved and host minor ore deposits of late Mesozoic age (Luedke and Burbank, 1981). But only one relatively small block of Paleozoic Leadville-Ouray limestone and one block of Hermosa/Molas formation are present in the relatively undisturbed area of Ironton Park, northwest of the caldera (Luedke and Burbank, 1964).

Because of its lithologic composition of thick tholeiitic and olivine basalts, ferruginous sediments, and iron-formation the Proterozoic basement is a favorable gold source. Tholeiites and iron-rich chemical sediments of Archean age have been shown to be gold enriched (Meyer and Saager, 1985). In addition, the great gold potential of Archean greenstone belts was pointed out by Hutchinson (1984). Therefore, the similar Proterozoic rocks of the Irving Formation to the south of the caldera and the Gunnison greenstone belt to the north represent likely sources for the gold, which probably was remobilized by the later Tertiary magmatism and related hydrothermal activity. Whatever the gold source, physico-chemical conditions combined to form a high grade, epithermal vein deposit. Hosted by relatively brittle Tertiary caldera fill above intensely fractured basement rocks the Gold King lode is close to the young Red Mountain porphyry stock. This stock probably supplied the heat source for meteoric water-dominated hydrothermal solutions. These solutions probably leached gold from Proterozic basement rock and carried it as bisulfide complexes.

A combination of Tilling, Gottfried, and Roe's (1973) view, emphasizing geologic setting and duration of hydrothermal systems over source rock enrichment to form an ore body, with Hutchinson's (1987) interpretation of hydrothermal leaching processes within basement rocks and with subsequent gold precipitation in overlying, less consolidated caldera fill provides an acceptable explanation for the geologic setting and origin of the Gold King deposit. During repeated magmatic and tectonic activity the caldera roof was intensely fractured. Thus a relatively large surface of Proterozoic basement rock was exposed to hydrothermal leaching and therefore a relatively short time period may have been sufficient for the formation of the Gold King veins. Chemically, acid sulfate (alunite) alteration follows the breakdown of bisulfide complexes and argillic (kaolinite) alteration. At decreasing pH gold solubility rapidly decreases as well. This also is true for silica which results in silica-flooding. Thus, the following alteration sequence is suggested for deposits like the Gold King. Argillic alteration with gold precipitation was followed by acid sulfate alteration, accompanied by only minor gold and this was followed inturn by silicification. A similar interpretation is possible for the alteration sequence at Summitville, described by Stoffregen (1987) and also for the Gold Quarry deposit, Nevada (Rota and Ekburg, 1988).

Although the latter is an epithermal disseminated gold deposit, a likewise alteration sequence may exist there. Thus, the sequence of alteration assemblages appears to be the same, although their morphology and distribution may be somewhat different.

The same principle of hydrothermal leaching from basement rocks probably also applies to the rich silver deposit of the Bulldog and OH-Amethyst vein system at Creede, eastern San Juan volcanic center (Bethke and Rye, 1979) and possibly to the Summitville gold mineralization in the southeast of the volcanic field. At Summitville, just as at Gold King, gold ore is hosted by quartz-latite flows and was accompanied by argillic and acid-sulfate alteration (Stoffregen, 1987). Because quartz-latite flows are relatively homogeneous and competent promoting the development of regular and continuous fault-fracture systems, these rocks may especially favor epithermal vein deposits in the San Juan volcanic field.

CONCLUSIONS

These results suggest that an earlier genetic model for the Gold King lode (Koch et al., 1987) should be modified as follows: The location and character of the rich Gold King lode is controlled by structural and lithologic properties of the host rock and by physico-chemical characteristics and changes of the metal-carrying hydrothermal solutions. The nearby Red Mountain porphyry stock served as the heat source for meteoric water-dominated solutions. These solutions probably leached gold from Proterozoic basement rock and carried it as bisulfide complexes. The flow path of hydrothermal solutions was outward from the Red Mountain stock along west-northwest fractures of the Silverton caldera. At the intersection of these fractures with faults of the Eureka graben fresh meteoric water mixed with these hydrothermal solutions which may also here have begun to boil due to upward decrease in confining pressure. This caused oxygenation and the breakdown of soluble gold-bisulfide complexes, re-precipitated gold and contemporaneous argillic alteration followed by acid-sulfate alteration and silification of the wall rock. Kaolinite and alunite represent the types of argillic and acid-sulfate alteration at Gold King, respectively. The argillic type accompanied the main (third) phase of gold-introducing mineralization. The acid-sulfate type followed this phase and was generated by strongly acidified solutions that resulted from oxygenation and possibly boiling that caused prior gold precipitation. Consequently the best gold values accompany argillic alteration, whereas veins in rocks with acid-sulfate alteration carry only low gold values. Alteration types are therefore useful exploration guides.

However precious metal deposits in the San Juan volcanic field may be interpreted, proximity to Precambrian basement rocks and the character of generative hydrothermal solutions seem to be key features. The alteration sequence recognized at Gold King probably also applies to other epithermal, vein type and/or disseminated gold deposits. These observations and suggestions may add to a better

understanding of future precious metal exploration at Gold King and
in the San Juan Mountains.

ACKNOWLEDGMENTS

Stimulating discussion and assistance with the determination of
telluride minerals was provided by Bruce Geller, University of
Colorado, Boulder, Colorado.   Richard Sanford USGS, Federal Center,
Denver, Colorado is acknowledged for discussion and suggestions for
this paper. Doug Taylor, geologist of Gold King Consolidated assisted
with helpful discussion and the selection of drill core.   Mark
Coolbaugh, Galactic Resources, receives thankful acknowledgment for
discussions and a very instructive tour of the Summitville deposit.
We thank Samuel Romberger for editorial work on parts of the
manuscript and we are very thankful to Georgia Parker for her
diligent typing of the manuscript.   Finally, we wish to thank the
organizing committee of this convention for its invitation to
participate.

REFERENCES

Barker, F., 1969, Precambrain geology of the Needle Mountains,
    Southwestern Colorado, United States Geological Survey,
    Professional Paper 644-A.

Bethke, P.M., and Rye, R. O., 1979, Environment of ore deposition
    in the Creede mining district, San Juan Mountains, Colorado; Part
    IV. Source of fluids from oxygen, hydrogen, and carbon isotope
    studies. Economic Geology, Vol. 74, pp. 1832-1851.

Black, E. D., 1987, The Gold King property San Juan County,
    Silverton, Colorado. Evaluation Report No. 2, MPH Consulting
    Inc., Denver, 65 p.

Burbank, W.S., United States, 1960, Pre-ore propylitization,
    Silverton   caldera, Colorado, Geological Survey Research 1960,
    B12-B13.

Burbank, W.S. and Luedke, R. G., 1964, Geology of the Ironton
    Quadrangle, Colorado, United States Geological   Survey,
    Map #GQ-291.

Burbank, W.S. and Luedke, R.G., 1969, Geology and ore deposits of
    the Eureka and adjoining districts, San Juan Mountains,
    Colorado, United States Geological Survey, Prof. Paper 535.

Casadevall, T., Ohmoto, H., and Rye, R.O., 1974, Sunnyside Mine,
    San Juan County, Colorado: Results of mineralogic, fluid
    inclusion, and stable isotope studies (abstr.); Geological
    Society of America, Annual Meeting., Abstr. Prog., No. 6, p.
    684.

Casadevall, T. and Ohmoto, H., 1977, Sunnyside Mine, Eureka Mining
    District, San Juan County Colorado: Geochemistry of gold and
    base metal ore deposition in a volcanic environment. Economic
    Geology, Vol. 72, pp. 1285-1320.

Eberl, D.D., Srodon, J., Lee, M., Nadeau, P.H., and Northrop,
    R. H., 1987, Sericite from the Silverton caldera, Colorado:
    Correlation among structure, composition, origin, and particle
    thickness. American Mineralogist, Vol. 72, pp. 914-934.

Gilzean, M.N., 1984, The nature of the deep hydrothermal system,
    Red Mountain district, Silverton, Colorado. Unpublished M.Sc.
    Thesis, University of California, Berkeley, 105 p.

Holland, H.D. and Malinin, S. D., 1979, The solubility and occurrence
    of non-ore minerals, in: Barnes, H. L., (editor), Geochemistry
    of Hydrothermal ore deposits, Second Edition, Chapter 9, John
    Wiley and Sons, 798 p.

Horlacher, C.F., 1988, Precambrian geology and gold mineralization
    in the vicinity of Ohio City, Gunnison County, Colorado.
    Unpublished M.Sc. Thesis, Colorado School of Mines, Golden,
    Colorado, 223 p.

Hutchinson, R. W., 1984, Archean Metallogeny: A synthesis and
    review. Journal of Geodynamics 1, pp. 339-358.

Hutchinson, R. W. , 1987, Metallogeny of Precambrian gold deposits:
    Space and time relationships. Economoic Geology, Vol. 82, pp.
    1993-2007.

Koch, B.C., Hutchinson, R. W., Free, B., 1987, The Gold King precious
    metal deposit, Silverton caldera, San Juan volcanic field,
    Southwest Colorado (abstr.), 40th Annual Meeting, Rocky Mountain
    Section Geological Society of America, Abstracts with Programs,
    Vol. 19, No. 5, p. 287.

Krauskopf, K.B., 1979, Introduction to geochemistry, Second Edition,
    Chapter 6-7, McGraw-Hill Book Company, 617 p.

Larson, P.S. and Taylor, H.P. Jr., 1986, An oygen-isotope study of
    water-rock interaction in the granite of Cataract Gulch,
    western San Juan Mountains, Colorado. Geological Society of
    America, Vol. 97, pp. 505-515.

Lipman, P.W., Fisher, F.S. Mehnert, H.H., Naeser, C.W., Luedke, R.
    G., and Steven, T.A., 1976, Multiple ages of Mid-Teritiary
    mineralization and alteration in the western San Juan Mountains,
    Colorado. 71, no. 3, p. 571-588.

Luedke, R. G. and Burbank, W.S., 1981, Geologic map of the Uncompahgre (Ouray) mining district, Southwestern Colorado. United States Geological Survey, Miscellaneous Investigation Series, Map #I-1247.

Meyer, N. and Saager, R., 1985, The gold content of some Archean rocks and their possible relationship to epigenetic gold-quartz vein deposits. Mineralium Deposita, 20, pp. 284-289.

Romberger, S.B., 1986, The solution chemistry of gold applied to the origin of hydrothermal deposits. Canadian Institute of Mining Special Volume 39, pp. 168-186.

Rota, J.C. and Ekburg, C.E. 1988, History and geology outlined for Newmont's Gold Quarry deposit in Nevada. Mining Engineering, Vol. No. 4, pp. 239-245.

Slack, J.F. 1980, Multistage vein ores of the Lake City district, western San Juan Mountains, Colorado. Economic Geology, Vol. 75, No. 7, pp. 963-991.

Steven, T.A. and Lipman, P., 1976, Calderas of the San Juan volcanic field, southwestern Colorado. United States Geological Survey, Profesional Paper 958, 35 p.

Stoffregen, R., 1987, Genesis of acid-sulfate alteration and Au-Cu-Ag mineralization at Summitville, Colorado. Economic Geology, Vol. 82, pp. 1575-1591.

Tilling, R.L., Gottfried, D., and Rowe, J., 1973, Gold abundance in igneous rocks: Bearing on gold mineralization. Economic Geology, Vol. 68, pp. 168-186.

Chapter 36

THE DISCOVERY OF THE WILD DOG VEIN GOLD DEPOSIT,
PAPUA NEW GUINEA - A CASE STUDY

I. D. Lindley

Principal, Kirakar Lindley Tamu & Associates
Rabaul, Papua New Guinea

ABSTRACT

The Wild Dog vein gold deposit in East New Britain Province,
Papua New Guinea, forms part of a recently discovered 26 km zone
of auriferous quartz vein mineralisation. The deposit, which was
discovered in July 1983, has no silt anomaly, and a panned
concentrate anomaly and float train restricted to within 500 m
and 800 m respectively, of the discovery outcrop. Discovery of
the Wild Dog deposit was a result of the methodical followup of
regional panning which, as followup progressed, was replaced in
importance by float sampling. Of the sampling methods used
during the exploration leading to discovery, panning and float
followup provided immediate indications of the presence of
mineralisation. This high degree of spontaneity of followup is
not provided by geochemical techniques, and has obvious
advantages in high cost exploration for precious metal deposits
in remote locations.

LOCATION AND PHYSIOGRAPHY

The Wild Dog gold deposit is located in the Baining Mountains, 50
km SSW of Rabaul, East New Britain Province, Papua New Guinea.
The deposit forms part of a larger 26 km zone of recently
discovered auriferous quartz vein mineralisation in the
uninhabited Nengmutka region (Figs. 1 and 2). Although it was
well known that colours of gold were to be found in many streams
draining the Baining Mountains, there are no surface indications
or records of any prospecting activity in the Nengmutka region
(Lindley, 1987).

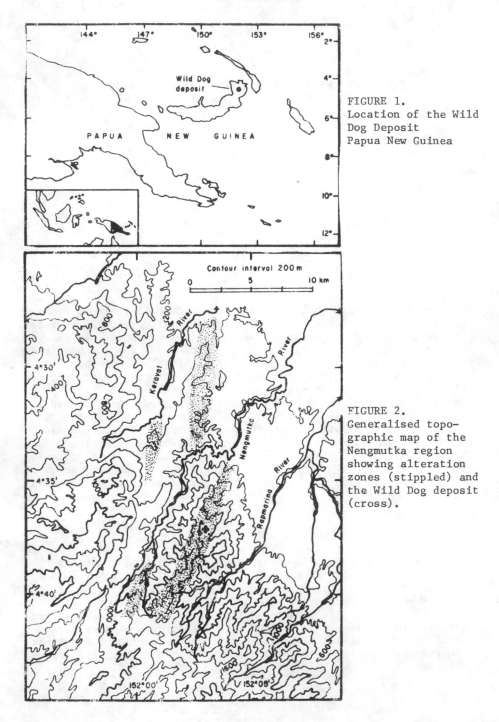

FIGURE 1.
Location of the Wild
Dog Deposit
Papua New Guinea

FIGURE 2.
Generalised topo-
graphic map of the
Nengmutka region
showing alteration
zones (stippled) and
the Wild Dog deposit
(cross).

The prevailing climate in the prospect vicinity is characterised by distinctive wet and dry seasons. Annual rainfall is in the order of 3000 to 3500 mm, with most of this falling between December to March. During this period, monthly rainfall commonly exceeds 450 mm. The dry season occurs from June to September and monthly falls of less than 100 mm are typical. Falls are generally very localised and characterised by extreme precipitation over a short time period.

The deposit, at 1000 m elevation, is situated in steep mountainous terrain varying between 200 and 1400 m elevation, with slopes of 20 - 45 degrees. Recent uplift has produced an incised drainage pattern. Because of the steep stream gradients and localised intense precipitation, rapid runoff occurs over short time periods. Streams experience high flow regime conditions very briefly before returning to normal flow conditions.

Good bedrock exposures throughout the region are a result of rapid erosion associated with the incised drainage network. The depth of weathering of bedrock is correspondingly relatively shallow with depths of 30 m typically encountered.

GEOLOGICAL SETTING

Local Geology

The Wild Dog deposit and associated zones of quartz veining are spatially related to Mio-Pliocene rhyodacitic volcanic centres, referred to as the Nengmuta and Keravat Calderas (Lindley, 1988). The mineralisation post-dates the caldera-forming events, and appears to have been controlled by structural elements of the caldera. The eruptive products of the caldera volcanism, referred to as the Nengmutka Volcanics, are intruded by a mid-Miocene tonalite, the Brabam Diorite. The tonalite is considered to be of a similar age to the volcanism, and may also be related to the auriferous quartz veining of the Nengmutka region.

Both the volcanic centres and the extensive areas of quartz veining are located within a major NNW trending extensional zone referred to as the Baining Mountain Horst and Graben Zone. The zone appears to have been active since the early Miocene.

Vein Structure

The deposit consists of three major veins and several minor
veins. The three major veins are referred to as the Eastern,
Central and Western veins. The Eastern vein has a width of 20 m
and a known strike of 350 m, the Central vein has a width of 30 m
and a known strike of 1350 m and the Western vein has a width of
20 m and a known strike of 550 m. Together they constitute a
vein system approximately 230 m wide which has a general strike
of 025 degrees and a westerly dip varying between 70 degrees and
80 degrees.

Two major zones of intense fracturing, cross-cutting the vein
system with an ESE strike, have allowed eastward flowing streams
to breach the vein system. They are referred to as the Wild Dog
Creek and Wet Creek fracture zones.

Alteration

The Wild Dog deposit is typical of other veins in the Nengmutka
region, in that it has three well developed wall rock alteration
types. An innermost envelope of replacement silification
surrounds the veins. This passes outward to an argillic wallrock
alteration in which plagioclase is completely replaced by
rectangular and irregular shaped masses of illite set in a ground
mass of fine-grained silica. An outer propylitic alteration is
characterised by the presence of epidote, pyrite and calcite.
Propylitisation of this nature is of regional extent.

Maximum development of the silification and argillic envelopes
occurs in the Wild Dog deposit between 920 to 940 m elevation.
The veins at this level pass vertically into replacement silica
and clay altered rocks, which form a bulbous shaped envelope with
a maximum lateral development of 100 m. The ground surface along
the deposit's strike length is generally above 920 m, and thus it
is only in the incised valleys of Wild Dog and Wet Creeks that
outcropping quartz veins are observed.

Mineralisation

Mineralisation occurs in a dark grey sulphidic silica and
includes chalcopyrite, chalcocite, bornite and telluride minerals
such as native tellurium, rickardite, hessite, calaverite, and
several gold-silver and bismuth tellurides. Gold is generally
restricted to calverite, petzite and/or sylvanite, however, rare
micron-sized free gold has been observed.

This episode of veining has been concentrated in vertical structures which cross-cut the westward dipping vein structures (Lindley, in press). These mineralised structures are referred to as ore shoots. Locally these cross-cutting vertical structures extend into the hanging wall and are referred to as hanging wall splits. These hanging wall splits can carry very high (bonanza) gold grades.

STREAM SAMPLING AND DISCOVERY

The Wild Dog deposit was discovered in early July 1983 (Lindley, 1987), fourteen months after the collection of the first reconnaissance samples from the 90 km^2 area between the headwaters of the Nengmuta and Rapmarina Rivers (Fig. 3). Regional sampling during this period indicated that auriferous quartz float was shedding from the main divide separating the two rivers. The followup work led to the discovery of two large isolated silica outcrops in Wild Dog and Wet Creeks, within the Rapmarina River catchment. The position of the outcrops indicated a possible NNE trending vein (now known as the Central vein) with a strike length of 750 m.

Sampling Methods and Analysis

During the definition of the Nengmutka mineralisation both stream sediment sampling and panning were completed. A sample density of one sample per 10-18 km^2 was sufficient to indicate that followup sampling was necessary.

Stream sediment sampling involved the collection of at least 2 kg of active sediment from a standardised location in the channel. For example, an effort was made to always collect sediment from the head of gravel bars where it was considered that maximum concentration of particle gold would occur. Large samples were obtained to permit the collection of at least 200 gm of -80# silt following sample drying and sieving. The entire 200 gm of material was analysed (in two 75 gm portions and one 50 gm portion) in an effort to overcome the particle sparsity effect. Analysis was by fire assay preparation and atomic absorbtion spectrophotometry, allowing a detection limit of 0.005 ppm gold. The three separate assay determinations were combined to provide a weighted average gold result for the sample.

Panned concentrates were submitted for analysis for both gold, base and other metals. In some cases the total sample was prepared for analysis or alternatively the concentrate was split into magnetic and non-magnetic fractions. Weighing of the total sample and its fractions was completed prior to grinding and

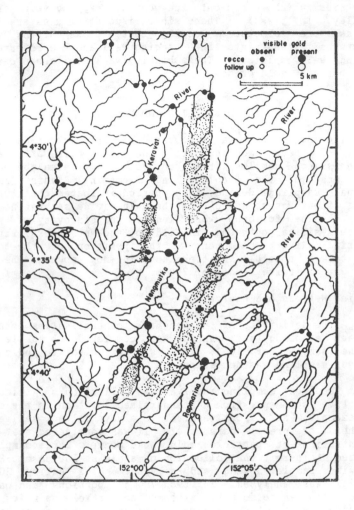

FIGURE 3. Panned concentrate distribution, showing
samples with colours of gold, zones of
alteration veining (stippled) and the
Wild Dog deposit (cross).

pulverising.   Subsampling of the concentrate was necessary to
obtain material for non-gold geochemistry.   Analysis was by
atomic absorbtion spectrophotometry, with a detection limit of
0.008 ppm gold.

Both   stream   sediments   and panned concentrates were analysed for
gold, silver, copper, lead, zinc, manganese, molybdenum and
arsenic.

Regional Reconnaissance

The   earliest   regional   indications of the gold mineralisation in
the Nengmutka region were obtained from the Nengmutka River and
its tributaries during May and August 1982.   During the May 1982
panning of the Sanakmat River (Fig. 3, 152 02'E, 4 35'S), one
angular colour of gold 0.1 mm in size was recovered from a
bedrock trapsite.   The concentrate assayed 1.08 ppm, and stream
sediment   0.04 ppm gold (Fig. 4).   Altered float and chalcedony
float were noted and sampled.   No significant gold (0.1 ppm)
results, however, were obtained.

Sampling resumed in the region in August 1982 at a site in the
Langmute River, another tributary of the Nengmutka River.   First
observations indicated the relative abundance of quartz float in
the stream (Fig. 3, 152 00'E, 4 39'S).   After the negative
results for 5 dishes of wash, a single small colour of gold was
observed   in wash obtained from a finely fractured bedrock
trapsite.   Less than a cupful of fine sand and mud was collected
from   the trap to yield the gold.   Analysis of both the
concentrate and stream sediment returned gold results below the
limits of detection.   Further field work was completed to
followup this result, prior to the receipt of assay results.   The
upper Nengmutka River may have been excluded from any further
followup if the field programme had been dictated by geochemical
results.   Why was the gold not detected in the concentrate?   The
gold may have been removed either by smearing during grinding and
pulverising of the sample, or during subsampling to obtain
sufficient material for base metals analysis. The results of
this sample serve as a striking illustration of the importance of
persistence in panning, based on the observation of favourable
stream float and the fallibility of geochemistry.

Immediate reconnaissance was completed upstream of this sample at
critical stream junctions and in the Nengmutka River upstream of
its junction with the Langmute. This reconnaissance indicated
that both the gold and float train persisted in the main Langmute
River, with negative results from its two major tributaries (Fig.
3).   The non-magnetic fraction of the panned concentrate assayed
3.43 ppm gold, with 0.008 ppm gold in the magnetic fraction. A

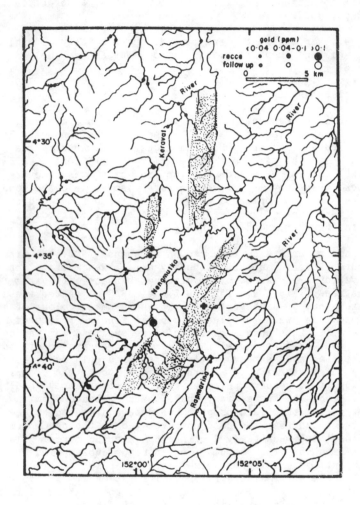

FIGURE 4. Stream sediment distribution, showing
samples equal to or exceeding 0.04 ppm
gold threshold, zones of alteration and
veining (stippled) and the Wild Dog
deposit (cross).

stream sediment from this point in the Langmute, where the channel is 3 m wide, returned a weighted average of 0.072 ppm gold.   This is anomalous when compared with the regional threshold of 0.04 ppm gold (Fig. 4).

From the Nengmutka River, upstream of its junction with the Langmute (Fig. 3), 152 01'E, 4 38'S), two colours of gold were panned from a bedrock trapsite on the outside of a right-angled bend.   The non-magnetic fraction of the concentrate assayed 2.24 ppm gold, and the magnetic fraction 0.008 ppm gold. Pannable gold was also obtained from moss, covering rocks on the outside of the bend.   The stream at this point varied between 10 - 15 m width.   Duplicate stream sediment samples returned erratic values of 0.364 ppm and 0.015 ppm gold when compared with the threshold of 0.04 ppm gold. Approximately 1% of the stream float consisted of vein quartz, with minor argillically altered float. Five specimens submitted for analysis returned no gold values in excess of 0.1 ppm.

The Nengmutka river upstream of this point was selected for followup ahead of the Langmute or Sanakmat Rivers, based on the presence of alluvial gold and the relative abundance of vein quartz float with respect to the stream's width.

Helicopter assisted reconnaissance allowed the completion of sampling in the Rapmarina, Nengmutka and Keravat Rivers during August 1982 and January 1983.   During this work pannable gold was obtained from the Rapmarina and Keravat Rivers (Fig. 3). This sampling defined an area of interest between the Rapmarina and Nengmutka rivers, shedding pannable gold and vein quartz float.

Additional reconnaissance completed as recently as 1986 in the Keravat River area resulted in the panning of 6 colours of gold from the mouth of the Mediva River (Fig. 3), 152 04'E, 4 28'S). The material was obtained from a fractured bedrock trapsite and the concentrate assayed 60.3 ppm gold.

Followup Sampling

Followup sampling commenced in September 1982 in the Nengmutka River upstream of its confluence with the Langmute River (Fig. 3, 152 01'E, 4 38'S).   This sampling in the Nengmutka River involved prospecting in all major tributaries in an attempt to determine the source of the alluvial gold. Both visible gold, stream sediment anomalies and vein float (Figs. 3 and 4) were obtained from almost all tributaries of the upper nengmutka River.   Panned concentrate and stream sediment results from the five major tributaries of the upper Nengmutka River are summarised in Table 1.

On the basis of the abundance of vein float, followup proceeded upstream in tributary C (Fig. 3, 150°00'E, 4°40'S). Emphasis during this work was to collect as many altered (argillic and vein) float specimens as possible for analysis. A total of 25 samples of dominantly vein material were submitted for assay. Results indicated that most vein samples were gold-bearing with four containing greater than 1 g/t gold (3.87, 1.93, 9.21 and 2.21 g/t). These samples also contained significant silver with concentrations up to 136 g/t.

During the period January to August 1983 this detailed followup was extended to cover the entire area of interest between the Nengmutka and Rapmarina Rivers. Panning as a prospecting technique, was replaced by float train followup. Almost every stream system was traversed in its entirety, as vein and argillic altered float was traced to its source. Float and outcrop material was methodically collected and submitted for analysis. This systematic work resulted in the location of many auriferous veins of 0.5 to 1.0 m width.

Assessment of gold results indicated that these veins were restricted to a northerly-trending zone of 1 - 1.5 km width, approximating the divide between the Nengmutka and Rapmarina Rivers (Fig. 2). Field work was subsequently extended northward towards the Nengmutka River and eventually, during limited field work completed in December 1983 and 1986, north of the Nengmutka River. Regional sampling in the area immediately south of the Nengmutka River indicated the presence of vein float, with an absence of pannable gold and no stream sediment anomaly. It was during this programme that the Wild Dog vein deposit was discovered 9 km upstream of a panned concentrate sample in Mangoman Creek, a tributary of the Rapmarina River.

The reconnaissance panned sample in Mangoman Creek (Fig. 3, 152° 05'E, 4°35'S) contained no visible gold but when assayed returned 0.243 ppm gold. This was well above the threshold of 0.1 ppm for panned concentrates and was therefore considered anomalous. The stream sediment assayed 0.018 ppm gold (duplicate 0.008 ppm gold). Two percent of the stream float comprised vein material, ranging from 1 to 15 cm diameter. A followup sample 1.1 km downstream of the discovery outcrop of the Wild Dog vein also contained no visible gold in the panned sample (assay 0.044 ppm), while the corresponding stream sediment sample assayed 0.026 ppm gold. Interestingly, vein float had decreased both in percentage (1/2%) and size (0.5 to 2 cm diameter).

In traversing the 800 m interval of Wild Dog Creek towards the discovery outcrop, the proportion of vein float rapidly increased to 100%. Nowhere along this interval was it possible to pan visible gold. A panned concentrate 230 m downstream from the discovery outcrop assayed 1.33 ppm gold, and the corresponding stream sediment 0.096 ppm gold. A panned concentrate collected 450 m downstream assayed 1.06 ppm gold, and the stream sediment 0.014 ppm gold. Five specimens of vein float were collected during float train followup and only two samples had greater than 0.1 g/t gold (0.19, 1.73 g/t gold).

At the discovery outcrop, three linear chip samples were collected across the 30 m wide interval of vein. These samples returned 0.058, 0.720 and 0.03 g/t gold respectively. It wasn't until the completion of a hand dug trench across the vein, permitting the collection of a representative channel sample, that an estimate of the vein's grade was obtained. Trenching indicated an average grade of 2.62 g/t gold over a linear distance of 40 m.

IMPLICATIONS OF SAMPLING RESULTS

Although the exploration methods used and the results obtained during the discovery of the Wild Dog deposit are not necessarily typical of all gold deposits, they illustrate several important implications for prospecting of gold deposits. These include implications for followup procedure, sampling methods and sample analysis.

Panning for the presence or absence of gold was found to be more useful as a regional reconnaissance method. As followup progressed, float sampling assumed greater importance, leading directly to the location of outcropping veins.

Of the sampling methods available, panning and float followup were the most useful, in that they provided immediate indications of the presence of mineralisation. The use of panning in the Nengmutka region has demonstrated the qualitative nature of the technique. A single colour of gold was as encouraging as many colours. the presence of alluvial gold with altered stream float directed much of the followup leading to the discovery of the Wild Dog deposit. Sediment sampling, apart from lacking the spontaneity of panning and float analysis, was of limited usefulness because of the erratic nature of results. This was well demonstrated by duplicate sampling in the upper Nengmutka River followup (Table 1).

TABLE 1:   FOLLOWUP RESULTS FROM THE UPPER NENGMUTKA RIVER

Tributary	Panned Concentrate (Assay Gold, ppm)	Stream Sediment Gold Assay, ppm (duplicate, ppm)
A	2 colours (2.90)	0.01 (0.021)
B	No colours (0.76)	0.045 (0.020)
C	1 colour (5.41)	0.039 (0.053, 0.036)
D	No colours (0.033)	0.024 (0.012, 0.091)
E	No colours (0.027)	0.013 (0.048, 0.278)

The laboratory results obtained from samples collected during the Nengmutka exploration also have implications for sample preparation and analysis. The analysis of samples of gold has always presented problems because of gold's particulate nature, malleability and low concentrations. All panned concentrates collected during the Nengmutka exploration were analyzed for base metals. This meant subsampling of the concentrate was necessary, and it is obvious that in several samples the one visible particle of gold was included in the subsample required for base metal analysis. If a gold analysis is required on a panned concentrate then the entire sample should be requested for assay. Sample preparation of the panned concentrate is also a problem. Samples are normally dried, and then crushed and disc pulverised prior to analysis. Because of the malleability of gold there is a possibility of coarse gold being lost from the sample during crushing and particularly pulverising, by the smearing of the gold on the crushing surfaces. This may

have been the case with the reconnaissance sample from the
Langmute River. The possibility of gold loss during sample
preparation reinforces the suggestion that it is most important
to thoroughly examine the concentrate in the dish prior to
bagging. It is also important to have the sample weighed (dry)
prior to analysis. Experience has shown that for sample weights
less the 10 gms the results of gold analysis are artificially
inflated compared with those from samples greater than, say, 70
gms.

Sediment samples suffer from the same particle sparsity problem
as panned concentrates. As mentioned above this was a major
problem with the sampling in the upper Nengmutka River.
Duplicate stream sediment samples from the Nengmutka River,
immediately upstream of its junction with the Langmute River,
returned values of 0.364 ppm and 0.015 ppm; one is highly
anomalous, and the other a background value. To help overcome
the problem of particle sparsity, large sediment samples are
required.

Stream dilution compounds the problems of stream sampling. In
the Nengmutka region, the panned concentrate and stream sediment
expression of the vein mineralisation is subtle, and is almost
negligible several kilometres downstream of mineralisation. This
is demonstrated by the rapid "tailing-off" of the stream sediment
and panned concentrate anomalies downstream from the Wild Dog
deposit. It must be noted, however, that in the case of the Wild
Dog deposit, the relatively recent exposure of the vein system
has also had an influence on the subtlety and rapid fallout of
stream anomalies. Other factors influencing stream dilution
include the stream gradient, amount of runoff, and stream channel
and valley profiles. Stream gradient and channel profile will
directly influence flow velocity and discharge rates, and runoff
and valley profile will influence the volume of water and
sediment entering the stream channel. Obviously an anomaly from a
stream in a steep, V-shaped valley, mountainous area will differ
from that of a stream in a broad, open area, assuming similar
rainfall conditions.

The influences of dilution also effect the float train, by
governing the rates of attrition and transportation. Within 1.1
km of the Wild Dog Prospect, the percentage of vein float
decreased from 100 to 1/2%, indicative of not only of its recent
exposure, but also attrition. Most of the material at this
distance ranged in size from 0.5 to 2 cm.

In general terms, methodical panning for the presence or absence of gold is a powerful tool. The technique permits a high degree of spontaneity of followup not provided by most modern day gold exploration techniques. So useful is the art of panning that the qualitative assessment of panning results, combined with float observations, will often override results obtained from other stream sampling techniques in directing the exploration effort.

ACKNOWLEDGEMENTS

The author wishes to thank City Resources Limited for their permission to publish this paper and financial support in attending this conference. Mrs. Josephine Kauli and Mrs. Tau Raka are thanked for typing of the manuscript.

REFERENCES

Lindley, I.D., 1987. The discovery and exploration of the Wild Dog gold-silver-copper deposit, east New Britain, P.N.G. Proceedings of Pacific Rim Congress 87, Gold Coast, pp. 283-286. The Australasian Institute of Mining & Metallurgy, Melbourne.

Lindley, I.D., 1988. Early Cainozoic stratigraphy and structure of the Gazelle Peninsula; east New Britain; An example of extensional tectonics in the New Britain arc-trench complex. Australian Journal of Earth Sciences, 35(2), pp. 231-244.

Lindley, I.D., in press. The Wild Dog vein gold deposit, in Geology of the Mineral Deposits of Australia and Papua New Guinea (Ed. F.E. Hughes). The Australasian Institute of Mining & Metallurgy, Melbourne.

10

Opportunities in Lesser Developed Countries

Chapter 37

COSTA RICA: GOLDEN OPPORTUNITY FOR 1989

Richard Keith Corbin

Mining Lawyer
MOTHER LODE MINING LAW CENTER
Sacramento, California

INTRODUCTION

With the world media full of predictions that the price of gold will quadruple in three years (Maybury, R.J., 1988), and with financial forecasters and strategists touting opportunities to make 400% in gold (Skousen, M., 1988), explorationists and entrepreneurs are scurrying to find enough gold to feed the global gold fever created.

At the same time, gold deposits in the United States and other developed and intermediate countries are becoming less attractive because of environmental restrictions, radical reductions in tax incentives, and depleting reserves, causing people in the mining industry to begin looking at opportunities for foreign investment in gold mining concessions in lesser developed countries (LDCs). Meanwhile, LDCs are responding with chamber-of-commerce aplomb in marketing their opportunities.

In the face of all this frenzy and hype, targeting gold mining concessions in LDCs for foreign investment may be done more objectively by employing a systematic approach. For this paper the SHOP rating system (Corbin, R.K., 1987), which considers factors of the LDC's stability, hospitality, ore occurrence, and profitability potential is used.

The result is that a tiny banana-growing republic in Central America, first discovered by Columbus and named the "Rich Coast" because of the many gold adornments worn by its Indians, has been rediscovered and emerges as a classic LDC gold target.

STABILITY

Throughout its history, the Republic of Costa Rica's political system has contrasted with that of its neighbors. The nation has steadily developed and maintained democratic institutions and an orderly, constitutional process of government succession. Several elements have influenced this advancement, among them educational opportunities, enlightened government leaders, comparative prosperity, flexible class lines, and the absence of a politically intrusive military.

During the quarter century ending in the mid-1970's, Costa Rica appeared to be a model developing country. Its gross domestic product grew by 7% per year in the period 1966 - 1970, 6% per year in the period 1976 - 1980. However, a combination of several factors, including accelerated and poorly structured external borrowing in the period 1978 - 1981, a sharp decrease in both the terms of trade and the level of trade and increasingly larger public deficits, slowed the Costa Rican economy.

Economically, Costa Rica is presently enjoying a resurgence. Although Costa Rica has traditionally had an agricultural-based economy, accession to the Central American common market has given a significant boost to the development of the local industrial base in Costa Rica. Moreover, because of Costa Rica's aggressive exploitation of its deposits of gold and other metals, mining is emerging significantly in terms of Costa Rica's production. In 1986, with the help of USAID, the Costa Rican Department of Geology, Mines and Petroleum (DGMH), in conjunction with the University of Costa Rica, brought in the United States Geological Survey. Fourteen man-months later, the three had teamed up to systematically examine all of the historic producing lode and placer gold mines, and had conducted airborne magnetic surveys, geochemical surveys, and other testing. The results were published in July 1987 and made available to foreign investors and operators.

Meanwhile, in 1986, state-owned Minera Nacional S.A. was created to promote investment in the mineral sector, particularly gold, and gold concessions were tendered for bid to both domestic and foreign investors. As a result, foreign investors and operators have arrived from as far away as Japan and Sweden. The resulting "jungle boom" has been a shot in the arm for the Costa Rican economy.

Other indicators bode well for the economy in Costa Rica. Unions there are relatively weak. Only about 15% of the workforce is unionized, mainly employees in the government bureacracy. The remaining unionization is in the agricultural industry.

While the rate of unemployment in Costa Rica was at an all-time high in 1982 at 9.5%, it gradually slowed to 5.7% in April of 1986, and presently it is near 4%.

The ultimate proof of the state of Costa Rica's economy is most readily observable by simply taking a taxi ride from the airport on the outskirts of the capital to downtown San Jose. There are no slums or "barrios," as they are referred to elsewhere in Latin America.

Politically Costa Rica is the oldest and most stable democracy in Latin America. In 1821 Costa Rica declared its independence from Spain and achieved its freedom without bloodshed.

Costa Rica has acted progressively even before that was a general trend. The early abolition of slavery during the colonial period, the establishment of free, obligatory education in 1869, the elimination of the death penalty in 1882, and the abolition of the army in 1949, all testify to its unique political character and vision.

Since the elimination of its army in 1949, the country has relied on a 4,500-member Civil Guard, and a 3,500-member Rural Guard to maintain internal security and public order, as well as emergency external defense, and on the OAS and the Rio treaty to guarantee the inviolability of its borders.

Costa Rica has a number of political parties; however, the two traditional, and strongest parties, are the National Liberation Party and the Social Christian Unity Party. The president is limited to a term of office of four years, and that office has pretty much alternated equally between the two strongest parties during the last 40 years.

Costa Ricans are very active in their politics, as voting is compulsory for every citizen over 18 years of age. There is no threat of a communist takeover in Costa Rica. Although there is a communist party, it only gained 3% of the popular vote in the last election.

A further measure of political stability in Costa Rica is the cordial relationship between Washington and San Jose. It is based on mutual respect for democratic traditions, common goals, and a relationship free from serious political disagreement.

Social stability in Costa Rica is helped by a relatively homogeneous racial composition and culture. Costa Ricans are Spanish-speaking with basic Catholic traditions persisting despite an influx of various western national groups -- French, Belgians, German planters and Americans. The values of Latin American culture are evident in the great importance attached to family ties, rather sedate, ritualized behavior, the yearly rounds of festivals, and an outwardly male-dominated and oriented society. The population of Costa Rica is approximately 2.5 million, with one-quarter centered in the metropolitan area of its capital, San Jose. Most Costa Ricans are Caucasian, although Blacks constitute a significant minority group of 4% and 2% of the population is Indian.

Education is a national passion of Costa Rica, which is reflected
in the vast array of schools, institutes and universities throughout
the country. Costa Rica has one of the highest literacy rates in
the world. 93% of the populace is literate, and the national government spends over 30% of its budget on education.

The country is well-known for its good medical care. Competent
surgeons, pediatricians, gynecologists, cardiologists, general prac-
titioners, dentists and opticians are to be found in San Jose. A
number of these have been trained in the U.S. or Europe, and speak
excellent English. Costa Rica has the best children's hospital in
Central America.

Costa Rica is an outspoken and active member of the international
community. It lobbied strenuously for the establishment of a United
Nations High Commissioner for Human Rights and was the first nation
to recognize the jurisdiction of the Inter-American Human Rights
Court, now based in San Jose. Also, the United Nations University
for Peace was founded there.

HOSPITABILITY

Costa Rica has the welcome mat out for foreign investors generally, and particularly for those interested in investing in its mineral
industry. Its primary target for mineral development has been, and
remains gold. The Republic realizes it needs infusions of foreign
capital, and equipment and technology to increase gold production,
and so it has made foreign investment in its gold concessions attractive. The most attractive feature is that foreign corporations or
individuals may own gold mining concessions outright, without the
typical LDC requirement that a national partner hold a controlling
interest.

Gold mining concessions are administered by the Ministry of
Industries, Energy and Mines, through its Department of Geology,
Mines and Petroleum. The Director of the Department, Jose Francisco
Castro Munoz, his deputy, Manuel Brenes Monge and staff are extremely
cordial and sincere in their assistance to foreign investors and
operators. The Department maintains a public room in its offices
in San Jose, where any interested party may requisition, read and
copy files regarding both lode and placer gold mining concessions.

Physical conditions within the interior of Costa Rica are also
quite hospitable. The climate is not bad for the tropics. There
are two seasons; rainy from May to November and dry from December
to April. The temperature ranges in the lowlands are from 80°F to
90°F while in the highlands the temperature ranges are from 73°F to
80°F throughout the whole year.

No unusual health problems or serious risks exist in San Jose.
The most serious health hazards are found outside the capital in
provincial areas. These include common diarrhea, amoebic dysentery,
bacillary dysentery, infectious hepatitis, and giardiasis. The common causes of intestinal diseases are polluted water, impure or un-
pasteurized milk, exposed fruits and vegetables, and unsanitary fruit
handling. Malaria cases have been reported in coastal areas at alti-
tudes below 2,000 feet. They are not common, although the incidence
rate has increased in recent years.

Most, if not all of these hazards can be alleviated by exercising
the simple precaution of utilizing commercially bottled water in the
field, or boiling it, and eating only fruits that you peel yourself
and avoiding fresh vegetables. Taking chloroquine tablets is a
recommended precaution to avoid malaria. Also a gamma globulin inno-
culation may help prevent hepatitis.

There should be no special concern for personal safety or welfare
while either in the metropolitan areas or in the provinces, as Costa
Rica has a very low crime rate. Moreover, Costa Rica's scrupulous
adherence to the protection of civil and human rights is world
renowned. Citizens and non-citizens alike receive a full panoply of
protections by the Republic's courts.

As far as security of the capital, equipment and technology inves-
ted in gold concessions, these will receive treatment at least equal
to that accorded other foreign and domestic investments. Costa Rican
law expressly prohibits investment expropriation, otherwise known as
"nationalization," unless effected within the parameters of prompt
and equitable compensation.

As an added hedge to secure overseas investments, the Overseas
Private Investment Corporation (OPIC) provides insurance to U.S.
investors seeking investment opportunities in Costa Rica, which
includes coverage for nationalization and a variety of other up-
heavals.

The logistics within Costa Rica are good. The Pan-American Highway runs from the Nicaraguan border in the north to the Panamanian
border in the south. The highway roughly parallels the lode deposits
in south-central Costa Rica, and is within an hours' drive of the
placer gold concessions on the Osa Peninsula. Costa Rica's neighbors
refer to Costa Ricans as "Ticos," which is a friendly term, referring
to their propensity to use diminutive forms such as momentico, for
momento. Ticos are warm and friendly people, especially towards
Americans.

Although the Costa Rican national language is Spanish, the population is very well educated and it is not difficult to find someone
who speaks English well, whether at offices such as DGMH, in hotels,
shops and restaurants, or on the streets. Many hotels feature menus

in both Spanish and English, and there is a weekly English language
newspaper available in San Jose.

There is no problem keeping in touch with the home office, as
Costa Rica has direct dialing telephone service to 65 countries,
including the United States. International airmail service can be
slow, but is generally reliable. First-class airmail and special
delivery from almost any point in the U.S. to Costa Rica usually
takes 2 - 8 days.

In summary, the Costa Rican government attitude towards foreign
investment is positive, and the favorable attitude has been shared
to greater or lesser extent by all major political parties likely
to play an active role in government in its long-standing democracy.
This attitude helped to encourage U.S. investors to invest $500 mil-
lion last year.

ORE OCCURRENCE

Christopher Columbus discovered Costa Rica in 1502. When he
arrived, the Indians on the Atlantic coast had already been mining
placer gold and rendering it into objects of art and personal adorn-
ment for hundreds of years. The gold these Indians possessed prompted Columbus to name Costa Rica the "Rich Coast."

A few years after Columbus' discovery, the Conquistadores returned
to the Rich Coast and hauled away $600,000 worth of gold artifacts in
their galleons. Today there is a collection of 1,800 pre-Columbian
gold artifacts, plus an 18-pound gold nugget on display at the
National Gold Exhibit in San Jose.

The first lode gold deposit in Costa Rica was discovered in 1815
by a Spanish bishop. This early discovery attracted miners and
investors to Costa Rica, resulting in two periods of intense mining
activity between 1824 and 1860 and between 1900 and 1950. After that
there was a quarter-century lull in gold mining activity. During
this lull, a noted explorationist wrote that Central America was one
area which the mineral explorer is justified in dismissing and specifically stated, "there is nothing in Costa Rica." (Bennett, R.H.,
1963) Accompanying his assessment was the disclaimer that at any
time it might be proved wrong, and it eventually was.

The current high price of gold has reactivated exploration, devel-
opment and mining activity in Costa Rica, and helped to prove up
world-class lode deposits, as well as first-rate placer deposits.
The principal lode gold deposits are in a region known as the "gold
belt," occurring in a zone about 110 km long and about 30 km wide.
The gold belt consists mainly of Sado epithermal vein deposits of the
Cordillera de Tilaran, and Montes del Aguacate ranges of south-central
Costa Rica. Also fueling Costa Rica's third "jungle boom" are

significant placer gold deposits on the Osa Peninsula in southwestern
Costa Rica, near the Panamanian border.

These deposits have been assessed in two 1987 reports, "The
Mineral Resource Assessment of the Republic of Costa Rica" ("the
Assessment"), and the "Geochemical Atlas." Both have been prepared
by the DGMH (Direccion General de Geologia, Minas e Hidrocarburos),
the U.S. Geological Survey and the University of Costa Rica. The
first publication is a 75-page report in both Spanish and English in
a color format with 20.5" x 32" size pages. The second is a 25" x
36" geochemical atlas of an equal number of pages, with colored maps,
duplicated on microfische, of the San Jose and Golfito Quadrangles.

These publications emphasize the Sado epithermal vein deposits of
the "gold belt," and the placer deposits on the Osa and Burica Peninsulas, as well as describing a number of silver and base metals deposits.

The Assessment states that the known gold deposits in the Cordil-
era de Tilaran and Montes del Aguacate ranges are principally of the
Sado epithermal vein deposit type. They were formed from upwelling
precious metal-rich hydrothermal fluids, derived in part from rhyolitic magmas but likely dominated by meteoric water, that were confined
to fractures and high-angle normal faults as the main fluids path-
ways. The rhyolitic intrusions may be responsible for some of the
faulting and fracturing. Precious metal-bearing veins formed as
fissure-fill along the main high-angle fault, along smaller faults
parallel and perpendicular to the main fault, and along shallow splay
faults that formed above the main fault. Precious metal content
decreases with depth whereas base metal content increases. Propylitic alteration is zoned farthest from the main fissure-fill vein
whereas sericitic alteration and silicic alteration are zoned near
the quartz veins and stockworks.

Parts of sheeted and stockwork quartz veins may have formed less
than 50 m below the surface beneath paleo hot-springs. The stockwork
quartz veins and high mercury contents at the Santa Clara and Rio
Chiquito mines may indicate that the gold-bearing epithermal veins
at these deposits were not as deeply emplaced as those in the Juntas
de Abangares and Montes del Aguacate areas.

In general, the median gold grade for Sado epithermal vein deposit
types is 0.6 grams per tonne and the median tonnage is 300,000 tonnes.
The Assessment reports that Costa Rican deposits are consistent with
the Sado type grade and tonnage distributions world-wide. It further
reports that chief among these deposits is the Santa Clara Mine, an
active open pit mine with heap leach gold recovery. Its estimated
pre-mining reserves are 5,200,000 tonnes at 2.06 grams Au per tonne.

The Bellavista-Montezuma Mine, according to another report, has
proven and possible gold reserves for open pit and underground mining

of 3,000,000 tons grading from 0.13 to 0.35 troy ounces Au per ton.
(Minerals Yearbook, 1986)

The premier placer gold deposits in Costa Rica are located on the
Osa Peninsula on the southwestern Pacific coast just north of the
Panamanian border. The Osa Peninsula is made up of a low range of
hills trending northwest and cresting at about 800 meters above sea
level. It is separated from the mainland by the Golfo Dulce, a bay
the size of Sand Francisco Bay.

Puerto Jimenez, population 1,000, on the Golfo Dulce is the only
real town on the peninsula. It serves farms on the narrow coastal
plain and the gold miners from the hills. It has an air strip, but
only Golfito has scheduled commercial flights.

The Osa Peninsula is underlain by Mesozoic and Early Tertiary
volcanic and sedimentary rocks. There is no recent volcanic activity
there, but the area is prone to earthquakes.

The best placers are in old river channel deposits that often but
do not always parallel the present drainage system. The coarsest and
most spectacular gold nuggets occur well back in the hills. Unfortunately, these coarse gold deposits are small and inconsistent. The
most consistent deposits with the largest volumes of material occur,
in classic placer fashion, just as the valleys open out onto the
coastal plain.

The Osa gravels are easily washed and disaggregated for sizing.
About 40 percent of the material is plus 1 centimeter and can be
screened out before concentrating. There is no clay problem to hin-
der the recovery of the gold.

Placer gold is generally 880 to 900 fine. However, the Osa gold
ranges between 910 and 960 fine. The long exposure of the gold in
the old channel deposits to the tropical climate has leached away
most of the silver and copper.

At this time, gold has only been found in placer deposits and
there are no good clues to its ultimate origin. The central part of
the peninsula contains residual placers of very coarse gold, so this
area is certainly the source. Whether the hardrock deposits that
supplied these are still there or whether they have been weathered
away has not been determined. (Bird, W.H., 1983)

The Assessment has also speculated on the source of the Osa placer
deposits. It states that placer gold deposits on the Osa Peninsula
appear to be from older gold-bearing deposits that occur at or above
the unconformity between the Pliocene Punta la Chancha Formation and
the older basaltic and limestone units. South-southeast- to south-
southwest-trending stream channels, 1-3 m deep, are filled with
shallow-water marine conglomerates and proximal turbidites which

grade upward into distal turbidites of the Punta la Chancha Formation. Lew (1983) suggested that the source of the gold was a vegetated volcanic land mass containing hydrothermal gold deposits and flanked by a shallow-marine environment. The source area was to the north and northeast of the Osa and Burica Peninsulas, as evidenced by mafic and felsic igneous rock fragments, littoral and neritic fauna, plant debris, and gold nuggets in the Pliocene sedimentary sequence.

The Osa Peninsula has recent alluvial and beach deposits as well as fossil placer deposits in Pliocene and younger sedimentary rocks.

Evidence for modern submergence of parts of the Osa Peninsula and adjacent coast areas indicates that the geologic environment permissive for this deposit type probably extends into the ocean. The Golfo Dulce did not develop before the Punta la Chancha Formation was deposited suggesting that the Golfo Dulce and adjacent coastal areas may contain marine placers.

The Assessment lists 45 known placer deposits on the Osa and Burica Peninsulas and indicates that the listing probably contains only a small fraction of all the placers that have been found there. It chronicles some production, and provides figures for reserves and ore grade for some of the Osa Peninsula gold concessions. The Isla Violin concession on the Rio Sierpe produced 69,145 grams of gold from 346,000 m^3 gravel in 1984.

In 1982 the Rio Tigre concessions, including concessions 698, 702 and five others, were reported to have 100,000 m^3 of gravel on the Bellida farm averaging 0.2 gm Au/m^3 and 687,000 m^3 of gravel on the Sanchez farm averaging 0.27 gm Au/m^3. (Castro M., J.F., and Vargas R., J.E., 1982) In 1983 the Corigold Partnership based in San Francisco, California, purchased concessions 698 and 702 and directed a systematic sampling program which resulted in a report of 825,000 m^3 of proven reserves on the Sanchez farm grading 0.4 gm Au/m^3 and 1,000,000 m^3 of possible reserves grading 0.4 gm Au/m^3. On the Bellida farm there were reported proven reserves of 100,000 m^3 at 0.27 gm Au/m^3 and 400,000 m^3 at 0.28 gm Au/m^3. (Bird, W.H., 1983)

According to the Assessment, the richest placer gold concession reported on the Osa Peninsula is Concession No. 1883 at the mouth of the Rio Carate, where it flows into the Pacific Ocean. The ore grade on Concession No. 1883 is reported at between 1-5 gm Au/m^3.

There may be placer gold deposits just as rich or richer within the Corcovado National Park, which occupies approximately one-third of the Osa Peninsula, including one-half of the Pacific coastline northward of Concession No. 1883. Because Corcovado is presently closed to mining, there are no legal concessions there; consequently the Assessment did not include data on reserves and grade in the park.

Nevertheless, there is independent evidence of the Corcovado's

riches. Until February 1986 illegal gold panners, "oreros," lived
in the jungles and sifted for their fortunes in the park's rivers
and Laguna Corcovado. Few struck it rich, but most at least made a
living. The Banco Central set up a special office in Puerto Jimenez
to buy the gold, which the country used to help pay the interest on
its huge foreign debt. But the gold panners' numbers grew so dis-
proportionately that their activity started causing real destruction.
The silt from their panning was filling up the lake and the park's
basin. So in February and March 1986, the Park Service and the
Costa Rican Civil Guard removed all the oreros, and now patrol regularly to prevent further mining. (Blake, B., and Becher, A., 1987)

In summary, the Assessment predicts there are even odds for 11 or
more undiscovered lode gold deposits in the range to 6,000,000 tonnes
of ore with a grade of approximately 2.2 to 18 grams of gold per
tonne.

These lode gold deposits are attracting major companies from
around the world. Freeport McMoRan is currently investigating several lode gold concessions, and other companies have already acquired
some. The Bellavista-Montezuma gold concession was taken by two
Canadian companies, Midland Energy Corp., and Westlake Resources Inc.
Another Canadian company, Rayrock-Yellowknife Resources Inc., gained
an option to obtain one-half of Midland's interest in the concession
through the expenditure of $1.5 million on exploration and feasibil-
ity studies. Rayrock plans to initiate work to upgrade the probable
and possible open pit reserves to the proven category. A $600,000
exploration program was to start in January 1987. (Minerals Yearbook, 1986)

Costa Rica's placer gold deposits are gaining the same high-caliber
attention as its lodes. Ralph L. Phelps, Jr., former owner-operator
of the New Almaden Mine near San Jose, California, once the largest
producing quicksilver mine in the United States, is presently directing Washington Gold Mining Company's negotiations with DGMH for a
placer gold concession on the Osa Peninsula. There are a number of
other companies from Canada and from as far away as Japan and Sweden
already operating concessions there.

PROFITABILITY POTENTIAL

The principal determinant of the profitability potential of an
LDC's gold concessions is to be found in its mining code. If the
mining code is too restrictive, no gold deposit can be mined at a
profit.

The principal legislation governing mining in Costa Rica is the
Mining Code (Codigo de Mineria) of October 4, 1982, Decree Law No.
6797, printed only in Spanish. Title II thereof governs concessions
in general, Title III, exploration permits, and Title IV, exploitation.

The Ministry of Industries, through its Department of Geology, Mines and Petroleum (DGMH) is entrusted with matters pertaining to the discovery of mines, exclusive exploration permits, denouncements and permits for the exploitation of minerals, as well as enforcement of obligations imposed by law.

Costa Rica's Mining Code treats mineral rights much as they are treated in the United States. Initially they are owned by the government. They are conveyed to private parties as concessions. These can be filed upon and held as long as a series of requirements are met. The requirements are reasonable, relatively inexpensive to achieve, and are simply designed to insure that the concession is mined. There are environmental requirements as well, but these are simple, and add no significant costs to the mining operation.

There are some important differences, however. In the United States claims are located on public lands after discovery of a valuable mineral deposit, such as gold, and the claimant's use of the surface for mining purposes takes priority over competing uses. The claimant may even obtain exclusive use of the surface by a process called patenting. By contrast in Costa Rica the surface is usually already privately held for farming, and the miner must pay the surface owner for any loss or damage he causes, or risk forfeiture of the concession.

Another distinctive feature is that in the United States a claimant must first be a citizen or declare his intention to become one in order to locate a claim. In Costa Rica any individual or company, foreign or national, may own a gold mining concession. In all but a few other Latin American countries, a national must own a controlling interest in the concession. This causes dilution of the investor's contribution in capital, equipment and technology. It also diminishes control of the operation, as well as security of the investor's interest in the concession.

Other important features of Costa Rica's laws applying to foreign investment are that there is no restriction on the repatriation of profits, no discriminatory import policies or license requirements, and no joint venture regulations which might represent barriers to direct foreign investment.

Presently it may be arranged for gold mining equipment to be brought into the country duty-free, as mining projects are exempt from import duties. Also the surcharges from the Central Bank are dropped. Although the Central Bank regulations levy no export tax on gold, the Mining Code exacts a severence tax that has varied from 2 to 6%. There is a 30% income tax on profits. However, mining companies can avoid much, if not all of this, by processing the gold in Costa Rica, for instance, into jewelry. (Valverte, A., 1988)

Aside from Costa Rica's favorable laws, it has a number of other features which are indicative of profitability. The Costa Rican labor force has been characterized as relatively well-educated, skilled and easily trainable, largely the result of an historical emphasis on state support of education. This effort has resulted in a literacy rate of over 90%. The average worker has demonstrated a willingness to seek, and an ability to absorb, additional special-ized training. Moreover, the cost of general labor is low. A $0.90 per hour minimum wage prevails, including fringe benefits. The med-ian Costa Rican income is only $300 per month.

Both lode and placer deposits are easily accessible. Miami, Florida is only 1,130 air miles from San Jose or 2.5 hours (non-stop) by air. Lacsa, the Costa Rican national airlines, has daily flights from Miami, as does Eastern Airlines. A number of other interna-tional airlines have service from Costa Rica to countries around the world. From San Jose it is only an hours ride to the lode gold deposits in the "gold belt." It is 3.5 hours via the Pan-American Highway to the Osa Peninsula, and another hour by paved and graveled roads to the placer gold concessions. In addition, there is commer-cial small-plane service to Golfito, and bush pilot service to sev-eral other points on the Osa Peninsula.

Food and housing costs in Costa Rica are relatively low, however fuel and transportation are not. Leaded gasoline is about $1.70 per gallon and diesel fuel is about $1.35 per gallon. Jeep and other four-wheel drive vehicles are available through rental agencies in San Jose, with monthly rates of approximately $1,000 to $1,500.

CONCLUSION

The stability of the Republic of Costa Rica in its socio-politico-economic sectors amply justifies its accolade of "the Switzerland of Latin America."

The hospitability of the "Rich Coast" jungle gold concession areas is a factor fairly characterized as inviting, and the welcomeness the Ticos extend to foreign investors is unmatched.

The ore occurrences of the Sado epithermal vein deposits of Costa Rica's "gold belt" are world-class, and the placer deposits on the Osa Peninsula are first-rate.

The profitability potential for foreign investment in Costa Rica's gold mining concessions is assured by the fair and reasonable provi-sions in the Ministry of Industries' Mining Code.

Based on the SHOP rating system, the opportunity for foreign investment in this LDC's gold mining concessions rates a 10, making Costa Rica the golden opportunity for 1989.

ACKNOWLEDGMENTS

"There is more adventure, romance, wonder,
excitement, success and failure in the profes-
sions concerned with metals and mining than in
any other calling." (Hoover, H., 1963)

It has been my distinct pleasure to share the adventure, romance,
wonder and excitement of the gold concessions in Costa Rica with my
esteemed colleagues and consultants, Dr. Howard L. Stensrud, Ph.D.,
Chairman, Dept. of Geology, California State University, Chico, CA,
and Dave McCracken, professional gold dredger, Happy Camp, CA. And
for sharing vicariously his five years' experience as managing partner of Corigold's gold concessions on the Osa Peninsula, I wish to
thank Fred Forster, San Francisco, CA.

Any measure of success I may gain from this experience, I owe
largely to the Republic of Costa Rica, its Ministry of Industries,
Mines and Energy, the Department of Geology, Mines and Petroleum
(DGMH), and to the kind indulgence and assistance of DGMH's Director,
Jose Francisco Castro Munoz, and Deputy, Manuel Brenes Monge.

The opportunity to share in turn my Costa Rican experience with
you could not have been possible without the efforts of C.O. "Chuck"
Brawner, and his Gold Mining 88 forum.

REFERENCES

American Embassy San Jose, 1988, "Costa Rica," Foreign Economic
    Trends, U.S. Dept. of Commerce.

American Embassy San Jose, 1986, "Costa Rica," Foreign Labor Trends,
    U.S. Dept. of Labor.

Bennett, R.H., 1963, "Quest for Ore," AIME, Star Press, T.S. Denison
    & Co., Inc., Minneapolis, pp. 128-129.

Bird, W.H., 1983, "Summary Report," Corigold Technical Operations

Blake, B., and Becher, A., 1987, "The New Key to Costa Rica," 7th
    ed., Publications in English, San Jose, C.R., pp. 184-185.

Castro M., J.F., and Vargas R., J.E., 1982, Mapa de recursos miner-
    ales de Costa Rica, 1:750,000:Costa Rica, Direccion de Geologia
    y Minas, 1 sheet.

Corbin, R.K., 1987, "Rating Opportunities for Foreign Investment in
    Gold Mining Concessions in LDCs," Gold Mining 87: First Interna-
    tional Conference on Gold Mining, Brawner, C.O., ed., AIME,
    Baltimore, pp. 539-553.

Hoover, H., 1963, "Quest for Ore," Foreword (paraphrased), AIME, Star
Press, T.S. Denison & Co., Inc., Minneapolis

Leyes, C.R., 1982, "Codiga de Mineria, Republica de Costa Rica,"
Imprenta Nacional, San Jose, C.R., 31 pgs.

Los Alamos National Laboratory, Ministerio de Industria, Energia, y
Minas, Central American School of Geology, 1987, "Geochemical
Atlas of the San Jose and Golfito Quadrangles, Costa Rica," U.S.
Dept. of Commerce, Los Alamos, N.M.

Maybury, R.J., 1988, "1987 Stock Crash Means Gold & Silver Will
Quadruple In Three Years," California Mining Journal, May 1988,
Harn, K.L., ed., Aptos, CA, pp. 41-56.

Minerals Yearbook, 1986, U.S. Dept. of Interior, "Costa Rica," The
Mineral Industry of Central American Countries," pp. 4-5.

Northcutt, E., 1970, "Costa Rica," Summary of Mining and Petroleum
Laws of the World, U.S. Dept. of Interior, Bureau of Mines, Infor-
mation Circular 8482, U.S. Gov't. Printing Office, Washington,
D.C., pp. 47-50.

Skousen, M., 1988, "Skousen's New Forecasts as of May, 1988," Forecasts & Strategies, Phillips Pub., Inc., Potomac, MD, p. 11.

U.S. Dept. of State, 1986, "Costa Rica," Post Report, Dept. of State
Publications 9363, U.S. Gov't. Printing Office, Washington, D.C.,
20 pgs.

U.S. Geological Survey, Direccion General de Geologia, Minas e
Hidrocarburos, University of Costa Rica, 1987, "Mineral Resource
Assessment of the Republic of Costa Rica," U.S. Dept. of Interior,
Reston, VA, Miscellaneous Investigations Series Map 1-1865, 75
pgs.

Valverte, A., July 5, 1988, CINDE Regional Director, personal communication.

Verity, V.H., et al., 1960, Comparative Mining Laws of Foreign
Countries," American Law of Mining, (1st edition), Rocky Mountain
Mineral Law Foundation, ed., Matthew Bender, Boulder, Colo., V. 2,
Title VII, §13.1 - 13.13, pp. 751-767.

Chapter 38

AN OFFICIAL APPROACH TO INVESTMENT IN ARGENTINE
MINING SECTOR - THE GOLD CASE -

Juan Eduardo BARRERA

Ph.D. Mining Engineer

Secretary of State of Mining
Buenos Aires, Argentina

SUMMARY AND CONCLUSIONS

Argentina presents to the foreign and local investors a mineral wealth that is largely unexplored, untested, underdeveloped and therefore incompletely evaluated.

In view of the above, the new administration of President Raul Alfonsín through its Department of Mines has designed and implemented a strategy (Plan de Expansión Minera P.E.M.), whose main objective aims at improving the "investment climate" in the sector.

It is the Government's basic intention to apply the Mining Promotion Law with responsibility but with generosity to encourage the attraction of investment (local and foreign) to the sector.

The establishment of a Data Bank system to evaluate possible new potential ore mining developments will help to connect the owners of potential projects with investors outside the sector.

Argentina has an interesting gold ore potential. Estimated reserves of individual projects already known amount to almost 300 tonnes of gold for potential production.

The government strongly believes and present information indicates that more systematic exploration has to be undertaken in the Jurassic areas that almost cover the whole of Patagonia. Two recent gold discoveries (Angela in the Chubut province, and Cerro Vanguardia in Santa Cruz province) indicate the potential of this area.

The government believes that most of the interest to develop new metallic mineral development projects will come from foreign investors through joint-ventures with local partners, from existing companies currently exploiting non-metallic minerals, and from construction firms that have idle equipment.

It is our intention that the establishment of a proper policy framework associated with a reliable inventory system will help to bring together the type of mining deposits and entrepreneurs needed to operate a sound financial mining operation.

INTRODUCTION

Although it is widely recognized that Argentina is a country with most of its vast mineral resources still untapped, it is also true that in the last twenty years the government has made a considerable effort in mining exploration. As a result of this effort a number of potential projects have been found, most of them of medium sized polymetallic deposits containing silver-zinc-lead or gold-silver-copper. To date, almost no follow-up investment has taken place.

One main cause that explains the underdevelopment of the mining sector in Argentina is the lack of local private entrepreneurs with relevant experience, and in particular, the lack of private initiative special in the pre-investment stage. The major exploration effort has been undertaken by the state in the recent past. Banco de Desarrollo (a national development bank) has had funds to finance the investment for the construction stage. However, in the phases between pre-feasibility and the final investment decision proposals have remained untouched by either the private sector or the government.

In view of this, since December 1983, the new administration of President Raúl Alfonsín through its department of mines (the Secretaría de Estado de Minería S.E.M.) has designed and implemented a strategy (Plan de Expansión Minero P.E.M.) the main objective of whole aims at improving the investment climate in the sector.

The present administration recognizes that improving the investment climate will result in the emergence of financially and economically feasible projects. Therefore, the Government is making a substantial effort to develop a rational policy framework so that these developments may take place.

Also, the new approach is to promote exploration of possible ore occurrences that present a good market prospective. Within this context, the Secretariat of Mines has identified two main priorities: a) to provide the private sector with important incentives vis-a-vis other economic activities, and b) to build-up a comprehensive Project Inventory System and to systematize and disseminate existing and future information.

The Government strongly believes that these two elements are critical and they constitute the basic engine for future mining growth.

This paper presents: i) a description of the major tools presently available to promote mining activities, and ii) an indication of investment opportunities in the gold sector in Argentina.

THE GENERAL POLICY

Institutional and Policy Framework

Incentive Administration: The present authorities have introduced radical changes in the administration of incentives affecting the sector. Thus, the incentives for investment in the sector stated in the Mining Promotional Law - deferred taxation, tax breaks and the Fondo de Fomento Minero (a mining finance fund administrated by S.E.M.) - have been reviewed to attract sound investment to the sector and to prevent their use as an exercise for financial purposes alone, as it was the case in the recent past. They are intended to make business attractive for small, medium and large firms in Argentina.

Mining in Argentina is promoted by a very favorable Mining Promotional Law (Law 22.095). It was passed in November 1979, favoring equally national and international investors. Beneficiaries are duly registered natural persons or legal entities, qualified to act in Argentina. Benefits and incentives are in no way linked to the origin of ownership or percentage of equity. Tax incentives may be granted to national and international companies. Benefits will depend upon the nature of the project and its coherence with the purpose of the P.E.M.

Benefits to promote mining activities fall into three categories: a) research, exploration, development, preparation and extraction of minerals; b) crushing, grinding, processing, pelletization, sintering, briqueting, roasting, smelting, refining and other ore treatments to be determined; and c) primary manufacture of products. There is a stipulated regional integration required for all processing operations. So, to qualify for these incentives, a firm must integrate processing in their mineral development phase.

General Benefits: The law permits a 100 percent income tax deduction for expenses involving investments made in research, prospecting and exploration, including mining license fees, construction of roads, railroad sidings, housing for personnel, etc.; and expenses for technical assistance to develop processes and products.

For the purchase of machinery, equipment, facilities, traction and loading gear, communications, power generation, vehicles, etc., the law allows a 100 percent deduction for Argentine manufactured goods, and a 90 percent deduction for imported goods.

Firms are exempt from income tax or capital gains tax on profits resulting from the transfer of mines and mining rights as corporate capital stock.

VAT tax is taxable on all mineral sales. However, the law allows the company, as an additional benefit, to retain the 100% value of VAT for the first seven years of operation. After that, there is a scaled reduction up to fifteenth year.

Ore reserve value may be capitalized up to 50 percent, and the non-capitalizable balance can constitute a reserve derived from assessment. This is for accounting purposes, and the capitalized half cannot be registered as foreign investment.

Benefits also include tax exemption on the issue and collection of fully-paid shares arising from capitalization of reserves, as well as from amendments to the memoranda and articles of association. The capitalization or allocation of shares received from another corporation as a result of capitalization is also tax exempt.

Special Benefits: These are obtained through the submission to the application authority a plans for concrete projects. These plans relate to new facilities or the expansion of existing ones; relocation of plants; and concentration, expansion, merger or reopening. It may include mining activities from research to refining and other processes determined by the Executive Branch.

Benefits granted by the authorities are special operating benefits from net worth and income tax for the first seven years. Taxes are then reduced in the eighth year (by 80 percent) and graduate to a 10 percent reduction by the fifteenth year. In some case, the application authority may shorten the time period (for example, to 10 years), but cannot change the corresponding percentages, which are written into the law.

A firm may obtain a stamp tax exemption for up to 15 years. These include stamp tax exemptions for contracts and extensions, capital increases and the issue of shares, plus any agreements entered into with national utilities for the supply of electric power, water and gas.

Another special benefit is that capital gains and net worth taxes may be exempt from the date a project is approved until the production process actually begins. Additionally, accelerated depreciation is available for specific time periods against income tax.

Finally, a company is exempt from all duty charges and taxes on capital goods, accessories and spare parts not produced in Argentina in equivalent conditions of quality, prices and delivery.

Deferred Taxation: Local companies, even those using foreign capital, can attract investment from third parties. The mining promotion system allows investors in local companies to defer payment of income tax, capital gains tax, net worth tax and value-added tax including advance payments which correspond to fiscal years with a general expiration date after the date of investment. The tax deferment is equal to 75 percent of the direct equity contribution, or 75 percent of the subscribed capital. Repayment consists of five equal and consecutive annual installments, beginning in the sixth fiscal year after the start-up of the project.

Also, investors in local companies may deduct the remaining 25 percent of the investment on the balance sheet for income tax purposes to correspond to the fiscal year in which the investment is actually accomplished.

In either case, the investment must remain the property of its shareholders for a period of not less than that for the repayment.

Conclusion: The Government has considerable powers to help develop the mineral sector. In recent years, the law was not been used to its full potential. (Since the law was promulgated in 1980, the Mining Secretariat had 4 different Secretarys in the brief period of 1980-1983). It is now the Government intention to apply the law with responsibility but with generosity to make possible the attraction of local and foreign investment to the sector.

The Project Inventory System

Setting-up an Information System: As a complement to the policies mentioned above, the Government is making significant efforts to improve both the status of information available within the Secretariat of Mines, as well as to upgrade of the capabilities of its technical staff, and the publication and dissemination of information already available. Within this context has been the development of initiatives in the general area of training, and the development of information systems and data processing.

The United Nations Department of Technical Co-operation for Development founded a consulting mission in 1984 which dealt with some of the overall requirements including those relating to the integrity of information. Members of the Secretariat, through a programme of Canadian bi-lateral assistance, have participated in study tours to the Quebec Provincial Department of Mines to study an excellent mineral data base system there, and to study the procedures used in Quebec for data compilation and analysis. Also, a close collaboration with the U.S.Bureau of Mines has continued since 1984. Currently, we are applying this knowledge to local requirements in Argentina, and two small data base files have been established.

Conclusion: As a consequence of the above developments, an embryonic Data Bank information system has been implemented. For the first time, the Mining Secretariat operate a Data Bank System that contains all the details of potential projects at different levels of knowledge (new discoveres, pre-feasibility, feasibility phase) that are known in Argentina.

At the present, the S.E.M. is issuing a monthly list of potentially available investment opportunities, as a way to inform investors of new developments in the sector.

We believe this is the first systematic approach to evaluate all the possible new potential ore mining developments in the sector, and the system will make it possible to connect the owner of the potential projects with investors outside the sector.

THE GOLD CASE

Geological Gold Potential: Argentina, with a large territory, has an interesting gold potential, due to the diversity of tectonic environments and therefore of metallogenic habitats. These conditions are similar to those found in gold producing countries, so we expect our potential will be recognized after an intensive exploration program.

The Argentine slope of the Andean Cordillera has some peculiarities shared with the Chilean slope, and the exploration and systematic search of porphyry copper type mineralization has yielded more than a dozen of these orebodies. In a similar way, the systematic search for gold deposits, so well known on the Chilean Andes, is revealing some interesting projects on the Argentine side.

We consider, with the information presently available, that two potential areas which belong to the Tertiary and Jurassic periods are the most interesting zones to begin a systematic programme of exploration and exploitation of gold mines in our country. Estimated reserves of individual deposits already known amount to almost 300 tonnes of gold for potential production. Below is presented the amount of reserves already estimated by geological period of origin.

Annexes 1 and 2, present a detailed break-down of the potential ore reserves given above. The table identifies each potential proposal by geological period, location, proven and probable reserves and type of ore deposit and provides an estimation of grades per tonne for each project. The locations are shown on a

map (see Annex 3). This information is all contained in our Data
Bank System.

TABLE I. ESTIMATED ORE RESERVES OF ALREADY KNOWN GOLD ORE
DEPOSITS

CYCLES	PROVEN RESOURCES	PROBABLE RESOURCES
TERTIARY	> 20 t.	~225 t.
JURASSIC	> 14 t.	~40 t.
PERMIAN TRIASSIC	~0,3 t.	~1,5 t.
OLD PALEOZOIC	~1,5 t.	7-8 t.
PLACER	~4-5 t.	~25-26 t.
TOTAL:	~41 t.	~ 300 t.

As you can see from Annexes 1 and 2, our main reserves are
located in deposits from the Tertiary period. The concentration of
occurrences is explained given the types of deposits searched for
in the past. Thus, Argentine has made large exploration efforts in
search of porphyry copper-type mineralization in its recent
history.

We strongly believe and our information indicates that more
systematic exploration efforts have to be undertaken in the
Jurassic Zones , that it almost covers the entire area of Patagonia.
Two recent gold ore discoveries Angela in Chubut province and
Cerro Vanguardia in Santa Cruz province indicate the potential. Up
to now, Patagonia has received little or no attention in
exploration terms.

Thus, the volcanic rocks of the Jurassic period in Patagonia
represent the most promising zones for exploration. The potential
of these ores is interesting, not only due to the vast areas still
unexplored but also because the types of deposits to be discovered
are generally larger than those in Tertiary formations. The
average type of deposit here is expected to contain; 1-10 million
tonnes of one, with grades of 5-8 g/t. Excellent examples are the
deposit of "Cerro Vanguardia" in Santa Cruz and "Huemules" in
Chubut.

We want to emphasize that the 300 tons of estimated gold
reserves are contained in potential production proposals in
different phases of the project development cycle. We believe that
additional gold ore reserves will be located if a systematic
exploration approach is undertaken in Patagonia.

DEVELOPMENT STRATEGY

The S.E.M.'s short-term strategy consists of concentrating the development effort on the pool of projects already known, starting with those most advanced and better known in order to quickly bring them into production and greatly reduce the exploration risks. By so doing, the success of the first projects would be better assured and this will have a "demonstration effect".

Thus, the success of the whole program depends on the ability of the Secretariat to make a convincing case that particular properties have economic potential. Once the studies are done and a proposal appears to indicate positive results, the Secretariat will use all the incentives provided for in the Mining Promotional Law, and the facilities of the Data Bank System to assure to bring together a responsible sponsor with the financial means needed to develop the proposal.

The one question critical to the present strategy success is: who is going to undertake the studies and develop the mines?. There are extremely few mining companies involved in the production of metallic minerals in Argentina and certainly not enough to take the new developments that the system expects to generate.

We believe that most of the interest in developing metallic minerals will come from foreign investors in particular through joint-ventures with local partners, from existing companies currently exploiting non-metallic minerals and from construction firms which have idle equipment.

The establishment of a proper policy framework associated with a reliable inventory system will help bring together the type of mining deposits and the entrepreneurs needed to operate financially sound mining operations.

REFERENCES

Ing. Juan Eduardo Barrera. Plan de Expansión Minera (P.E.M.)-
Versión revisada 1988 - Secretaría de Minería

"Oro y Plata en Argentina" 1983, Centro Internacional Información
Empresaria.

Ivosevic, Stanley W. 1984. "Gold and Silver Handbook", Pub. by
author, Denver, Colorado, USA.

Cox Denis P. y Singer Donald A. 1986. "Mineral Deposit Models",
U.S. Geological Survey, Bull. 1693, Washington, USA.

Bache Jean Jaques, 1980. "Les gisements d'or dans le monde", Mém.
B.R.G.M. nro. 118, Orleans, FRANCE.

Angelelli, V. et al, 1984. "Yacimientos Metalíferos de la
República Argentina", Vol II, C.I.C. La Plata ARGENTINA

Annex I

ORO EN YACIMIENTOS DE "ROCA DURA"

AGE	DISTRICT	PROVINCE	EVALUATED RESERVES (t)	POTENTIAL RESERVES (t)	SURFACE INTEREST FOR PROSPECTS (km2)	TOPOGRAPHY OF LAYERS	TYPES OF POTENTIAL RESERVE	MAXIMUM Au (g/t)	POTENTIAL (t/Au)
	RINCONADA	JUJUY	---	1	500	2	0,1 - 0,2	> 10	< 2
PALEO-	INCAHUASI	CATAMARCA	---	3,1	100	2	0,1 - 0,2	> 10	< 2
ZOICO	CULAMPAJA	CATAMARCA	0,072	0,5	200	3	< 0,1	> 15	< 2
INFERIOR	CANDELARIA	CORDOBA	1,2	2	240	3	< 0,1	> 15	< 2
	MORADO	SAN JUAN	0,14	1	100	3	0,1	> 15	< 2
PERMO-	CASTAÑO NUEVO	SAN JUAN	0,09	DESC (*)	100	6C	< 1	> 5	< 5
	GUANDACOL	LA RIOJA		DESC.	500	5C	< 10	2 - 9	< 90
TRIASICO	PUNILLA	SAN JUAN	0,196	DESC.	100	5C	< 10	2 - 9	< 90
	HUEMULES	CHUBUT	3,5	26,5	50	6C	< 1	> 5	< 90
JURASICO	VANGUARDIA	SANTA CRUZ	7,12	13	500	6C	< 1	> 5	< 90
	ANGELA	CHUBUT	3,25	DESC.	100	6B	< 1	> 5	< 5
	A. DIONISIO	CATAMARCA	15	> 200 (*)	200	4B/6C/6D	> 100	< 1	> 100
	MOG. R.BLANCO	LA RIOJA	1,3	DESC.	240	6C	< 1	> 5	< 5
TER-	GUALILAN	SAN JUAN	1,7	DESC.	100	5C	< 10	2 - 9	< 90
	MARAYES	SAN JUAN	0,43	DESC.	120	6C	< 1	> 5	< 5
CIA-	ORO DEL NORTE	MENDOZA	0,6	DESC.	100	6C	< 1	> 5	< 5
	TUPUNGATO	MENDOZA	0,054	DESC.	100	6C	< 1	> 5	< 5
RIO	ANDACOLLO	NEUQUEN	1,6	DESC. (*)	250	6C	< 1	> 5	< 5
	VALLE D. CURA	SAN JUAN	---	DESC.	1000	6C	< 1	> 5	< 5
TOTALES:			~36 (1,2% TOTAL)	~265 (88% TOTAL)	4600				

Annex II

ORO EN PLACERES

DISTRICT	PROVINCE	REASONABLE VOLUME (M m3)	CONTENTS (mg/m3)	GOLD CONTENT (t)	SIZE OF PARTICLE	ACTUAL MINING	PRECEDING	INFRA-STRUCTURE	OWNER
FAMATINA	LA RIOJA	> 20	200 - 250	4 - 5 (*)	FINO	ARTESANAL	SI	BUENA	PRIVADA
A.LOS MINEROS	NEUQUEN	>20	300 - 600	6 - 12	MED-GRUESO	ARTESANAL	SI	BUENA	ESTATAL
TOMOLASTA	SAN LUIS	<15	150 - 250	2 - 4	MED-GRUESO				
					FINO	INDUSTRIAL	SI	BUENA	PRIVADA
EPUYEN	CHUBUT	10 - 25	100 - 200?	1 - 5	FINO	NO	NO	MALA	ESTATAL
RIO PERCEY	CHUBUT	10 - 15	~100?	1 - 2	MED-FINO	NO	NO	MALA	ESTATAL
AMANAO (1)	CATAMARCA	>25	~100?	2 - 4	MED-GRUESO				
					FINO-MUYFINO	NO	NO	BUENA	ESTATAL
RIO PINTO	CORDOBA	>25	200 - 250	5 - 7	FINO-MUYFINO	NO	NO	BUENA	PRIVADA
RINCONADA -									
STA. CATALINA	JUJUY	>>40	50 - 100	2 - 4	FINO-MUYFINO	ARTESANAL	SI	ACEPTABLE	ESTATAL
GUALCAMAYO	SAN JUAN	>5	<150	<1 (*)	FINO-MUYFINO	NO	SI	MALA	ESTATAL
TOTAL:			175 - 200	140 - 225	23 - 45	FINO>>GRUESO			

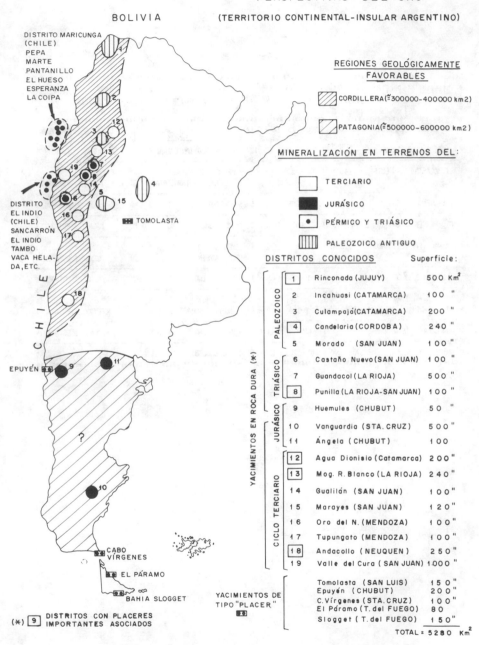

Annex III

PERSPECTIVAS DEL ORO

(TERRITORIO CONTINENTAL-INSULAR ARGENTINO)

BOLIVIA

DISTRITO MARICUNGA
(CHILE)
PEPA
MARTE
PANTANILLO
EL HUESO
ESPERANZA
LA COIPA

DISTRITO
EL INDIO
(CHILE)
SANCARRON
EL INDIO
TAMBO
VACA HELA-
DA, ETC.

CHILE

EPUYÉN

CABO
VÍRGENES

EL PÁRAMO

BAHIA SLOGGET

(∗) [9] DISTRITOS CON PLACERES
IMPORTANTES ASOCIADOS

REGIONES GEOLÓGICAMENTE
FAVORABLES

CORDILLERA (≃300000-400000 km2)

PATAGONIA (≃500000-600000 km2)

MINERALIZACIÓN EN TERRENOS DEL:

TERCIARIO

JURÁSICO

PÉRMICO Y TRIÁSICO

PALEOZOICO ANTIGUO

DISTRITOS CONOCIDOS   Superficie:

PALEOZOICO	1	Rinconada (JUJUY)	500 Km²
	2	Incahuasi (CATAMARCA)	100 "
	3	Culampajá (CATAMARCA)	200 "
	4	Candelaria (CORDOBA)	240 "
	5	Morado (SAN JUAN)	100 "
TRIÁSICO	6	Castaño Nuevo (SAN JUAN)	100 "
	7	Guandacol (LA RIOJA)	500 "
	8	Punilla (LA RIOJA-SAN JUAN)	100 "
JURÁSICO	9	Huemules (CHUBUT)	50 "
	10	Vanguardia (STA. CRUZ)	500 "
	11	Ángela (CHUBUT)	100
CICLO TERCIARIO	12	Agua Dionisio (Catamarca)	200 "
	13	Mog. R. Blanco (LA RIOJA)	240 "
	14	Gualilán (SAN JUAN)	100 "
	15	Marayes (SAN JUAN)	120 "
	16	Oro del N. (MENDOZA)	100 "
	17	Tupungato (MENDOZA)	100 "
	18	Andacollo (NEUQUEN)	250 "
	19	Valle del Cura (SAN JUAN)	1.000 "

YACIMIENTOS EN ROCA DURA (∗)

TOMOLASTA

YACIMIENTOS DE TIPO "PLACER"		
Tomolasta (SAN LUIS)	150 "
Epuyén (CHUBUT)	200 "
C. Vírgenes (STA. CRUZ)	100 "
El Páramo (T. del FUEGO)	80 "
Slogget (T. del FUEGO)	150 "

TOTAL = 5280 Km²

Chapter 39

EXPLORATIONS AND THE STUDY PHASE OF EXPLOITATION OF THE MACEDONIAN GOLD
DEPOSITS IN YUGOSLAVIA

by V.Pavlović , A. Davaliev , K. Hrković

Professor of Open pit mining on Mining and Geology Faculty,
Belgrade University , Yugoslavia
Geological engineer , Head of the Geological Bureau,
Gevgelija, Yugoslavia
Lecturer of Mineral Deposits on Mining and Geology Faculty,
Belgrade University , Yugoslavia

INTRODUCTION

On the territory of Macedonia in South Yugoslavia primary and secondary gold occurences were
known since ancient times. Remnants of numerous goafs and remainders of washaries on river
terraces and alluviums date right back to the times of Phillip II and his son Alexander of
Macedonia and from Roman times, as well as from the Middle Ages witness exploitation of
primary and secondary gold.
More recent explorations on primary and secondary gold occurences date from the perod
between two World Wars. Over that period a number of private companies, each on their own
jointly, carried out gold exploration, particulary of alluvial occurences on rivers emptying
into River Vardar, starting from Skopje to the Greek frontier.
The latest gold explorations carried out in the alluvium of River Konjska date back in 1976,
and in 1978 old terraces of River Konjska were explored.
The Visibility Study was prepared on possibilities of exploitation of gold from River Konjska
alluvium. In the paper are collected the basic   data an all known primary and secondary
occurences of gold, the results of the latest explorations, and the study itself.
To-date geological explorations and analyses show very interesting potentionality of River
Konjska and River Vardar alluvium, which proves continuing of the further explorations. For
the moment, it is necessary to re-explore more complete view of the deposit. After that it will
be possible to make complete techno-economical estimation of the auriferious alluviums and
the conditions of the evaluations of the alluvial gold.

1. REVIEW OF PRIMARY AND SECONDARY GOLD DEPOSITS IN MACEDONIA,YUGOSLAVIA

1.1. Primary deposits

Scarse data are available for primary gold deposits in Macedonia, being unadequate for clear indication of gold-bearing explored centres. The most important primary gold occurences are located in the region of Demir kapija, as well as in the Strumica-Valandovo region, surroundings of Bitola (eastern Pelister hillside), and surroundings of Gevgelija (eastern mountain Kožuf) (Fig. 1).

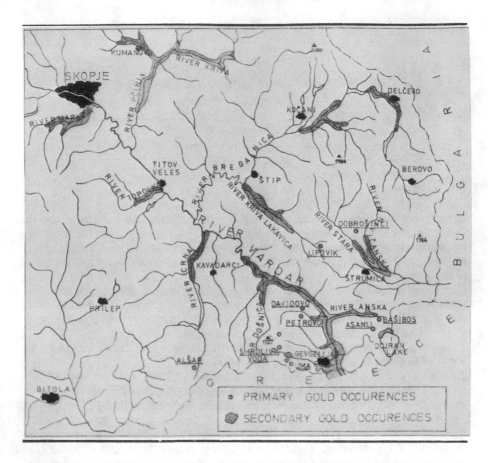

FIG. 1 – THE MAP OF MACEDONIAN GOLD OCCURENCES EXPLORATION IN 1976

1.1.1. Demir Kapija Region

There are several gold occurences in the Demir Kapija region : Bukovi, Iberli, Košarka, Korešnica-Tropište, Došnica near village Dren. Gold occurences are found on both River Vardar banks. The occurences are on the contact between limestones and the basic rocks, where mineralization occurs in quartz veins, represented by Pb-Zn-Cu mineralization containing certain quantities of gold.

1.1.2. Strumica - Valandovo Region

In this region there are numerous gold goafs dating back to the times of Alexander of Macedonia (336-323 B.C.) and Roman times around year 146.

a) Gold was exploited from quartzite on Abadaz (800-1000m above sea) 3km northeast of villages Kočuli and Jumru Koja.
b) The second occurence is in spring section of River Lakavica, near village Lipovil, i.e. in the stream Volontar, which intersects, in this part thick layers of quartzite considered gold-bearing, because downstream in stream Volontar alluvial deposits, over a length of 7 km and width of 100 to 800 meters, remnants of numerous gold washing goafs are found.
c) The third gold occurence is near village Dobrošinci, 13km North-Northeast of Strumica, on the southwestern slope of mountain Novomalska (1204 m a.s.). The terrain is built of quartzites, gneisses, argiloshist, and phyllite. Gold was found in streams formed on this terrain and gravitate towards River Strumica. In the deposit, the gold originates from quartzite, which also contains magnetite, while copper-bearing veins with pyrite occur in argiloshists and phyllites.

According to the information afforted by T. Vitorović , on the areas of villages Suševo - Dobrašinci - Novo Malo (year 1935) gold exploitation were performed in alluviums of River Suva , River Suševa, River Novčanska and stream Aktoprak, but due to the high groundwater table (1.00m) exploratory shafts caved, so that detritus was explored only in surface section, yielding only modest results.

1.1.3. Bitola Surroundings

There are two gold occurrences in the vicinity of town Bitola :

a) Occurence near village Lavci on the contact between gneisses and mica shists, representing the land base, and younger parashists overlaying them, contain pyrite with 4.5 g of gold per ton.
b) Occurence near Velušin and Barešanj (10-15 km) to the south of Bitola in dip ravines of streams falling from mountain Pelister, in pegmatites penetrating through gneisses and other crystalline shales, visible gold grains are found according to T. Vitorović. Quartz in pegmatitic veins contains native copper. Samples of this material also contain gold. Unfortunately, this area remained unexplored up to the present.

1.1.4. Gevgelija Surroundings (Eastern slope of Kožuf mountain)

The basin of River Konjska is one of the most important gold deposits covering the eastern hillside of Kožuf massif. In the times of Alexander of Macedonia, from old river terraces of River Konjska gold was mined over a distance of some 15 km, where huge volumes of alluvial materials were treated. Primary gold occurences are located in the spring section of River Konjska, where the land structure is represented by Precambrian gneisses and facies of old Paleozoic methamorphized rocks, including chlorite-sericitic shales, phylites, quartz porphyries and other rocks intersected by younger magmatic rocks such as metarhyolytes, rhyolytes, and andesites , associated with intesely hydrothermally altered zones in which kaolinizations and silicifications with Pb–Zn and Sb–As mineralizations are observed. the present ore minerals include antimonite, realgar, arsenopyrite, pyrite, galenite, and other minerals.

In addition, a fair amount of quartz veins and lenses occur in Precambrian and Old Paleozoic products.

On the basis of hitherto gold explorations completed from 1976 onwards three types of gold occurences were defined :

a) The zone with hydrothermally altered kaolinized phyllites and sericitis shales with arsenopyrite and pyrite impregnations. The zone is represented by three orebodies several hundred meters long each and 20 to 100m wide.

b) Silicified-Kaolinized zone near Smrdliva Voda, located in the area of Smiljevo Bačilo and Javor Reka, represented by antimony, arsenic, lead and zinc mineralizations, while the present minerals include antimonite, realgar, auripigment, galenite, markasite and pyryte. During antimony explorations gold was found in samples.

c) Quartz rif located to the Northwest of Smrdliva Voda, occuring in the zone of phyllites chloritic-sericitic shales. During past explorations gold was found in quartz samples.

In addition to the River Konjska basin , there are another two ore occurences containing gold :

a) The ore occurence Petrovo, located on the bank of River Petruška. Mineralization occurs in gabbrodiabases in the form of veins. Mineralization is represented by chalcopyrites and pyrites containing gold aswell.

b) The second gold occurance is Davidovo, on the contact between gabbrodiabase and granites. On the surface, the gold occurence is highly limonitized, and depthwise it is represented by pyrytic veins up to 2.5 m thick and according to some old data they contain up to 34 g of gold per ton.

2. SECONDARY GOLD DEPOSITS

Secondary alluvial and deluvial gold deposits on the Macedonian territory are numerous, and among them as the most important the following alluviums may be emphasized : River Konjska, River Lakavica, River Anska, River Bregalnica, River Pčinja, River Kriva, River Topolska and whole southern part of River Vardar. (See Fig. 2)

In a number of above alluviums gold was exploited in ancient times, so a large portion of gold has been exctracted. In the nearer past, alluvial deposits were explored between the two World Wars, and results are found, as already mentioned, in the Report of Professor B. Milovanović. After the war, more important and considerable explorations were carried out in River Konjska alluvium, particulary after 1976. This brief review of gold alluvial deposits in Macedonia includes a short description of some alluviums on the basis of pre-war data, while a specific review is given for the River Konjska alluvium and southern part of River Vardar.

2.1. Alluvial deposits in the River Vardar lower stream

River Vardar alluvium from Demir Kapija, and particulary from Udovo to the Greek frontier, may be fairly interesting regarding gold content. This section of River Vardar stream receives a number of gold-bearing rivers such as : River Došnica, River Stara, River Anska, River

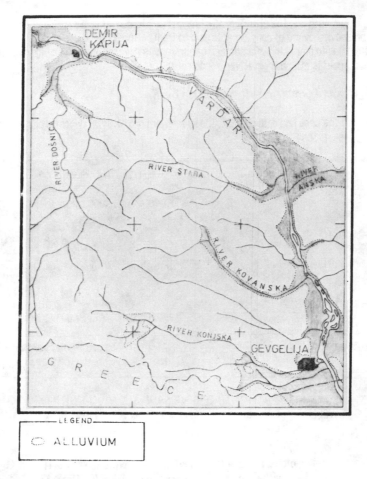

LEGEND

ALLUVIUM

FIG. 2 - SUMMARY MAP OF ALLUVIUMS OF RIVER KONJSKA AND RIVER VARDAR FROM DEMIR KAPIJA TO GEVGELIJA

Kovanska, Rivar Konjska, and some smaller tributaries located in this section.
The River Vardar lower stream alluvium has a length of 50 km from Demir Kapija, i.e. 27 km from Udovo to the border near Gevgelija, being one kilometer and even more wide, with an average deposit thickness of about 10m. According to the information obtained by drilling (for other purposes), the deposit thickness in River Vardar alluvium near Gevgelija reaches 90m. According to its structure, the alluvium consists of sand and gravel which occur alternatively. Information on gold presence in the alluvium, as well as about exploratory activities is found in Professor Milovanović Report, where is stated that some samples contain right up to 5 g per cu.m of alluvium material (oral announcement by M.Popović in 1940).
According to the information of a Belgium company which operated to the east of Gevgelija, an increased gold content in the detritus is found at the depth of 16m.
Over the post-war period no exploration were carried out in River Vardar alluvium for gold.

According to our opinion, the alluvium of River Vardar lower stream may be of interest for exploration and discovery of considerable gold concentrations, particulary in the bedrock (paleorelief) due to emptying of a number of auriferous rivers, and according to unconfirmed information , gold is mined from River Vardar alluvium.

2.2. Alluvial gold deposits in the River Konjska basin

In regard with the area and gold content, the River Konjska alluvial deposits are the most interesting ones. (See Fig.3)

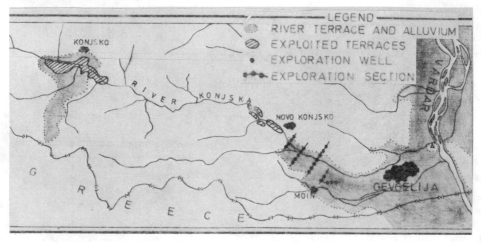

FIG. 3 – SUMMARY MAP OF RIVER TERRACES AND ALLUVIUMS WITH EXPLORATION WELLS FOR GOLD ON RIVER KONJSKA NEAR GEVGELIJA

As already stated above, old terraces of this alluvium were exploited in the time of Alexander of Macedonia, particulary in the area from village Novo Konjsko, to village Konjsko, were huge rock lumps exist even today, witnessing exploitation. At that time, the richest deposit sections were exploited, and it is assumed that large amounts of gold were extracted.
The River Konjska auriferous deposits are represented by a Recent alluvium and Old terraces, so they wil be presented separately.

2.1. Recent alluvium

The recent alluvium belongs to the Gevgelija valley and extends from the village Novo Konjsko to the river Konjska mouth into River Vardar, over a length of abut 8 km and width up to 3km, while the depth , according to the geophysical data, ranges up to 90m. The alluvium is composed of sand and course gravel with clayey partings, and also there is a large number of gabbro boulders with sizes exceeding 1cu.m .
Before the war fair activities were devoted to gold exploration in the River Konjska alluvum, in the immediate proximity of village Moin.
Explorations were performed by a Belgium company over the period 1937-1938, and to the

East of village Moin 45 shafts of varying depths were sunk in which gold presence was determined in concentrations ranging from 0.12 to 0.28 g per cu.m . To the west of the village where the river leaves canyon and flows into the valley 180 shafts were found and four workings where opened with a size over 2,000 cu.m . The obtained results are much better, the average gold content being 0.48 g per cu.m . In individual zones 2 to 4 g per cu. m were extracted (Data by M. Jovanović).

The latest alluvium explorations date back to 1976. At that time 21 exploratory wells were drilled in the alluvium, arranged in 4 parallel profiles transversally on the alluvium with the objective of defining gold ore reserves. Later on in 1981 another 3 wells were drilled. Washing was carried out by the method of gravity concentration of shaking tables (1976) and a "Oold-Sever" machine for gravity concentration.
In 1976 drilling reached the depth of 35m and in 1981 up to 49m , and only in the peripherial section of the alluvium individual wells reached the paleorelief.

Explorations in 1976 yielded the following results :

- Total reserves in the explored section are 60,000,000 cu.m of alluvium material
- Average gold content is 0.45 g per cu.m of alluvium material
- Totally 2.5 tons of gold in the reserves

In addition to this subsurface horizon, there is another hanging horizon at the depth between 25 and 35m with an increased gold content, but the reserves were not calculated for it.
In 1981 identical results were obtained. It should be emphasized that gold occurs in the alluvium in a flaky and finegrain form, indicating long transport route.

2.2.2. Old terraces of River Konjska

The terraces are located near village Konjsko and extend southwards towards village Uma over a length of about 4 km and width of 1 to 2 km. They consist of sands and gravels,containing gneiss, shale, and quartz boulders and clayey partings. The average detritus thickness in the terraces ranges between 10 and 15m and occasionaly even 20 and 25m.
Most extensive mining works in ancient times took place in this area and millions of cubic meters of detritus were processed. The richest gold sections on areas with a smaller detritus thickness were exploited, this being proved by rock lumps extracted during mining. In deeper terraces mining was carried out by adits, because major gold concentration occur in the paleorelief.
The terraces were explored in 1978 , and 16 wells were drilled, most of them down to the paleorelief .
Due to the appereance of groundwaters in the southern part of explored area, the exploratory wells were not drilled down to the paleorelief, so this section remained unexplored.
The samples were treated in "Gold-Sever" machine. The amount of gold extracted from the samples was small, excluding samples taken directly from the paleorelief, where gold content ranges between 0.3 and 0.4 g per cu.m . According to the size and the shape of grains, samples with coarse grain gold are in question with grain sizes as large as wheat grains. Since this gold is concentrated in the paleorelief and the upper and the upper layer is waste, this amount of gold has no practical value because it comes from a depth of 15 to 20m .
It is necessary to explore the southern parts of the terraces to obtain a clear image of auriferous terraces in this section.

3. OTHER AURIFEROUS ALLUVIUMS IN MACEDONIA

Data on auriferous alluviums in Macedonia mainly date back before the War, so they will be outlined as such.

a) River Anska alluvium - It is located near Valandovo, and extends over 25 km , being gold-bearing over the whole length. At the depth of 2 to 3m there is an enriched gold horizon. The richest alluvium section is the branch reaching village Prsten. Individual samples contain as much as 1.5 g of gold per ton of detritus.

b) In Dojran Lake terraces and "deep conglomerates", the gold content ranges, according to data by M. Popović and S. Pavlović from 1939 , between 3 and 4 g per cu.m .

c) River Bregalnica alluvium - At Delčevo, the Bregalnica alluvium reaches a length of 10 km , width of 200m, and thickness of about 5m . According to T.Vitorović data near village Mitrešinci (altitude 756), gold was washed, and coarse grain gold was found, of which individual grains were as heavy as 2 grams.
Unfortunately, were not conducted in this region, so there is no more detailed knowledge about the gold-bearing potential of the River Bregalnica.

d) River Kriva Lakavica alluvium - It extends over about 12 km , from Mareško Polje to Štip. Detailed explorations were performed in this region regarding gold with 494 boreholes, of which 51 with depths from 9 to 14m . Measurable amounts of gold were obtained on the mouth of River Majdanska into River Kriva Lakavica, were gold most probably originates from the andensite massif near village Šapur. Gold content ranges from 0.005 to 0.13 g per cu.m .
Old washaries exist between 17. and 19.th kilometer of the road Štip-Strumica on terraces above the river which were exploited very cleverly.

e) River Pčinja alluvium - The gold is located in old terraces and in a recent alluvium. The terraces on the north of the village Novi Čitluk have a volume of approximately 2,000,000 cu.m. The gold is nearly completely contained in the paleorelief, and occurs in small amounts. Richer portions are rare, and of small sizes, and the lowest gold content is 0.018, medium 0.16 , and highest 0.256 g per cu.m .
The recent alluvium enriched with gold covers a considerable area. Gold content varies from to 0.10 to 2.94 g per cu.m , but immediately above the paleorelief at a depth from 6 to 7 m .

f) River Kriva alluvium - This is a River Pčinja tributary. Gold occurs in this alluvium from its mouth into River Pčinja to River Kratovska. Most probably, the gold is carried by River Kratovska and River Povištica. There is no information on gold exploration in River Kriva, but it is characteristic that River Pčinja is gold-bearing from its mouth downstream, and there is no gold at all in River Pčinja alluvium.

g) River Topolska alluvium - River Topolska is a right hand tributary of Vardar, and empties to the south of Titov Veles. Gold washing was carried out near villages Gorno

Jabolčište, Gorno Vranovce and Melnica. According to T.Vitorović, gold occurs in here in grains characterized by courseness and massiveness, individual grains having a weight of 2 to 3 grams. There are no finegrains and flakes because they are transported further away.

In addition to foregoing auriferous alluviums, there are also others not explored at all, and gold was found only by a few samples. Such alluviums include those of River Markova, River Došnica, River Stara, River Petruška, Rivere Kovanska, etc. which deserve exporation in the future.

4. PRELIMINARY MINE FEASIBILITY STUDY – RESERVES OF SOUTHERN PART OF RIVER KONJSKA ALLUVIUM (ABSTRACT)

According to the data obtained by exploratory operations in 1976, the Feasibility study was made about exploitation of the southern part of the alluvium where the amounts of gold-bearing alluvium were calculated. This whole area is devided by vertical sections and the reserves were determined. The main gold seam was the one of 6.3m thick and with the gold content of 0.1 g per cu.m . This seam lays in the upper part of the alluvial strata. Using the method of vertical cross-sections (profiles), the following results are obtained :

- total volume of encountoured alluvium : 8,000,000 cu.m
- average gold content : 0.1 g per cu.m
- total amount of gold   : 0.84 t

All calculations were made for the thickness of some 8 to 10 meters, but having in mind the possibility that even better gold content may be found on the contact between alluvium and the bedrock (paleorelief), it is planned that final estimations of the depth of the pit should be performed after a number of additional explorations before and during the exploitation.

The calculaton of the gold reserves is given on table 1 :

TABLE 1

section mark	ponderized content	average thickness	section area	distance	volume	gold total
	(mg/cu.m)	(m)	(sq.m)	(m)	(cu.m)	(g)
block O-I	107.62	6.30	1,575	250	393,750	42,375
block I-II	103.97	6.30	4,031	800	3,224,800	333,348
block II-III	96.42	8.65	4,651	800	3,720,800	358,760
block III-IV	93.71	3.91	2,815	100	1,126,000	105,517
AVERAGE : 100.15 mg/cu.m				TOTAL :	8,000,000	840,000

According to determined volumes of alluvial material (direct evaluation of those elements "in situ") total amount of excavating masses of overburden and underlying auriferous seam are established.

The overburden in this auriferous alluvium is represented by topsoil (humus) seam of varying thickness. This topmost overburden seam should be removed separately for recultivation purposes, in order to provide present agricultural land surface in post-exploitation period. Besides this topsoil seam, the most of the overburden is consisting of alluvial material, made of sands and gravels, which are also auriferous, but considerably below the exploitable limits.

According to land surveying data and observations taken "in situ", the following overburden masses are calculated :

- without topsoil (humus) $Q1 = 528,000$ sq.m x 1.45 m   = 756,000 cu.m
- with topsoil (humus) $Q2 = 228,000$ sq.m x 0.80 m   = 182,400 cu.m

                                                              TOTAL :   948,000 cu.m

4.1. Estimation of exploitabile reserves

The estimation of River Konjska alluvium quantities was performed in second, geological part of this study, according to performed explorations during 1976.
As the total gold quantities of the deposit are not officially approved due to its insufficient exploration, only volumes of the auriferous alluvium are determined.
Taking into account the geographical positon of the deposit, positon of urban areas

- without topsoil (humus) $Q1 = 528,000$ sq.m x 1.45 m   = 756,000 cu.m
- with topsoil (humus) $Q2 = 228,000$ sq.m x 0.80 m   = 182,400 cu.m

                                                              TOTAL :   948,000 cu.m

4.1. Estimation of exploitabile reserves

The estimation of River Konjska alluvium quantities was performed in second, geological part of this study, according to performed explorations during 1976.
As the total gold quantities of the deposit are not officially approved due to its insufficient exploration, only volumes of the auriferous alluvium are determined.
Taking into account the geographical positon of the deposit, positon of urban areas and some important buildings in the immediate vicinity of the deposit, the total geological reserves are reduced for the part of remaining massif of village Gorničet. The total quantities of this massif, which are not to be taken into account in calculations are 393,750 cu.m , i.e. the 42.375 kg of gold will remain unexcavated :

        8,000,000 cu.m   840   kg
    -     393,750 cu.m   - 42.4 kg
        _____   _____

        7,606,250 cu.m   797.6 kg

Which can be taken as 7,606,000 cu.m of alluvial material and 800kg of gold for further calculations.
Overburden to ore ratio will be determined only in volume-units according to the total amounts of auriferous alluvium. Estimating in that way, this ratio will be :

$$\frac{948,000 \text{ cu.m}}{7,606,000 \text{ cu.m}} = 0.12$$

Relatively small thickness of topsoil seam and favorable overburden to ore ratio shows that topsoil masses can be removed quite simply by using bulldozers.

4.2. DETERMINING OF APPROPRIATE OUTPUT

First of all, many specific characteristics of this deposit should be underlined, for both geologically-tecnological and systematical reasons. Between all those groups of characteristics stated, there is a considerable conditional relationship.
A number of considerable indicators of the processing system, particulary a yearly output is determined most according to the empirical reasons in exploitating these types of alluvial deposits, and the technology itself.
It necessary to underline that River Konjska is completely dry most of the time, when in fact the stream does not exist at all, but some ponds here and there. Only in rainy periods the weak flow is formed in the riverbed.The divergence ditch is planned along western side of the site for providing security of exploitation during that periods of the year.

The output of this opencast facility is determined on 760,000 cu.m per year. This will be provided by one Dragline excavator of standard type, operating with 5 cu.m bucket.
The transportation of the excavated material and also the dumping of the waste will be performed by using 50 t dumpers of standard type.
The alluvium handling facilities are planned on the eastern side of the site, situated to provide easy connection with main road (See Fig. 4).
Exploitation starts with opening cut nearby the plant, in the very middle of the deposit and the developing of the workings is planned both on the northwest and southeast direction simultaneously, perfoming the dumping of the waste alluvial material into the excavated space backwards.
For the waste handling on the plant the wheel loader with 5 cu.m bucket is planned, in accordance with the dragline bucket size.
Two bulldozers are planned for topsoil stripping and some ancillary operations required.
The considerable part of the excavated alluvium waste is consisting of good sands and gravels, very convenient for bulding purposes, which are planned for the market. This also improves the ecconomic effect of the whole exploitating process itself.

Atthe present time the Feasibility study on gold exploitation from River Vardar alluvium is in progress. It is expected to obtain more favourable results, having in mind that the alluvium thickness is considerably less, which enables easier reaching the bedrock contact. Not only Yugoslavian firms and institutions are interested in realization of this project, but also there is a very considerable interest by foreign capital.

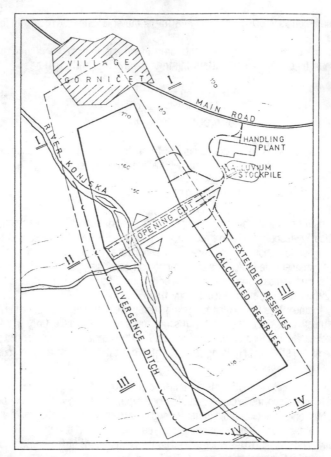

FIG. 4 – DISPLACEMENT MAP OF OPEN PIT ON THE RIVER KONJSKA AND ITS
SURROUNDINGS, ACCORDING TO THE PRELIMINARY MINE FEASIBILITY STUDY

REFERENCES

Arsovski, M. 1976, "Geomorfologija sliva Konjske Reke" , Stručni fond Geološkog zavoda SR
   Makedonije, Skopje

Džalev , I. 1987 , "Kratak pregled zlatonosnih pojava u SR Makedoniji", Sovet za
   finansiranje na studii i istragi, Gevgelija, Yugoslavia

Hrković, K. 1978. "Zlatonosni aluvion Konjske reke i mogućnost njegove valorizacije"
   MSC Thesis, Rudarsko-Geološki fakultet , Beograd , Yugoslavia

Chapter 40

PROSPECTS FOR GOLD IN NEW ZEALAND

Colin J Douch

Senior Minerals Geologist
Ministry of Energy, New Zealand

Abstract

New Zealand has been a significant producer of gold
for over a century. Total recorded production is over
850 tonnes. There were three major periods of
production: the first and biggest was between 1860 and
1880, the second was between 1895 and 1920 and the
third was between 1930 and 1945. After 1945 gold
production fell continually until 1980 when the major
upward adjustment in gold price led to renewed
interest.

There have been important new developments in each
of the known gold-rich areas: Westland, Otago and
Coromandel and exploration for gold in these areas and
in previously unprospected regions has reached
unprecedented levels.

In Westland over 140 gold processing plants are
each producing between two and six ounces a day from
low-grade placers. A major gold dredging operation
will soon be producing over a tonne of gold annually
and other new dredges are proposed. There are still
vast reserves of gold-bearing gravels in Westland and
prospecting is continuing to reveal new resources. In
the Reefton area there is renewed interest in
gold-bearing quartz reefs which were the source of
over a million ounces of gold between 1900 and 1951.
New large opencast reserves are now defined.

In Otago there is intense prospecting and mining
activity in the placer deposits along all the major
river systems with particular emphasis on the Clutha
River, part of which is due to be flooded in a major
hydro-electricity project during 1989. At Macraes
Flat, north of Dunedin a large gold resource has been
proved in a quartz reef environment and there is
growing interest in prospecting in other areas with a
similar geological environment in Otago.

In the Coromandel peninsula several epithermal gold
deposits are being appraised and developed. Large
low-grade opencast mines are being developed at the
site of the historic Martha mine and at Golden Cross,
nearby while several small-tonnage high-grade deposits
are awaiting mining consents. There has been intense
interest in exploration on the Coromandel peninsular
because of its volcanic epithermal geological
environment. Exploration has more recently been
extended into other areas, particularly active
geothermal and hydrothermal environments, in the hope
of locating recently extinct epithermal systems which
have not been eroded. There is growing interest in
prospecting for Carlin-type mineralisation in the
almost unexplored axial mountain ranges of the North
Island.

New Zealand has a lot of the right types of geology
and experience both here and overseas suggests there
is a lot more gold to be found.

Historical perspective

Figure 1 shows location of major gold-bearing
localities in New Zealand.

New Zealand has a history of major gold production.
During the century 1860 to 1960 over 850 tonnes of
gold, currently valued at nearly 20 billion New
Zealand dollars, were produced (1 $NZ = 0.66 $US
August 1988). This production of an average 3.5 kg of
gold per square kilometre from New Zealand compares
with less than 0.5 kg per square kilometre from
Western Australia (mainly from the Kalgoorlie Golden
Mile) and 29 kg per square kilometre from South Africa
(mainly from Witwatersrand). There were three main
periods of gold production in New Zealand (fig. 2),
each successively smaller. The first, and greatest,
was between 1860 and 1880. During that period gold was
New Zealand's biggest export commodity accounting for

MAJOR
GOLD BEARING LOCALITIES
IN NEW ZEALAND

COROMANDEL PENINSULA

Auckland

Thames • Golden Cross
Karangahake • Martha

Taupo

MAIN AXIAL RANGE

Sams Creek •

Buller R

WESTLAND

Grey R • Reefton
Taramakau R

SOUTHERN ALPS

Wellington

Christchurch

Shotover R

Cromwell
OTAGO
Kawarau R Macraes •
Flat

Clutha R

Dunedin

Invercargill

• Gold Bearing Localities

0   100   200   300
KILOMETRES

Figure 1

three-quarters of foreign exchange earnings. It
financed the development of the agriculture and
forestry industries and transport and communications
systems. The second was between 1895 and 1920
resulting from a huge inflow of risk capital, a demand
for gold from Britain to pay for wars and to pay for
continuing massive imports particularly of machinery
for agriculture and infrastructure development. The
third, and smallest, was between 1930 and 1945 when
economic depression and consequent lack of private
investment finance resulted in large-scale, Government
funded scientifically planned exploration programs in
the South Island placer fields.

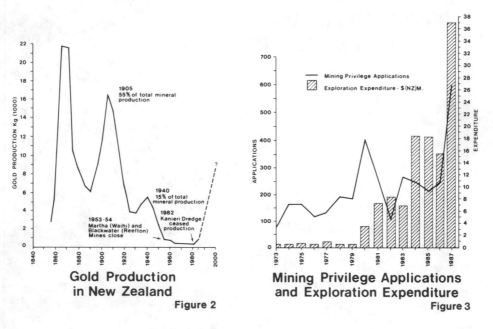

**Gold Production
in New Zealand**

Figure 2

**Mining Privilege Applications
and Exploration Expenditure**

Figure 3

Until 1934 the price of gold was 4/4/0 sterling
per ounce at which level it had been since 1717. In
January of 1934 the United States returned to the Gold
Standard and the price of gold was fixed at US$35
where it remained until the introduction of the
two-tier gold pricing system during 1973. During the
economic crises of the 1930's inflation continuously
debased the value of the pound. Prices, particularly
of labour, escalated until gold production could no
longer pay for necessary development costs. Investment
capital from Britain, the traditional source, was no
longer available. Gold mining in New Zealand almost

ceased. The two biggest hard rock gold producers, Blackwater mine at Waiuta and Martha mine at Waihi closed during 1951 and 1952 respectively. During 1952 there were nine bucket-ladder dredges still operating on the South Island placers, reduced from 201 during the peak year, 1903. This was reduced to five in 1955, three in 1960 and one in 1963. This sole survivor, the Kanieri dredge, continued to operate with steadily declining returns until 1982.

The political and economic events of the 1970's culminated in a huge increase, to US$850, in the free market price of gold in January 1980. Following that initial rise the gold price settled back to between US$400 and $450 per ounce with occasional, usually brief, increases towards $500 per ounce. That sustained price level, which has been even more dramatic in New Zealand dollar terms resulted in a massive re-investment into the New Zealand gold mining industry. This is reflected in the increase in number of licence applications and in the level of expenditure on exploration (fig. 3).

Summary geology of New Zealand gold deposits

Past gold production in New Zealand has been from four main geological environments(1):

1 Cenozoic placers of Southland and Otago (240,000 kg) and Nelson and Westland (237,000 kg)

2 Tertiary epithermal deposits of the Hauraki Goldfield   (270,000 kg)

3 Schist-hosted lodes of Otago (9,450 kg)

4 Metagreywacke-hosted lodes of Nelson and Westland (94,000 kg)

Cenozoic placer deposits. The Cenozoic placer deposits of Southland, Otago, Nelson and Westland were formed during alternating glacial and interglacial periods. During glacial periods glaciers advanced from the Southern Alps bringing with them large quantities of rock which had been eroded from the basement schists. This rock included quartz from gold-bearing veins. Gravel deposits accumulated because rivers were unable to transport the material as fast as it was being

supplied. During the following interglacial period,
temperatures warmed, precipitation increased and
rivers cut into the glacial deposits until sea level
rose and drowned the system. During this phase large
scale particle-size and density grading took place in
each river system. This whole process occurred at
least four times. Each cycle added complexity to the
gravel stratigraphy and further concentrated heavy
minerals, including gold. During these cycles several
different deposit types were formed:

> -buried river channels
> -fluvial deposits
> -glacial meltwater channels
> -degraded terraces

The cycles are represented by six main morraines
and associated outwash terraces which are preserved in
the main valleys(2). It is these, and recent river
gravels, that were worked and are currently being
re-appraised.

The Grey, Buller and Hokitika river valleys and
their tributaries in Westland host what is probably
among the largest of remaining gold placer deposits in
the world with resources conservatively estimated at
over ten million ounces (312,500 kg) of gold. In
Otago, the Shotover, Kawarau and Clutha rivers
likewise host large resources, estimated at over three
million ounces (93,750 kg). In Nelson the Aorere,
Waingaro and Anatoki valleys and in Southland the
Waiau and Aparima valleys contain largely untested
resources.

Tertiary epithermal deposits. The Hauraki goldfield
was the first major gold discovery in New Zealand(3).
Gold was discovered at Coromandel in 1852 but little
development took place for nearly ten years because of
the discovery of more easily won gold in Otago. Gold
in the Hauraki goldfield is associated with silver and
base-metal sulphides in quartz veins. It is now
recognised that the mineral association including the
metallic and host rock alteration mineral suites are
characteristic of the near-surface hydrothermal
(epithermal) zone observed throughout the
calc-alkaline volcanic environment of the Pacific rim,
including New Zealand's North Island. In such a system
gold is dissolved and carried as a bisulphide complex
$(Au(HS)_2')$ in near neutral to acid water at
temperatures in the order of 250oC. Gold and

base-metal sulphides precipitate in the boiling zone
within a few hundred metres of the surface. Such
systems which are actively depositing gold and
sulphide mineralisation have been extensively studied
in the Taupo volcanic zone(4,5). The Hauraki
gold-bearing quartz vein systems show all the
characteristics of epithermal deposits. At least 47
separate deposits were mined between 1852 and 1952 but
more than 80 percent of gold produced came from the
Martha mine at Waihi.

Schist-hosted lodes of Otago. Quartz lodes containing
gold, scheelite and stibnite crop out over a wide area
in greywacke-derived chlorite schist zones of Otago
and Marlborough. The lodes are localised along single
or multiple parallel shear zones. They are generally
discordant to the schist foliation and dip
steeply(1,6). The largest known lode, at Macraes Flat,
is the exception in that it shows a gentle dip and is
sub-concordant to the foliation. The lodes are thought
to have been deposited from hydrothermal fluids
generated by dehydration of the sedimentary pile
during metamorphism(1,6).

Metagreywacke-hosted lodes of Nelson and Westland.
Quartz lodes containing only gold, pyrite and
arsenopyrite occur within lower Paleozoic Greywacke
which crops out in two separate zones in southern
Nelson and Westland provinces. The most important lode
systems are those in the Reefton area where over
67,000 kg of gold were won from 84 mines within a belt
of weakly metamorhosed greywacke 34 km long and 10 km
wide. The lodes are lensoid quartz bodies which were
emplaced along shear zones parallel to fold axes(7,8).
Ore shoots have limited lateral extent but persist
down-plunge for up to 1200m. A hydrothermal,
metamorphic origin for the lodes has been
suggested(6).

New developments in exploration and mining

   Since 1982 there have been important new
developments in each of the historically important
regions and discoveries in areas previously regarded
as unprospective. Except in the placer goldfields the
most significant single factor has been a breakthrough

in understanding of the geology of deposits. In the
Hauraki goldfield it was the realisation that the
deposits had a near-surface hydrothermal origin and
were formed by processes similar to those observed
operating presently in the central North Island. In
Otago it was the understanding of structural controls.
In Reefton it was the understanding of structure and
genetic controls. In the placer areas the breakthrough
was in engineering with the development of small-scale
equipment which could achieve a steady optimum
throughput combined with high gold recovery, and the
engineering of a large dredge which could operate
within environmentally acceptable limits.

The present mining industry

   In early 1988 production of gold began, once more,
at the Martha mine at Waihi. This is the first of
several large new hard rock mining projects in the
epithermal environment of the Coromandel Peninsula.
The original Martha mine was New Zealand's biggest
mine. It was worked almost entirely underground from
15 levels over a depth range of almost 600 metres for
a return of 187,000kg of gold and 625,000kg of
silver(9). Ore originally averaged half an ounce (16g)
of gold and nearly two ounces of silver per tonne. The
new mine will be entirely opencast. Ore reserves of
over 10 million tonnes with an average grade of 2.6g/t
equivalent (based on 1:63 silver to gold value ratio)
using a cut-off grade of 1 gram of gold equivalent are
proved. The final pit will be 140 metres deep. It is
expected the mine life will be at least 13 years
(Waihi Gold Company Ltd. reports to Ministry of
Energy). This one mine will, during 1988, more than
double New Zealand's current gold production.

   The New Zealand gold mining industry has achieved a
considerable psychological boost by the opening of the
Martha mine. There was a 35 year period prior to the
re-opening when New Zealand had no hard rock gold
mines and with the decline of the industry there was
also a loss of the institutions which accompany it.
With the return of hard rock mining, institutions such
as the Mining and Exploration Association (the
fledgling local equivalent of a Chamber of Mines)
Coromandel Miners Association and West Coast
Commercial Goldminers Association have sprung to life.

The section of the industry which has shown the greatest sustained development since 1982 has been that of the small alluvial (placer) operation. After the closure of the Kaniere dredge gold production fell to a low of about 300 kg during 1983. Almost all of this gold was produced by about forty units each winning less than an ounce of gold a day. Since then the number of plants operating has grown to over one hundred and forty and gold production has increased accordingly to almost two tonnes per year.

Gold content of typical ore-grade gold-bearing gravels along the Grey River and its tributaries ranges between 65mg and 500mg, with an average of 150mg, per cubic metre. Size range distribution varies by locality, formation and level. Mostly the gold is coarse with 80 percent in the range 1 to 7mm and 18 percent in the range 0.1 to 1mm. Only 2 percent is larger than 7mm or smaller than 0.1mm(2).

A typical processing unit comprises a pontoon base on which is mounted a forward-sloping, 1.2m diametre rotating tubular screen (trommel) with a mesh size of 19mm or 25mm. For primary recovery a riffle table system is mounted beneath the screen. Riffle units are of variable size and style. Hydraulic riffles in which there is an upward jet of water behind the bar are now being widely used, as are sophisticated new rubber mat designs. Knudson bowls and Knelson concentrators are widely used for secondary gold recovery. The whole unit is floated in an enclosed pond. It is fed by an hydraulic excavator with a bucket capacity of one cubic metre and an arm length of six metres. Oversize material is either conveyed or bulldozed away from the end of the trommel. Optimum throughput is about 60 cubic metres per hour.

Currently the largest plants operating are three floating units in Otago. They use primary and secondary jigs and Knudson bowls for primary and secondary gold recovery. The largest is fed by a 165 tonne Hitachi UH801 excavator with a bucket capacity of 1.5 cubic metres and an effective arm reach of 14 metres. The three plants produced a total of 275kg of fine gold during 1985.

The Clutha river which drains through central Otago was historically a rich source of alluvial gold(10). It is currently the site of a major hydro-electricity dam development, at Clyde, ten kilometres south of Cromwell. The terraces adjacent to the river north of

Clyde are the scene of frenetic gold recovery activity
because the hydro lake is scheduled to be formed in
late 1989 with resultant flooding of much of the
gold-bearing ground. The town of Cromwell has already
been half mined away. Clutha river gold is typically
finer than west coast river gold with 50 percent in
the range 0.05 to 1mm. The extremely fine gold is
difficult to recover because the finest flakes either
float because of surface tension or plane in the water
stream. Much experimental work is currently being
carried out to develop a circuit capable of recovering
the very fine gold. Most of this work is being done on
gold-bearing heavy mineral beach leads in Westland
where almost all the gold is in the size range 0.01 to
0.1mm. Some of the leads contain zones of sand
containing up to 10g of very fine gold per cubic
metre. The best recovery has come from combinations of
spiral classifiers, Reichert cones, Knelson
concentrators and Knudson bowls.

Next year

   Currently there are two major mining projects
awaiting the granting of mining licences and one on
which a licence has been granted, but operations have
not yet commenced.

   The Golden Cross deposit at Waitekauri, eight
kilometres northwest of Waihi is expected to be the
next major mine to become operational. The reef system
at Golden Cross was worked underground on seven levels
from 1892 to 1903 and 1906 to 1920. Three
quartz-calcite reefs yielded 2,897kg of gold and
9,175kg of silver(3). Mining stopped when ore reserves
ran out and drilling failed to reveal further reserves
at depth. Recent intensive drilling on geophysical
anomalies has revealed further ore in the vicinity of
the old workings and in a newly discovered
north-to-south trending vein system south of the old
workings. Like the ore at Martha mine most of the ore
at Golden Cross is confined to the uppermost few
hundred metres from the surface. During mining the
larger, southern, system will be worked by both
underground and opencast methods while the northern
system will be entirely underground. Opencast
reserves of 2.3 million tonnes of ore with a grade of
2.7g/t gold and 13g/t silver are proven. Underground
reserves of over three million tonnes with a grade of
7.3g/t gold and 18.5g/t silver are probable. The vein
system is now interpreted, from the presence of a
characteristic metallic and alteration mineral suite,

as having been emplaced in a geothermal boiling zone
during a Miocene volcanic phase. Geological mapping
and photo-interpretation indicates it was emplaced
near the margins of a volcanic collapse structure, as
are many of the existing geothermal zones in the Taupo
region. (Cyprus Minerals Ltd. reports to Ministry of
Energy)

Macraes Flat deposit 60 km north of Dunedin, is at
a similar stage of development. Lodes at Macraes were
discovered by alluvial prospectors in 1866. Little
mining took place until 1890 when the main lode system
at Round Hill was discovered. The lodes were worked
underground from 1890 to 1930 during which an
estimated 100,000 tonnes of ore yielded approximately
469kg of gold and 100 tons of scheelite. A second
phase of mining between 1932 and 1936 produced 3,374
tons of ore for a return of 31kg of gold(6).

Recent exploration began at Macraes in 1984 and by
1987 proven and probable ore reserves of 6.1 million
tonnes containing 2.3g/t gold using a 1g/t cutoff had
been established. Known lode systems outside of the
main zone remain largely untested as does the main
zone at depth but no estimate of a possible resources
has been published. It is proposed to mine the
existing reserves entirely by opencast methods but to
continue an active exploration program during mining
(BHP Ltd. reports to Ministry of Energy). The lodes at
Macraes Flat have been shown to occur as pinch and
swell structures within a northwest-trending shear
zone which is about 125m thick and dips to the
northeast at about 20o. The shear zone has been traced
in outcrop for at least six kilometres. Interpretation
from aerial photographs indicates a minimum 25km
extension. The lodes are thought to have been formed
during an early Cretaceous phase of mineralisation in
response to the rapid uplift of the Otago schist and
the escape of metamorphic fluids into higher
structural levels(1,11).

A major dredging operation on the Grey river is
expected to become operation during late 1988 or early
1989. This project is a successor to the Kaniere
dredge which operated on the Taramakau river for many
years. The old company failed because it had entered a
region of gravels containing sub-economic gold grade
and thus faced the expense of building a new dredge on
the Grey river where it held mining licences. Despite

the increase in price of gold during this period the
company failed after it had built the pontoon part of
a large new dredge(12).

A new company has taken over the old licences and
designed, and almost completed, a technically advanced
dredge incorporating the earlier pontoon. The new
design combines both a bucket wheel and bucketline on
the one dredge to simultaneously mine both overburden
and auriferous gravels at the rate of 12 million cubic
metres per year. Tailings will be automatically
re-spread using conveyor systems to grade the material
with coarse at the base and fines on the surface. The
dredge will be the largest bucketline dredge in the
world. The company currently has under mining licence
an area containing proved reserves of over 100 million
cubic metres of gravels with an average grade of less
than 130mg of gold per cubic metre. It has prospecting
licences over surrounding river terraces within which
there is potential to quadruple this reserve. The
innovative new design should achieve greatly reduced
production costs and high annual throughput and the
time required to achieve satisfactory restoration is
greatly reduced (Grey River Gold Ltd. reports to
Ministry of Energy). Possibly most important, the
innovative new design has potentially opened up large
areas which have been previously considered uneconomic
because of thick overburden or low grade.

The next five years

Exploration is being carried out which should
result in the development of at least two and possibly
four big hard rock gold mines within the next five
years. All but one are in areas of past mining
activity and have resulted from re-appraisal of old
workings, detailed mapping and geochemical sampling
and re-interpretation of structure.

Sam's creek prospect in Nelson Province was discovered
as a result of regional exploration. The prospect is a
series of complex microgranite sills and dikes which
crop out discontinuously and have been traced for
nearly fifteen kilometres. Detailed drilling over a
600m length of the microgranite has identified a
resource of approximately five million tonnes
averaging 2 g/t of gold within which there are high
grade lenses. Within the same zone a further 10
million tonnes is possible. There is potential for
discovery of similar zones over the remaining strike
length.

Reefton in Westland has been the site of intense
exploration activity over the past several years. The
quartz lodes of the Reefton goldfield consist of
steeply- to moderately-dipping lenses and shoots
containing an average 15g/t gold. The ore shoots are
located within shear zones. The host sequence of
sandstones and shales is moderately folded and the
lodes are located in the more tightly folded zones,
particularly on the limbs of folds.
A major geochemical sampling and geological mapping
program identified one major prospect - the Globe
prospect - and several large drill targets. In the
Globe prospect diamond drilling has indicated a
mineralised unit of over ten million tonnes with a
grade between 2.5 and 3 g/t gold. Of this amount at
least four million tonnes is opencastable. Drilling is
continuing to further define the resource. Drilling
the other large targets is proceeding and results are
encouraging with the possibility of several orebodies
being defined. Drilling is also being carried out in
the vicinity of the old Blackwater mine with the
immediate aim of delineating near-surface
(opencastable) reserves and the longer term aim of
identifying sufficient deep ore to justify reopening
the mine (CRA Ltd reports to Ministry of Energy and
personal communications).

Talisman mine at Karangahake, south of Thames, was one
of the big mines with a recorded yield of 109,709kg of
bullion (22% Au and 78% Ag). The mine was worked on 15
levels vertically over nearly 700 metres during the
period 1895 to 1918. The first cyanidation plant in
the world was installed at Karangahake by the Cassells
Company in 1890 and was used to treat ore from all the
local mines including that from Martha. The Talisman
mine worked two of four shoots within a single reef
which was the most persistent of several reefs. The
mine is currently being re-appraised and drilling from
underground into the main reef system has returned
core containing bonanza grades of gold. It is highly
probable that sufficient ore will be defined to
justify re-opening this mine (NZ Goldfields Ltd
reports to Ministry of Energy)

Numerous small underground mines on the Coromandel
peninsula are currently being re-appraised. Many of
those in the area of Thames contained extremely rich
(bonanza) ore. A typical example was the Caledonian
Mine near Thames which yielded 8,661kg of bullion (70%

Au and 30% Ag) from 55,529 tonnes of quartz. Modern geophysical surveying, geochemical sampling and geological mapping in the light of the full understanding of the epithermal origin of the Coromandel deposits has resulted in discovery of many new drill targets in the vicinity of the old mines. Mining licences have been applied for on two of the old mines in which ore reserves have been clearly defined.

In Otago numerous old mines are being actively re-appraised. A mine on the Bullendale reefs, which crop out in the hills west of the Shotover river and are considered as the source of the fabulous quantities of alluvial gold in the Shotover, is being re-opened for exploration. Bendigo mine north of Cromwell, yielded nearly 1,000kg of gold from sporadic mining of high grade oxide ore between 1876 and 1937. Primary ore averaged 10g/t but could not be worked profitably. The reef system is currently being drilled from the surface.

Prospecting and exploration

Intensive prospecting and exploration is being carried out in all prospective parts of New Zealand. Activity is directed dominantly at the historically important gold producing localities although areas previously considered of little prospectivity are receiving considerable attention.

Epithermal environments. A much more thorough understanding of ore-forming processes in the epithermal environment has lead to a new approach to prospecting. In particular a whole range of minerals diagnostic of epithermal acid sulphate centres is now recognised. These include adularia, albite and cristobalite, the clay minerals illite, kaolinite and halloysite, the zeolites mordenite, laumontite and wairakite (calcium analogue of analcime), which are temperature indicators, and the sulphides pyrite and pyrrhotite. Geophysical properties of such systems are now also better understood. Hydrothermal alteration of feldspars to clays strongly alters the electrical properties of the rocks. Quartz veins are high resistivity features within zones of low resistivity clay alteration. Clays also give a moderate IP response while quartz gives a weak response. Such

centres also show very weak magnetism. In a typical
geophysical survey areas of poor aeromagnetic response
are selected for ground follow up. Such areas are then
tested for resistivity and IP properties. A method
which is gaining wide acceptance is Controlled Source
Audio Magnetotelluric surveying (CSAMT) which is able
to measure resistivity at varying depths by changing
the signal frequency. Such a system is able to locate
epithermal systems buried by later volcanic deposits.

Careful geological mapping and sampling and
geophysical surveying is being applied throughout the
Coromandel Peninsular to locate quartz vein systems
which are buried beneath a shallow regolith. In the
Taupo area the same geophysical methods are being
applied, with considerable success, in defining
extinct epithermal systems under thick tuff or ash
layers. Two such areas are currently being drilled.
Samples from both areas contain potentially ore grade
gold.

Placer environments. Prospecting and exploration using
classical methods is continuing throughout all the
main river valleys where auriferous gravel terraces
and river alluvium are present. Hydraulic excavators
are most widely used to dig pits of up to $100m^3$ for
sampling in gravels up to 12 metres thick. In units
thicker than 12 metres, becker or keystone drills are
most commonly used. Magnetic surveys are commonly
carried out to locate concentrations of black
(magnetite) sand with which gold is associated.
Shallow seismic surveys are commonly carried out to
define bottom profiles particularly to determine the
existance of deep leads (or gutters) or faults. Large
areas in central Otago are being re-appraised using
seismic profiling. In the Manuherekia and Ida valleys
a basal auriferous conglomerate horizon has been
thrust up along marginal faults where it has been
worked for gold historically, by hydraulic sluicing.
It is likely this same conglomerate horizon underlies
both valleys at shallow depth. Seismic profiling is
being used to closely define the bottom profile with a
view to tunnelling into rich deep leads.

Schist environment. All the gold producing areas in
the Otago schist belt are being re-appraised using
classical exploration techniques. Such methods are now
assisted by interpretation of structures revealed from
satellite imagery. In particular, low grade,
opencastable resources are sought.

Metagreywacke environment. Metagreywacke in the region
of Reefton is being closely examined using classical
techniques assisted by structural interpretation from
high resolution aerial photographic imagery. Low grade
opencastable resources are primarily sought but the
possibility of deep, high grade reserves in the
vicinity of previously-worked mines is also being
tested.

Sedimentary environment. gold mineralisation with some
similarity to the Carlin type mineral association has
been discovered at one locality in the North Island
main axial range. This discovery has caused
considerable interest in what is, in New Zealand, a
new gold environment. Almost the whole of the ranges
has been pegged for exploration.

The offshore environment. Shallow sampling in deep
water offshore from the Grey river mouth has revealed
the existence of some $75km^2$ in which gold grade
averages $200mg/m^3$. Within this area there is an zone
of some $33km^2$ in which gold grade averages $300mg/m^3$.
Because of the deep water, even sampling is at the
limits of current technology. Further exploration and
possibly mining will be dependent on technological
developments.

Licensing

   New Zealand has a licensing process which derives
from the Mining Act 1971. This act is administered by
the Resource Allocation Group of the Ministry of
Energy.

There are basically three types of licence:

Exploration licence. This allows low impact
prospecting over a maximum area of $500 \ km^2$. It is
granted for a maximum term of two years and is
non-renewable. It gives priority for any subsequent
application

Prospecting licence. This allows exploration up to
mining feasibility stage over a maximum area of 40
hectares. It allows drilling, pitting, bulk sampling,
re-opening of old workings and sampling and
geophysical and geochemical surveying. It is granted
for a maximum term of three years and is renewable for
a further three years. It gives priority for a
subsequent mining licence application.

Mining licence. This allows mining of all types over a maximum area of 400 hectares. It is granted for a maximum term of 42 years. No royalties are currently charged for gold production. An environmental impact report is normally required, for major projects, before a licence is granted

Because of the requirement for public involvement in the planning procedure and for recommendations from agencies such as local and regional authorities, catchment authorities and other Government departments to be considered, granting of a licence can be a lengthy process. On average, exploration licences take 12 months, prospecting licences take 18 months and mining licences take two years to grant.

Environmental aspects and opposition to mining

Environmental factors are of utmost importance in New Zealand when any development is proposed. In the past, as in most countries, environmental considerations were secondary to development because of the urgency to develop agricultural land and forestry and basic mining operations such as aggregate, limestone and building stone quarries. New Zealand had gold and it was a wise use of that resource to mine it to help pay for development. Today, the equation is not that simple. There is a far greater understanding of ecological balance, sensitivity of some ecosystems to damage and the difficulty of restoration. This necessarily leads to multiple land use considerations when any development is proposed. All development proposals, including applications for exploration or mining licences, are open to objection from interested parties. Such objections are heard by the Planning Tribunal who may permit or refuse an application or may recommend modifications to a proposal.

New Zealand has a National Park system and numerous forest parks and reserves of various status which are controlled by the new Department of Conservation. A consent to explore or mine on Conservation land (under the Mining Act 1971) is required from the Minister of Conservation before an exploration or mining licence may be granted. Consents are normally granted, commonly with conditions. In the year since it's formation only one mining proposal has been blocked because the Minister of Conservation considered that conservation values of the area concerned outweighed the benefits from a small mine.

There is a vigorous anti-mining lobby, with
interests mainly in the Coromandel peninsula and in
Nelson Province, whose able questioning of prospecting
and mining proposals provides serious debate about
land use alternatives. Although such groups rarely
stop prospecting or mining projects their obvious
committment and beliefs often cause mining companies
and Government departments to consider modifications
to proposals. Other groups such as Native Forest
Action Council and Forest and Bird Protection Society
provide keen opposition to development interests.

REFERENCES

1 Brathwaite, R.L., and others, 1986. The geological background to a resurgence of gold exploration in New Zealand. Vol 2 Geology and exploration, 13th Congress of the Council of Mining and Metallurgical Institutions.

2 Douch, C.J., 1988. Procedures for prospecting alluvial gold deposits in Westland. Ministry of Energy report, in prep.

3 Downey, J.F., 1935. Gold mines of the Hauraki district. Government Printer, Wellington. (out of print)

4 Henley, R.W., and Hoffman, C.F., 1987. Gold: sources to resources. Proceedings of the Pacific Rim Congress 1987. Australasian Institute of Mining and Metallurgy.

5 Brown, K.L., 1986. Gold deposition from geothermal discharges in New Zealand. Economic Geology 81.

6 Williams, G.J., 1974. Economic geology of New Zealand. Monograph Series 4, Australasian Institute of Mining and Metallurgy.

7 Suggate, R.P., (ed) 1978. The Geology of New Zealand. Vol 1. New Zealand Geological Survey publication.

8 Downey, J.F., 1928. Quartz reefs of the West Coast mining district. Government Printer, Wellington. (out of print)

9 McAra, J.B., 1978. Gold mining at Waihi 1878-1952. Waihi Historical Society, Pegasus Press, Christchurch.

10 Hearn, T.J., and Hargreaves, R.P., 1985. The speculators dream: gold dredging in southern New Zealand. Allied Press, Dunedin.

11 Brathwaite, R.L., and Piranjo, F., 1985. Metallogenic epochs and tectonic cycles in New Zealand. Geologicky Zbornik Geologica Carpathica 36.

Author's Index